Targeted Therapeutic Systems

Targeted Diagnosis and Therapy

Editor

John D. Rodwell
Vice President, Research and Development
CYTOGEN Corporation
Princeton, New Jersey

1. Antibody-Mediated Delivery Systems, *edited by John D. Rodwell*
2. Covalently Modified Antigens and Antibodies in Diagnosis and Therapy, *edited by Gerard A. Quash and John D. Rodwell*
3. Targeted Therapeutic Systems, *edited by Praveen Tyle and Bhanu P. Ram*

Targeted Therapeutic Systems

edited by

Praveen Tyle
Sandoz Pharmaceuticals Corporation
Lincoln, Nebraska

Bhanu P. Ram
Idetek, Inc.
San Bruno, California

Marcel Dekker, Inc. **New York • Basel**

ISBN 0-8247-8181-3

This book is printed on acid-free paper.

MARCEL DEKKER, INC.
270 Madison Avenue, New York, New York 10016

Current printing (last digit):
10 9 8 7 6 5 4 3 2 1

PRINTED IN THE UNITED STATES OF AMERICA

Introduction to the Series

Targeted Diagnosis and Therapy is a series intended to collect new knowledge generated in the research and development of self-directed diagnostic and therapeutic agents. The powerful tools of recombinant DNA and monoclonal antibody technologies have contributed immensely to our understanding of the concept of molecular recognition as well as protein structure-function relationships. This has yielded a view of the future that includes the use of a variety of new pharmaceutical products. These products will have the property of localizing to a predetermined site, with a consequent diagnostic or therapeutic effect. The clinical use of these products will include the treatment in vivo of malignant organs or tissues as well as elimination of specific cell types in ex vivo bone marrow-purging procedures. Each volume will focus on one product or strategy, and contain the relevant preclinical and/or clinical experience. The list of near-term subjects includes a variety of antibody conjugates, the interferons, the interleukins, tissue plasminogen activator, and gene therapy. Other volumes will deal with the next generation of agents, such as genetically engineered toxins and fusion proteins. It is expected that the series will be useful for basic researchers and clinicians alike.

New frontiers lie ahead. The opportunities for research and development of important new pharmaceutical products are considerable. It is hoped that Targeted Diagnosis and Therapy will assist in the efforts to achieve this goal.

John D. Rodwell
Editor

Preface

Since the discovery of monoclonal antibodies, researchers have a new weapon, "the magic bullet" in their arsenal for devising novel systems for delivering drugs to a target site. Although perhaps an idealistic image, it is true that the concept of targeted therapeutic systems is gaining more and more interest in the pharmaceutical and medical community. An important aspect of the material discussed in this book is that, for targeting to be successful, not only are site selection and access important, but so is the persistence of therapeutic drug levels at the target below toxic levels.

Toxicity is reduced by delivering a drug to its target in higher concentrations, thereby reducing harmful systemic effects by a drug that can be administered in a smaller quantity to produce a desired effect. Toxicity can also be reduced by administering the drug in a nontoxic form that is activated only at the site of action, or in other ways. The growing dissatisfaction with currently available drugs, and particularly with the cytotoxic agents used in cancer therapy, has caused many scientists to consider the possibility of using antibody-directed therapy. This includes monoclonal antibodies, their conjugates, and the manipulation of the liposomal surface. Targeted drug delivery therapy of tumor using monoclonals or their conjugates has been reported by many investigators, and the results are quite promising.

Exogenous antibodies have been administered to both laboratory animals and humans as anticancer agents. The antibodies have been directed against cell-surface determinant present on the tumor. Antibodies can act as their own therapeutic systems when directed against receptors that are involved in the uptake of an essential component for the cell or when they are directed against receptors that provide signaling events. A variety of drugs and toxic agents have been linked to antibodies for both imaging purposes and as therapeutic systems in an attempt to destroy unwanted cell populations, particularly malignant tumors. The added advantage of using drugs is the vast body of information on mode of action, side effects, and toxicity that has been established and is available to the designers of drug conjugates. The argument for the use of toxins such as diphtheria toxin and ricin is that they act as a catalyst and are many times more effective than drugs with stoichiometric mechanisms. Thus dosages are minimized while the chances of a successful outcome are maximized, particularly in those instances where a target antigen is not strongly expressed.

Liposomes have been used to deliver drugs to a diseased area by using antibodies as homing probes. More relevant to this volume are the therapeutic systems in which the antibody is the actual drug to be targeted rather than its use as a "homing device." One of the largest advantage of using liposomes as a carrier is that a large amount of drug can be packed in the liposomes without losing the immunoreactivity of the linked antibody. It is anticipated that the use of antibody as a means of conferring specificity to liposomes would markedly increase their usefulness in the future in designing targeted therapeutic systems.

Considering the number of recently published papers dealing with targeted therapeutic systems, it is clear that the interest in the development of such systems is increasing rapidly. Nevertheless, as far as we know, there is no comprehensive book covering the subject. The aim of this work is to partially fill this gap and act as a stimulus and basis for research in this most interesting and needed area. The book describes the principles of monoclonal antibodies, immunoconjugates, and specialized delivery systems, and it examines the various targeting possibilities suggested by these systems. The chemistry and biology of these systems range from very simple reaction to highly complex formulation development.

The book begins with a chapter that is an introduction to monoclonal antibodies immunoconjugates, and liposomes as targeted therapeutic systems. Part Two of the book specifically covers monoclonal antibodies and contains chapters on large-scale production, purification, regulatory agency concerns, and therapeutic applications of monoclonal antibodies. The immunoconjugates are gaining increasing popularity in targeted drug delivery for cancer therapy. These are presented in Part Three of the book. The chapters in this part include immunoconjugates in drug delivery, evaluation of drug monoclonal conjugates for carcinoma, and radioimmunoconjugates for targeted delivery. This part also covers specific immunoconjugates currently being investigated by various research laboratories. New product performance standards to ensure safety and efficacy will have to be developed for these systems. The final section, Part Four, presents various specialized systems currently being utilized to achieve targeted therapy. These include liposomes, nanospheres, and LDLs.

This volume is designed for individuals of diverse backgrounds who are interested in the design, development, use, and/or optimization of targeted therapeutic systems, and in carrying the product from discovery phase to the marketplace. Scientists and management involved in clinical practice and in research and development should find this book useful, as should personnel responsible for innovative developments in the therapeutic systems and their use in medicine.

We are indebted to specialists in several areas who prepared chapters and persevered through revisions and deadlines. Special appreciation is extended to Sandra Beberman and Carol Mayhew of Marcel Dekker, Inc., for their expert assistance in the preparation of this book.

Praveen Tyle
Bhanu P. Ram

Contents

Contributors

Donald A. Baker, Ph.D. Vice President of Research and Director of Regulatory Affairs, Regulatory Affairs Department, Bio Response, Inc., Hayward, California

Byron Ballou Assistant Research Professor, Departments of Surgery and Pathology, University of Pittsburgh School of Medicine, Pittsburgh, Pennsylvania

Martin K. Bijsterbosch, Ph.D. Center for Bio-Pharmaceutical Sciences, Sylvius Laboratories, University of Leiden, Leiden, The Netherlands

Dominique Bourel, Ph.D., Pharm. D. Laboratoire des Réactifs de Groupages Sanguins, Centre Régional de Transfusion Sanguine, Rennes, France

Donald J. Buchsbaum, Ph.D. Assistant Professor, Department of Radiation Oncology, The University of Michigan Medical Center, Ann Arbor, Michigan

John B. Cannon, Ph.D. Senior Pharmaceutical Scientist, Pharmaceutical Product Development, Abbott Laboratories, North Chicago, Illinois

Passchier C. de Smidt Center for Bio-Pharmaceutical Sciences, Sylvius Laboratories, University of Leiden, Leiden, The Netherlands

Pedro H. di Rocco Research Associate, Department of Pharmaceutical Technology, School of Pharmacy and Biochemistry, University of Buenos Aires, Buenos Aires, Argentina

Thomas L. Evans Scientist, Manufacturing Department, IDEC Pharmaceuticals Corporation, Mountain View, California

Vipin K. Garg, Ph.D. Manager, Protein Purification, Bio Response, Inc., Hayward, California

Mridul K. Ghosh, Ph.D. Research Associate, Department of Industrial and Physical Pharmacy, School of Pharmacy and Pharmaceutical Sciences, Purdue University, West Lafayette, Indiana

Thomas R. Hakala Division of Urological Surgery, University of Pittsburgh School of Medicine, Pittsburgh, Pennsylvania

W. Scott Harkonen, M.D. Associate Medical Director, Becton Dickinson Monoclonal Center, San Jose, California

Ho Wah Hui, Ph.D. Senior Research Scientist, Pharmaceutical Product Development, Abbott Laboratories, North Chicago, Illinois

Theodore S. Lawrence, M.D., Ph.D. Assistant Professor, Radiation Oncology Department, The University of Michigan Medical Center, Ann Arbor, Michigan

Richard A. Miller IDEC Pharmaceuticals Corporation, Mountain View, California

Ashim K. Mitra, Ph.D. Assistant Professor, Department of Industrial and Physical Pharmacy, School of Pharmacy and Pharmaceutical Sciences, Purdue University, West Lafayette, Indiana

Marcelo C. Nacucchio, Ph.D. Adjunct Professor, Department of Microbiology, Parasitology, and Immunology, School of Medicine, University of Buenos Aires, Buenos Aires, Argentina

Karen Kashmanian Oates Assistant Professor, Department of Biology, George Mason University, Fairfax, Virginia

Stefano Persiani, Ph.D. Research Associate, Division of Pharmaceutics, University of Southern California School of Pharmacy, Los Angeles, California

Bahnu P. Ram, Ph.D. Manager, Research and Development, Idetek, Inc., San Bruno, California

S. Ramakrishnan, Ph.D.* Assistant Medical Research Professor, Department of Medicine, Division of Hematology/Oncology, Duke University Medical Center, Durham, North Carolina

---

**Current affiliation*: Assistant Professor, Department of Pharmacology, University of Minnesota Medical School, Minneapolis, Minnesota

Alain P. Rolland, Ph.D., Pharm. D.* Visiting Scientist, Advanced Drug Delivery Research Center, Ciba-Geigy Pharmaceuticals, Horsham, England

Wei-Chiang Shen, Ph.D. Associate Professor, Pharmaceutics Division, University of Southern California School of Pharmacy, Los Angeles, California

Daniel O. Sordelli, Ph.D. Assistant Professor, Department of Microbiology, Parasitology and Immunology, School of Medicine, University of Buenos Aires, Buenos Aires, Argentina

Praveen Tyle, Ph.D. Senior Scientist B, Sandoz Research Institute, Sandoz Pharmaceuticals Corporation, Lincoln, Nebraska

Theo J. C. van Berkel, Ph.D. Professor, Center for Bio-Pharmaceutical Sciences, Sylvius Laboratories, University of Leiden, Leiden, The Netherlands

Alan L. Weiner, Ph.D.† Director, Novel Drug Delivery Systems, The Liposome Company, Inc., Princeton, New Jersey

---

**Current affiliation*: Formulation Group Leader, Department of Pharmaceutics, Centre International de Recherches Dermatologiques, Valbonne, France

†*Current affiliation*: Director, Research and Development, Escalon Ophthalmics, Inc., Rocky Hill, New Jersey

Targeted Therapeutic Systems

part one
INTRODUCTION

1

Monoclonal Antibodies, Immunoconjugates, and Liposomes as Targeted Therapeutic Systems

PRAVEEN TYLE
Sandoz Pharmaceuticals Corporation, Lincoln, Nebraska

BHANU P. RAM
Idetek, Inc., San Bruno, California

I. INTRODUCTION

Since the discovery of monoclonal antibodies (MoAbs) in 1975 [5], researchers have a new weapon, "the magic bullet," in their arsenal to use in devising novel methods of delivering drugs to the target site. Antibodies including monoclonals have found use in sensitive immunodiagnostic tests [1–4]. However, the use of MoAbs and their conjugates for therapeutic purposes is still in its infancy. The purpose of this chapter is to introduce monoclonal antibodies, immunoconjugates, and liposomes and their role in targeted drug delivery and immunotherapy of cancer.

In order to appreciate the impact of MoAbs, it is necessary to understand the limitations of conventional polyclonal antisera/antibodies. The production of highly specific antisera is difficult and unreliable; further, it requires highly purified antigen, and different lots of antisera have different specificities and affinities toward the antigen. On the other hand, MoAbs are highly specific. Unlimited quantities of such antibodies can be produced against virtually any molecule, regardless of the purity of immunizing antigens.

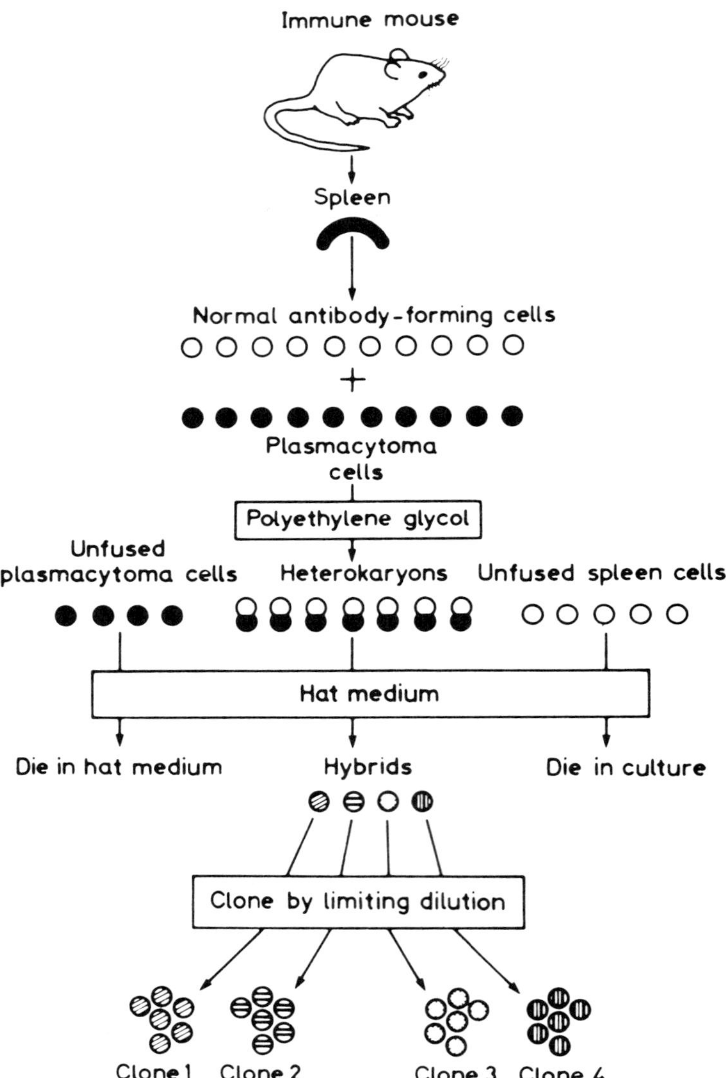

Figure 1 General scheme for the production of monoclonal antibodies. (From Ref. 95.)

II. MONOCLONAL ANTIBODY PRODUCTION

Each B lymphocyte in an animal expresses an antibody of only one specificity on its membrane. Once driven to differentiate, the B cell becomes a plasma cell with the cytoplasmic machinery to produce and secrete large amounts of its own unique immunoglobulin. For the production of monoclonals, usually a mouse is immunized with the antigen of interest (Figure 1). Once an immune response ensues, B lymphocytes from the immunized animal's spleen or lymph nodes are harvested in a single cell suspension. These cells are then fused with myeloma cells for the same species. Myeloma cells are immortal and have the cytoplasmic machinery to produce large quantities of immunoglobulin. They also contain an enzyme deletion, hypoxanthine guanine phosphoribosyl transferase (HGPRT), which is required for their survival in the presence of the folic acid antagonist, aminopterine. Fusion of cells is generally triggered by Sendai virus, polyethylene glycol, or electric current. The cell suspension is then distributed into the wells of a microliter plate in a section medium such as hypoxanthine-aminopterine-thymidine (HAT), where only hybridomas that have acquired the HGPRT from lymphocytes via cell fusion survive. The hybrids are cloned by limited dilution to one cell per well, which makes it easy to identify an antibody of the desired specificity, titer, and avidity for propagation in mass culture. Alternatively, the antibody-producing cells can be injected into the peritoneal cavity of mice for the production of ascites fluid. Cell supernatant or ascites can later be purified by column or affinity chromatography to obtain the pure antibody (See Chapter 3 for more details).

Since the inception of hybridoma technology in 1975 [5], many improvements in the technique have been made. The use of genetically altered mice for immunization [6], fusing cells in the presence of electric current, hybridoma selection by flow cytometry [7], and in vitro immunization [8,9] are some examples. Evaluation of hollow fiber bioreactor systems for large-scale production of murine monoclonal antibodies are discussed in Chapter 2. Therapeutic and drug delivery applications of monoclonal antibodies are discussed in Chapter 5.

III. IMMUNOCONJUGATES

Conjugates of small molecules (drugs, some toxins) and large molecules (proteins, enzymes, bacterial and plant toxins) with monoclonal or polyclonal antibodies are used in a myriad of immunodiagnostic assays [10], in immunohistochemistry [11], and as therapeutic agents. Various methods are described in the literature for the preparation of protein conjugates [12,13]. Antibodies (immunoglobulins) carry several reactive groups, such as the carboxyl groups of the C-terminal and of aspartic and glutamic acid residues, the amino groups of the N-terminal and of lysine residues, the imidazol of histidine and phenolic

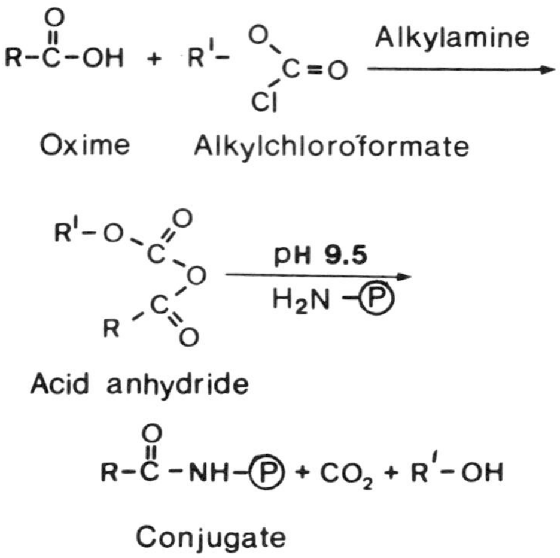

$$R-\overset{O}{\overset{\|}{C}}-OH + R'-O-\underset{Cl}{C}=O \xrightarrow{\text{Alkylamine}}$$

Oxime Alkylchloroformate

$$R'-O-\overset{O}{\overset{\|}{C}}-O-\overset{O}{\overset{\|}{C}}-R \xrightarrow[H_2N-(P)]{pH\ 9.5}$$

Acid anhydride

$$R-\overset{O}{\overset{\|}{C}}-NH-(P) + CO_2 + R'-OH$$

Conjugate

Figure 2 Mixed anhydride method for conjugation of carboxyl group-containing haptens.

function of tyrosine, the sulfhydryl group of cysteine, and the guanidino group of arginine.

Immunoconjugates used in cancer therapy constitute both haptens, with a molecular weight below 1000 daltons (doxorubicin, daunorubicin, vindesine), and large molecules (plant and bacterial toxins such as ricin and diphtheria toxin) [14–16]. Ligands with carboxylic groups can be directly coupled to proteins after their activation with a mixed anhydride, a carbodiimide, or N-hydroxysuccinimide. The mixed anhydride method requires the use of alkylchloroformate, trialkylamine, and acid in equimolar quantities to form a mixed anhydride of acids at low temperature in an inert organic solvent (Figure 2). The mixed anhydride-activated acid is then added to protein solid to obtain the desired product. Some of the drugs conjugated by this method are testosterone-17-hemisuccinate [19] and synthetic estrogens [23]. In recent years, the use of N-hydroxysuccinimide (NHS) in the presence of water-soluble carbodiimide to form esters of the acids has become popular. The NHS ester of the acid reacts quickly with amino groups on proteins to form amide or peptide bonds. The reaction shown in Figure 3 can be performed in a controlled manner. Protein conjugates of aflatoxin-B_1-oxime [2], thyroxine, and progesterone prepared by this method are some good examples. Many compounds have reactive groups to which a carboxyl group can be attached, for example, by alkylation of oxygen or nitrogen substituents with halo esters, followed by ester hydrolysis [17, 18], and by formation of hemisuccinate esters [12] and carboxymethyloximes

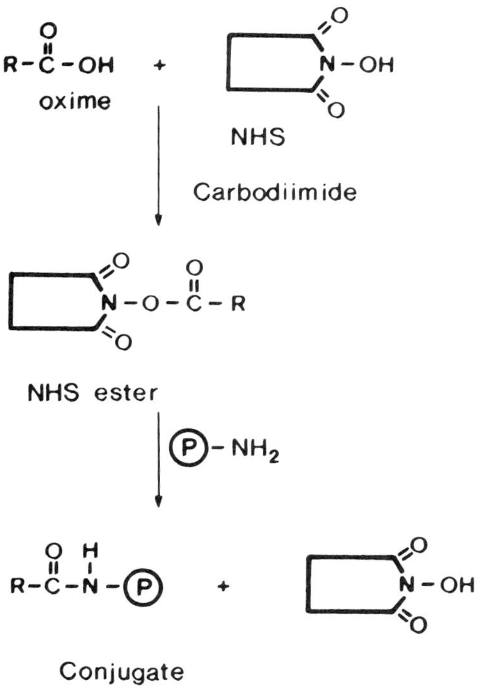

R–C(=O)–OH (oxime) + NHS

Carbodiimide

NHS ester

Ⓟ–NH_2

R–C(=O)–NH–Ⓟ (Conjugate) + NHS

Figure 3 N-hydroxysuccinimide method for conjugation of carboxyl groups-containing haptens.

[19]. Likewise, procedures are available for coupling ligands with amino groups [20,21], hydroxyl groups [22], and carboxyl groups [23]. Multiple methods are available for the production of protein-protein conjugates, for example, using glutaraldehyde as the coupling agent [24,25] or by periodate oxidation [26]. Some of these methods are outlined in Chapters 7 and 12.

Both haptens and large molecules can be coupled to proteins using homo- or heterobifunctional reagents. A general disadvantage of common cross-linking with homobifunctional reagents is a lack of control over the incorporation of protein and hapten molecules into the resulting conjugate. Some of the homobifunctional reagents used for coupling are N,N′-O-phenylenediamaleimide [27], 4,4′-di-fluoro 2,2′-dinitrophenyl sulfone [28], toluene-2,4-diisocyanate [29], benzoquinone [30], and bis-succinic acid N-hydroxysuccinimide ester [31]. Of these N,N-O-phenylenedimaleimide has been used extensively to prepare conjugates of β-galactosidase. The method is suitable for proteins contain-

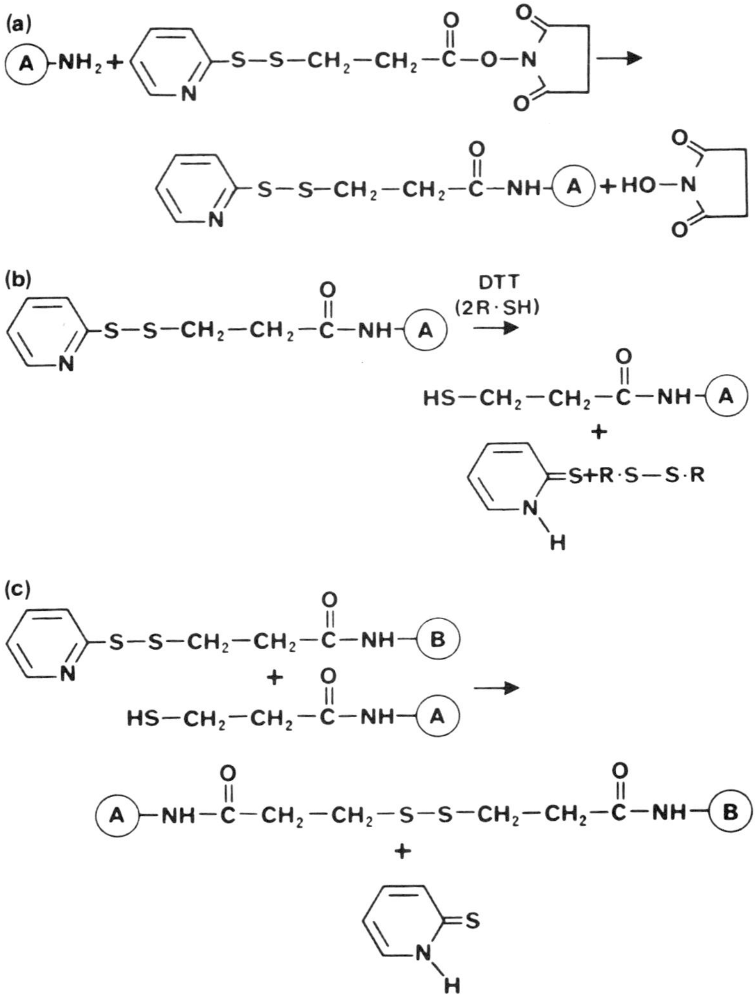

(a)

$$\text{(A)}-NH_2 + \text{(2-pyridyl)}-S-S-CH_2-CH_2-\overset{O}{\overset{\|}{C}}-O-N\text{(succinimide)} \longrightarrow$$

$$\text{(2-pyridyl)}-S-S-CH_2-CH_2-\overset{O}{\overset{\|}{C}}-NH-\text{(A)} + HO-N\text{(succinimide)}$$

(b)

$$\text{(2-pyridyl)}-S-S-CH_2-CH_2-\overset{O}{\overset{\|}{C}}-NH-\text{(A)} \xrightarrow[(2R\cdot SH)]{DTT}$$

$$HS-CH_2-CH_2-\overset{O}{\overset{\|}{C}}-NH-\text{(A)}$$

$$+$$

$$\text{(pyridine-2-thione, N-H)}=S + R\cdot S-S\cdot R$$

(c)

$$\text{(2-pyridyl)}-S-S-CH_2-CH_2-\overset{O}{\overset{\|}{C}}-NH-\text{(B)}$$

$$+ \; HS-CH_2-CH_2-\overset{O}{\overset{\|}{C}}-NH-\text{(A)} \longrightarrow$$

$$\text{(A)}-NH-\overset{O}{\overset{\|}{C}}-CH_2-CH_2-S-S-CH_2-CH_2-\overset{O}{\overset{\|}{C}}-NH-\text{(B)}$$

$$+$$

$$\text{(pyridine-2-thione, N-H)}=S$$

Figure 4 Scheme for protein:protein conjugation. (a) Introduction of 2-pyridyl sulfide structures with a protein using SPDP. (b) Thiolation of the modified protein. (c) Conjugation of proteins A and B. (From Ref. 96.)

ing sulfhydryl groups, which can also be introduced into the proteins with S-acetylmercaptosuccinic anhydride [32], 5,6,5-methylmercaptobutyrimidate [33], or disulfide reduction.

Heterobifunctional reagents [13] are becoming increasingly popular for the preparation of protein-protein and protein-hapten conjugates, as the reactions leading to conjugate formation can be performed in a controlled fashion. The heterobifunctional reagent, N-succinimidyl 3-(2-pyridyldithio)propionate (SPDP), has been used [34] to conjugate IgG to toxins [16,35], liposomes [35], tuberculin [37], and enzymes [38,39]. The conjugation scheme using SPDP is presented in Figure 4. Some examples of related heterobifunctional reagents used in conjugation are m-maleimidobenzoic acid N-hydroxysuccinimide ester (MBSE) [40] and N-(4-carboxycyclohexylmethyl)maleimide N-hydroxysuccinimide ester [41].

A. Assessment of Immunoconjugates for Therapeutic Use

The following criteria must be considered if an immunoconjugate is to be used in a clinical setting: purity, binding properties, and pharmacological activity. For the purification of protein conjugates of drugs of low molecular weight (MW), the free drug can be removed by dialysis or gel filtration on Sephadex G-10 or G-25 [42]. Purification of conjugates of two high-MW components is achieved by a protein fractionation technique using Sephadex G-150 [43] and G-200 [44], Ultrogel AcA-44 [45], ion-exchange chromatography [46], and affinity chromatography [47].

Antibodies may undergo conformational changes in their structure after conjugation, which may affect their binding characteristics and thus their efficacy as a carrier vehicle. However, many antibodies do not lose their specificity and binding characteristics after conjugation [48]. For example, of several monoclonal antibodies against carcinoembryonic antigen (CEA) [49], human osteogenic sarcoma [50], p97 emlanoma-associated antigen [51], and neuroblastoma cells [52], only the CEA antibody lost some antigen-binding affinity when conjugated to the anticancer drug vindesine (VDS).

Loss of antibody activity may also be related to the drug/antibody ratio in the conjugate. Increasing the ratio of drug on the conjugate in general decreases the antibody activity. Three different VDS-antibody ratios gave the following antibody activities relative to the unconjugated antibody: 3:1, 84%, 7:1, 76%; and 9:1, 70% [48].

The pharmacological activity of anticancer drugs in a monoclonal immunoconjugate may decrease or increase in comparison with the free drug. The antimitotic drug VDS, when conjugated to a MoAb raised against the human osteogenic sarcoma cell line, 791T, showed less toxicity on cultured 791T cells than the free drug [53]. On the other hand, the conjugate of the drug daunorubicin with melanotropin had three times more toxicity on cultures of melanoma cells than free daunorubicin [54]. In addition, the conjugate showed selective toxicity to mouse melanoma cells but not to mouse 3T3 fibroblasts, whereas unconjugated daunorubicin was equally toxic to both cell lines [55]. The enhanced selective toxicity of the conjugate may be due to the recognition of conjugate by melanotropic receptors present on melanoma cells [56] but absent from 3T3 cells as hypothesized by the authors [54].

IV. DRUG DELIVERY AND TARGETED THERAPY OF CANCER

Monoclonal antibodies may, by themselves, kill tumor target cells and may therefore serve as anticancer agents. For example, Centocor (Malvern, Penn.) has introduced monoclonals, 17-1A and CA-125, in clinical trials to treat gastrointestinal and ovarian cancers, respectively. Biotherapy Systems (BTS, Mountain View, Calif.) and Biotherapeutics, Inc. (Franklin, Tenn.), are developing monoclonals tailored to each individual's tumor. A number of complete and partial remissions have been achieved in clinical trials using this antibody approach [57].

Most researchers believe that this approach to treating cancer will not prove feasible in most cases due to several severe limitations of the antibody: (1) Cancer cells are heterogeneous, so those cells that are not recognized by the monoclonal antibody can escape and proliferate; (2) many cancer cells can interact with the antibody, rendering it harmless; (3) some tumors contain semidead cores that cannot be reached by antibodies; and (4) the need for readministration will increase the likelihood of immunogenic reactions by patients.

A. The Immunoconjugate Approach

The use of MoAbs for targeting therapeutic agents to tumors is currently a major area of interest in tumor immunology [58]. The aim is to produce conjugates of antibody and a toxic agent that will localize selectively at the tumor site, causing maximum damage to tumor cells but not to normal cells. Much attention has been focused on the use of plant or bacterial toxins such as ricin, the A chain of abrin, gelonin, and the A chain of diphtheria toxin [16,59-62]. These molecules alone are relatively nontoxic, but when combined with an antibody they become bound to target cells, followed by their internalization, which leads to death of the tumor cells. The toxic agents or drugs in the conjugates ride piggyback on monoclonal antibodies, which carry them to tumor-specific sites. Some of these systems are discussed in Chapters 8-11.

B. Homing of Liposomes

Many unwanted reactions arising from the use of drugs, enzymes, and proteins in cancer treatment or prevention of disease could be minimized by the entrapment of such agents in liposomes [63,64]. Injected liposome-entrapped agents do not come in contact with blood, and their clearance from plasma and tissue distribution is controlled by their carriers [65,66]. The endocytolic mode [67] of liposome uptake by cells warrants the entrance of agents in otherwise inaccessible cells and also provides a convenient mechanism for the release of entrapped agents by the disruption of liposomes in the lysosomal milieu [64]. However, the localization of injected liposomes mainly in the fixed macrophages of the liver and spleen restricts their use [64]. It has been suggested that direction of liposomes to alternative targets could be attained by appropriate manipulation of the liposomal surface. Liposomes have been used to deliver drug to the diseased area by using antibodies as homing probes. More relevant to this discussion are drug delivery systems in which the antibody is the actual drug to be delivered rather than being used as a "homing device" [68]. The cytotoxic drug bleomycin, used in cancer chemotherapy, was entrapped in liposomes with antibody [69]. It appeared that following its attachment to the cell surface, liposomal IgG effected the uptake of the associated liposomal moiety itself and the drug bleomycin. It also appeared that only in a limited population of such liposomes could anti-cell IgG, available on the surface, effect the association of the liposomal carrier with cells.

Investigation of the subcellular fate of liposomal bleomycin taken up by cells (HeLa) via cells-specific IgG has shown that only 20.5% of the cellular ^{111}In radioactivity is bound to the plasma membranes, the remainder, presumably interiorized through endocytosis, being recovered in the lysosome-rich particulate fraction [70].

Studies by Gregoriadis have shown that the liposome-associated antibody is able to get to its binding sites on the cell surface resulting from the internalization of complex in the cell's lysosomes, where disruption of liposomes and liberation of entrapped drug capable of reaching the cellular target occur [69,71]. Recently, liposomes have been coupled with monoclonal antibody through covalent linkage using the heterobifunctional reagent SPDP, discussed earlier in this chapter [36]. The coupling method resulted in efficient binding of protein to the liposomes without aggregation and denaturation of the coupled ligand. Further, liposomes did not leak encapsulated carboxyfluorescein as a consequence of the reaction [36]. In another application, a secondary antibody, raised against a primary antibody against a tumor antigen, was encapsulated into liposomes. Injection of the radiolabeled primary antibody into a patient followed by the second antibody in liposomes led to clearance of the double antibody-liposome complex and aided in tumor detection and localization [72].

One of the biggest advantages of using liposomes as a drug carrier is that a large amount of drug can be packed in the liposomes without losing the immunoreactivity of the linked antibody. It is anticipated that the use of antibody as a means of conferring specificity to liposomes would markedly increase their usefulness in the future (see Chapter 12 for details).

C. Toxin Immunoconjugates

Koprowski et al. [73] have isolated hybridoma clones that secrete antibody to epitopes on colorectal carcinoma cells; these antigens have not been detected on normal tissues or other tumor cell lines tested. Conjugates of MoAB 1083-17-1A (designated 17-1A) with A chain of diphtheria-toxin (DTA)-SS-(17-1A) or ricin-toxic (RTA)-SS-(17-1A) were evaluated for their binding to human cell lines.

Specificity was demonstrated by measuring conjugate binding to cell lines that lack antigen. As shown in Figure 5, (DTA)-SS-(17-1A) bound to both SW948 and SW1116 but did not bind to four other cell lines that do not express antigen. These results correlate with the known binding specificity of 17-1A antibody [74]. The higher level of binding of the conjugate to SW 948 cells relative to SW 1116 cells reflects a difference in the number of cells in the assay, not in the amount of antigen expressed by SW 948 cells.

The same conjugates were further studied in a separate experiment to see the inhibition of protein synthesis (an indicator of cell growth inhibition) of human cell lines, which provided evidence of 100% protein synthesis inhibition in colorectal carcinoma cell lines but not in other human cell lines.

Allogenic bone transplantation is an effective therapy for acute nonlymphocytic leukemia [75] and has a role in the therapy of acute lymphoma [76]. However, the number of patients treated with allogenic transplants is small and limited by the availability of compatible donors and patient age. Recently, autologous transplantation has been proposed as an alternative therapy for leukemia and lymphoma patients. In the procedure, marrow is removed from patients and incubated in vitro with drugs [77] or specific antibodies [78] to remove residual leukemia cells. The feasibility of this experiment has been demonstrated in experimental animal systems and clinical trials [78,79].

MoAbs are appropriate reagents for purging marrow of malignant cells. Preliminary trials using monoclonal antibodies and complement for autologous transplantation have recently been reported [78]. However, many monoclonals do not fix complements or require high antibody concentrations for complement-mediated cell killing [80]. Conjugates of ricin with murine monoclonal antibodies have been used successfully in animal transplantation models [62, 81]. Recently, Leonard et al. [35] synthesized the conjugates of whole ricin with the pan-T-cell monoclonal antibodies T101 and 3A1 and studied their

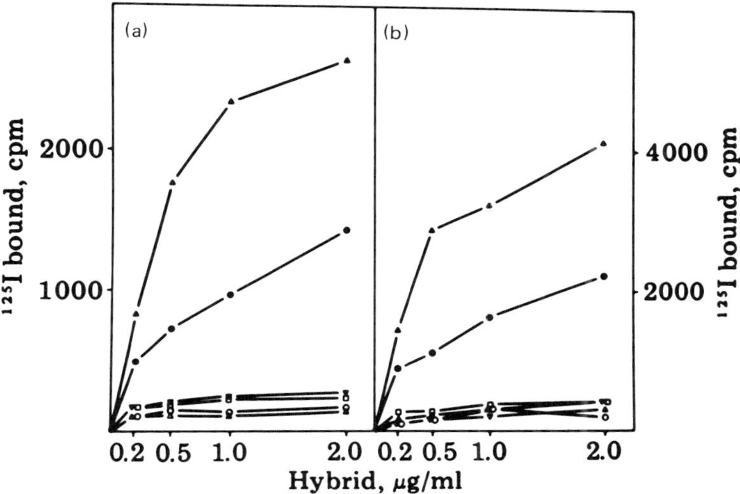

Figure 5 Specificity of conjugate binding to human cell lines: ▲, SE 948, colorectal carcinoma; ●, SW 1116, colorectal carcinoma; ▽, MRC-5, normal fibroblast; □, WM 56, melanomas; △, MBA-9812, lung carcinoma; ○, HS-0853, lung fibroblasts. (a) (DTA)-SS-(17-1A); (b) (RTA)-SS-(17-1A). (From Ref. 16.)

toxicity for cell lines, peripheral blood T lymphocytes, and normal cells and cell lines exhibited the following sensitivities to ricin: 8392 (human malignant B-cell line) < E rosette-positive lymphocytes < bone marrow progenitors < 8402 (human T ALL) < CEM (human T ALL). Ricin sensitivities correlated with ricin binding in the presence of lactose; peripheral blood T cells were resistant to 0.1-nM ricin, but a similar concentration of T101-ricin inhibited normal and malignant T-colony formation by 98%. 3A1-ricin was slightly less effective. The authors concluded that (1) normal blood cells and malignant cell lines exhibit varying degrees of ricin sensitivity in the presence of lactose; (2) T101-ricin is at least 10-fold more toxic to T lymphocytes than to bone marrow progenitor cells and is effective in mixtures of normal and malignant cells; and (3) treatment of unfiltered marrow with anti-T-cell immunotoxins should safely remove target T cells without excessively damaging normal progenitors of producing excessive free ricin [35]. Thus T101 whole-ricin immunotoxins merit clinical trials for removing malignant T cells from human bone marrow.

D. Drug Immunoconjugates

An alternative to plant or bacterial toxins is to use conventional anticancer drugs that are already used in clinical practice. Thus, doxorubicin coupled to

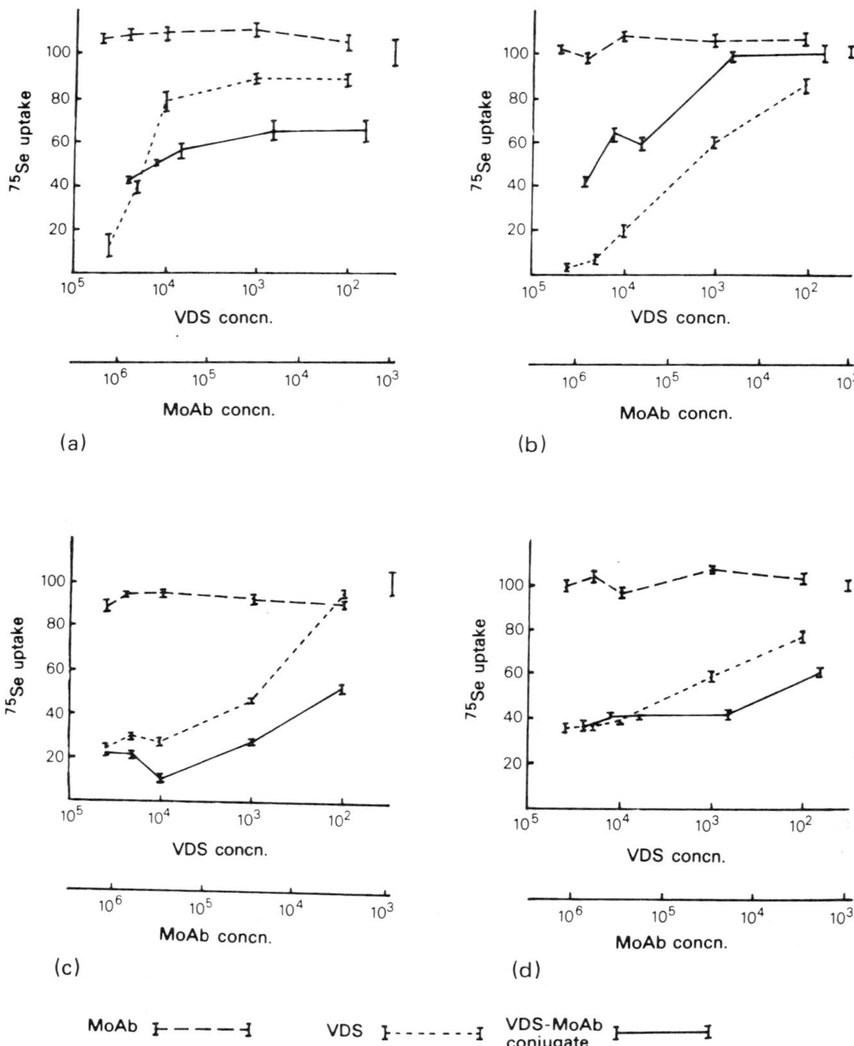

Figure 6 Relative effects of VDS. VDS-α791T/36 conjugate (VDS-MoAb) and α791T/36 (MoAb) on osteogenic sarcoma cell lines. (a) 791T target cells; (b) 788T cells; (c) 2 OS cells; (d) T278 cells. ^{75}Se update is expressed as a percentage relative to that in PBS controls. Vertical bars indicate the SE. The SE at the far right (100%) is that obtained in PBS controls. The concentrations of VDS and MoAb indicated are in nanograms per milliliter. (From Ref. 53.)

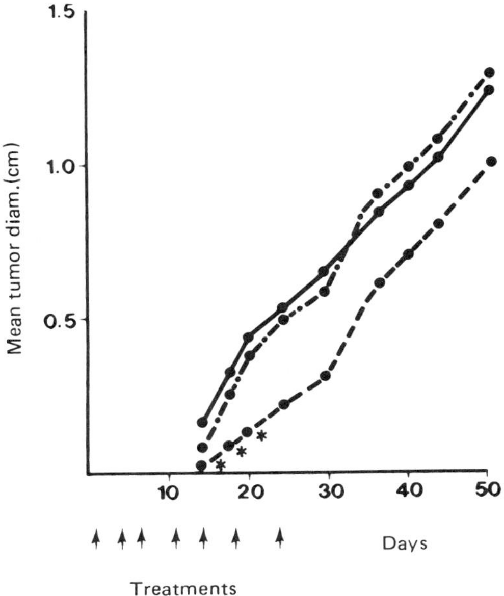

Figure 7 Effect of VDS-96.5 conjugate or antibody 96.5 alone on growth of human melanoma H2169 xenografts in groups of five athymic nu/nu mice. Mice were given tumor on day 0 and treated on days 1, 4, 7, 11, 14, 18, and 24 and PBS (arrowheads). 96.5 alone at 124 mg/kg (-●-) or VDS-96.5 at 5.1 mg/kg VDS and 124 mg/kg 96.5 (— — —). The conjugate-treated group showed a significant difference ($P < 0.02$) from the PBS group at the time points indicated (*) (Student's t test). For clarity, standard error bars are omitted. (From Ref. 15.)

MoAbs has been reported to have therapeutic effects against a rat mammary carcinoma [14].

Many drug-antibody conjugates for suppression of tumor growth have been reported [82-84]. It has been suggested [85] that linkage of cytotoxic drugs to antibody through an inert intermediate carrier offers a wide scope of improved cancer chemotherapy. Rowland et al. [85] conjugated p-phenylenediamine mustard (PDM) to an antibody against mouse lymphoma cells (EL4) through inert intermediates, e.g., polyglutamic acid (PGA). The PDM-PGA-antibody conjugate showed greater effectiveness in vivo as an antitumor agent (data not shown). The antimitotic drug VDS has been conjugated to MoAbs against various carcinomas by many investigators [15,48,53] and shown to have selective toxicity against tumor cells.

Embleton et al. [53] use a MoAb originally against the human hosteogenic sarcoma cell line 791T to study the cytotoxic effect of VDS immunoconjugates in vitro on 791T sarcoma cell lines and other antigenically cross-reactive osteogenic sarcoma cell lines, and, also, on tumor cell lines that have no detectable reaction with MoAb. Continuous exposure to cultured 791T cells indicated that the VDS was partially inactivated following conjugations, since the conjugate was less toxic than the free drug. However, MoAb binding activity was essentially preserved following conjugation. Despite diminished drug activity in the conjugate, assays designed to mimic antibody binding to tumor, in which target cells were treated with conjugate and washed before culture, showed selective cytotoxicity for four osteogenic sarcoma lines, with little or no effect on non-cross-reactive control cells (Figure 6). In the case of 791T, 2 OS, and T278, the immunoconjugate was more toxic than VDS alone, and with 788T it was less toxic, but in all cases cytotoxicity was highly significant at doses of 10 mg/ml VDS or greater ($P < 0.001$ by Student's t test). Complete cytotoxicity was not achieved, presumably because not all target cells entered mitosis during the assay. In comparison, free VDS was equally toxic to all cell lines and free antibody was nontoxic.

Conjugates of VDS with MoAbs 96.5, 11.286.14, and 14.95.55 (antimelanoma, IgG2a, anti-CEA, IgGI, and IgG2a, respectively) were evaluated by Rowland et al. [15]. Conjugate VDS-96.5 was markedly cytotoxic for cells bearing a high concentration of p97 antigen but not for cells expressing low levels of p97. Cells expressing an intermediate p97 level were moderately susceptible to the conjugate.

The effect of conjugate VDS-96.5 was also tested in vivo on xenografts of melanoma H2169 in nude mice (Figure 7). The xenograft initially grew at a lower rate in mice treated with conjugate than in mice treated with free 96.5 antibody or PBS; but from 27 days after implantation (3 days after treatment had ceased), growth rates became similar in all three groups. During the initial phase of retardation (up to 25 days after implantation), the difference between VDS-96.5-treated mice and controls was statistically significant. Antibody alone did not affect tumor growth. Similar results were obtained using VDS conjugated to anti-CEA monoclonal antibodies. However, anti-CEA MoAbs alone were able to suppress the growth of a human colorectal tumor implanted in athymicmice [15,86]. Conjugates with the drugs methotrexate, chlorambucil, antibiotics, radionuclides, and some alkylating agents have been explored [97]. Thus, there seems to exist a potential for application of immunoconjugates in targeted tumor chemotherapy and drug delivery.

V. MONOCLONAL ANTIBODIES, IMMUNOCONJUGATES, AND CANCER: SOME PROBLEMS AND PROBABLE SOLUTIONS

Although tumor therapy using monoclonal antibodies (MoAbs) and their conjugates seems to be quite promising, their use as a therapeutic agent has been hampered by many problems and unanswered questions listed below [1].

1. Most of the MoAbs developed against human tumor cells have been made in mice by immunizing the mice with human tumor cells or extracts thereof; hence, these MoAbs are mouse immunoglobulins and represent the way a mouse spleen cell sees a human tumor cell. These MoAbs may cause unwanted side effects in humans by serving as antigens and stimulating an immune response in the patient. This would make the MoAb therapy impossible. This problem may be circumvented by using human MoAbs, which are much less likely to generate an immune response. However, there are two major obstacles to this approach. First, there is no adequate, genetically marked human myeloma cell line to fuse with immune cells producing MoAb. This makes it very difficult to find a MoAb-producing cell after fusion between the immune cells and the parental myeloma cell line has been performed. Second, obtaining immune cells has proved difficult [87-89].

2. MoAbs in direct human therapy may have the potential for generating antigen-antibody complexes. Such complexes have long been known to damage the kidneys, as well as to act as antigens themselves. In a recent study, the induction of human anti-mouse IgG antibody by repeated administration of mouse antibody has been circumvented by the construction of chimeric antibody. Since human antibodies of appropriate specificity are not available, chimeric antibodies with variable regions identical to those of mouse hybridoma and human constant regions may provide antibodies of appropriate specificity that are less immunogenic that the complete mouse antibodies. Furthermore, the ability to genetically engineer changes in the DNA segments enables one to produce antibody molecules with "tailor-made" effector functions [90-92].

Following the above approach, Ghrayeb [90] recently described the construction of immunoglobulin genes in which the DNA segments encoding the variable regions from the heavy and light chains of the mouse MoAb 17-1A were joined to the DNA segments encoding human γ^3 and κ constant regions. The transfection of expression vectors containing these chimeric Ig genes into mouse myeloma cells resulted in the production of functional chimeric IgG with the same binding specificity as the original hybridoma antibody. The early results were very promising using the chimeric IgG.

3. For both drugs and toxins, a major problem may be the need for the antibody to attach and translate the agent into each cell. Antigen modulation may play a role as a mechanism by which this translation can occur [87]. Where antigenic heterogeneity exists, these conjugates may be limited by heterogeneity. In

contrast, radioisotopes will have a certain "field" effect and may circumvent the problem of heterogeneity. Unfortunately, this field effect and toxicity, without intracellular translation, will cause greater toxicity to normal organs where the antibody conjugates might be retained, generally in liver and spleen. An additional feature of isotope-labeled antibody is the potential for imaging in addition to therapy [93,94].

4. Each immunoconjugate has the potential for enhanced therapeutic specificity given the conjugation to the antibody molecule and its inherent specificity for the antigenic site on the tumor cell. However, many problems remain to be defined as to the class of immunoglobulin, the purification of antibody, the route and schedule of administration, and the use of immunoconjugates as cancer-specific reagents.

5. Most of the work on the use of immunoconjugates has been done using in vitro systems. The results of in vivo studies are limited and do not warrant definitive conclusions.

6. Worst of all, an immunoconjugate effective for one type of cancer in one patient may not be active against similar cancers in other patients. This means that it may be necessary to raise monoclonal antibody against tumor from each patient to prepare an effective immunoconjugate. Tailoring of MoAb to individual lymphoma patients would present additional practical difficulties if additional work bears out the initial success. In view of this problem, a few companies are producing custom-tailored MoAbs to each patient's lymphoma.

7. The goal is to use an antibody that recognizes a tumor-specific antigen that distinguishes absolutely between tumor cells and normal cells; however, with the notable exception of idiotypes on B-cell tumors, no such tumor-specific antigen has yet been found. Rather, a number of tumor-related antigens have been defined that are more or less restricted to tumor cells and their tissue of origin.

8. Binding of some antibodies to cells causes the target antigen to disappear from the cell surface. This antigenic modulation may decrease the effectiveness of repeat loss of antibody [87,88]. However, not all antibodies have this effect.

9. The human pharmacology on antibody-drug conjugate has not been studied at all. Carefully planned clinical trials are needed to learn how to best administer drug-antibody conjugates with the goal of improving drug delivery over the conventional therapy with drug alone [98].

The early reports of tumor therapy with monoclonal immunoconjugates are promising [99]. The improvement in in vitro immunization of human lymphocytes, in constructing chimeric antibodies by genetic engineering techniques, and in finding optimal conditions for producing human-human hybrids and, especially, the promise of transfected stable cell lines and of better conjugate synthesis give hope that MoAbs and their immunoconjugates can be used as therapeutic agents in the future.

VI. ACKNOWLEDGMENT

The authors are grateful to Sue Kendrick for typing the manuscript.

VII. REFERENCES

1. B. P. Ram and P. Tyle, *Pharm. Res. 4*(3):181–188 (1987).
2. B. P. Ram, L. P. Hart, O. L. Shotwell, and J. J. Pestka, *J. Assoc. Offic. Anal. Chem. 69*:904–907 (1986); B. P. Ram, L. P. Hart, R. J. Cole, and J. J. Pestka, *J. Food Prot. 49*:792–795 (1986); R. Warner, B. P. Ram, L. P. Hart, and J. J. Pestka, *J. Agr. Food Chem. 34*:714–717 (1986).
3. D. Dixon, L. P. Hart, B. P. Ram, and J. J. Pestka, *J. Agr. Food Chem. 35*: 122–126 (1987).
4. M. T. Liu, B. P. Ram, L. P. Hart, and J. J. Pestka, *Appl. Environ. Microbiol. 50*:332–336 (1985).
5. G. Kohler and C. Milstein, *Nature 256*:494–497 (1975).
6. R. T. Taggart and I. M. Samloff, *Science 219*:1228–1230 (1983).
7. M. Andreeff, A. Bartal, C. Feit, and Y. Hirshaut, *Hybridoma 4*:277–287 (1985).
8. C. L. Reading, *J. Immunol. Methods 53*:261–291 (1982).
9. S. D. Wolpe, in *Mammalian Cell Culture: The Use of Serum Free Hormone-Supplemental Media*, Plenum Press, New York, 1984, p. 103.
10. T. T. Ngo and H. M. Lenhoff, in *Enzyme Mediated Immunoassay*, Plenum Press, New York, 1983.
11. I. Linnoila and P. Petrusz, *Environ. Health Perspect. 56*:131–148 (1984).
12. B. F. Erlander, *Pharmacol. Rev. 25*:271–280 (1973).
13. E. Ishikawa, *J. Immunoassay 4*:209–327 (1983).
14. M. V. Pimm, J. A. Jones, M. R. Price, J. G. Middle, M. J. Embleton, and R. B. Baldwin, *Cancer Immunol. Immunother. 12*:125–129 (1982).
15. G. F. Rowland, C. A. Axton, R. W. Baldwin, J. P. Brown, J. R. F. Corvalan, M. J. Embleton, V. A. Gore, I. Hellstrom, K. E. Hellstrom, E. Jacobs, C. H. Marsden, and M. V. Pimm, *Cancer Immunol. Immunother. 19*:1–7 (1985).
16. D. G. Gilliland, Z. Steplewiski, R. J. Collier, K. F. Mitchell, T. H. Chang, and H. Koprowski, *Proc. Natl. Acad. Sci. USA 77*:4539–4543 (1980).
17. S. Spectro and C. W. Parker, *Science 168*:1347–1348 (1970).
18. J. W. A. Findlay, R. F. Butz, and R. M. Welch, *Res. Commun. Chem. Pathol. Pharmacol. 17*:595–603 (1977).
19. B. F. Erlanger, F. Borek, S. M. Beiser, and S. Lieberman, *J. Biol. Chem. 228*:713–727 (1957).
20. F. A. Anderer, *Biochem. Biophys. Acta 71*:246–248 (1963).
21. K. Landsteiner, in *The Specificity of Serological Reactions*, Harvard University Press, Cambridge, Mass., 1945.
22. R. Bredehorst, A. M. Ferro, and H. Hilz, *Eur. J. Biochem. 82*:105–113 (1978).

23. R. J. Warren and K. Fotherby, *J. Endocrinol. 62*:605–618 (1974).
24. S. Avrameas, *Immunochemistry 6*:43–52 (1969).
25. S. Avrameas and T. Ternynck, *Immunochemistry 8*:1175–1179 (1971).
26. A. Tsuji, J. Maeda, H. Arakawa, K. Matsuoka, N. Kato, H. Naruse, and M. Irie, in *Enzyme Labelled Immunoassays of Hormones and Drugs* (S. B. Pal, ed.), Walter de Gruyter, Berlin, 1978, p. 327.
27. K. Kato, Y. Hamaguchi, H. Fukui, and E. I. Ishikawa, *J. Biochem.* (Tokyo) *78*:235–237 (1975).
28. P. K. Nakane and G. B. Pierce, Jr., *J. Histochem. Cytochem. 14*:929–931 (1967).
29. A. Schick and S. J. Singer, *J. Biol. Chem. 236*:2477–2485 (1961).
30. T. Ternynck and S. Avrameas, *Immunochemistry 14*:767–774 (1977).
31. J. W. Ryan, A. R. Day, D. R. Schultz, U. S. Ryan, A. Chung, D. I. Marborough, and F. E. Dorer, *Tissue Cell 8*:111–124 (1976).
32. I. M. Klotz and R. E. Heiney, *Arch. Biochem. Biophys. 96*:605–612 (1962).
33. E. Gnemmi, M. J. O'Sullivan, G. Chieregatti, M. Simmons, A. Simmond, J. W. Bridges, and V. Marks, in *Enzyme Labelled Immunoassay of Hormones and Drugs* (S. D. Pal, ed.), Walter de Gruyter, Berlin, 1978, p. 29.
34. J. Carlsson, H. Drevin, and R. Axen, *Biochem. J. 173*:723–737 (1978).
35. J. E. Leonard, R. Taetle, D. To., and K. Rhyner, *Blood 65*:1149–1157 (1985).
36. L. D. Leserman, J. Barket, F. Kourilsky, and J. N. Weinstein, *Nature 288*: 602–604 (1980).
37. P. J. Lachmann, A. Vyakarnam, and K. Sikora, *Immunology 42*:329–336 (1981).
38. D. Pain and A. Surolia, *J. Immunol. Methods 40*:219–230 (1981).
39. P. Nilsson, N. R. Berquist, and M. S. Grundy, *J. Immunol. Methods 41*: 81–93 (1981).
40. T. Kitigawa and T. Aikawa, *J. Biochem. 79*:233–236 (1976).
41. S. Yoshitake, Y. Yamada, E. Ishikawa, and R. Masseyeff, *Eur. J. Biochem. 101*:395–399 (1979).
42. B. G. Joyce, G. R. Read, and D. R. Fahmey, *Steroids 29*:761–770 (1977).
43. W. H. Stimson and J. M. Sinclair, *FEBS Lett. 47*:190–192 (1974).
44. D. M. Boorsma and G. L. Kalsbeck, *J. Histochem. Cytochem. 23*:200–207 (1975).
45. D. M. Boorma and J. G. Steefkerk, *J. Histochem. Cytochem. 24*:481–486 (1976).
46. S. Yamashita, N. Yamamoto, and K. Yasuda, *Acta Histochem. Cytochem. 9*:227–233 (1976).
47. P. D. G. Dean, W. S. Johnson, and F. A. Moddle, *Affinity Chromatography*, IRL Press, Oxford, Washington, D.C., 1985.
48. G. F. Rowland, R. G. Simmonds, J. R. F. Corvalan, R. W. Baldwin, J. P. Brown, J. J. Embleton, C. H. J. Ford, K. E. Hellstrom, I. Hellstrom, J. T. Kemshead, C. E. Newman, and C. S. Woodhouse, *Protides Biol. Fluids 30*: 375–379 (1983).

49. J. R. F. Corvalan, C. A. Axton, D. R. Brandon, W. Smith, and C. S. Woodhouse, *Protides Biol. Fluids 31*:921-924 (1984).
50. M. J. Embleton, B. Gunn, V. S. Byers, and R. W. Baldwin, *Br. J. Cancer 43*: 582-587 (1981).
51. J. P. Brown, K. Nishiyama, I. Hellstrom, and K. E. Hellstrom, *J. Immunol. 127*:539-546 (1981).
52. J. T. Kemshead, J. Fritschy, V. Asser, R. Sutherland, and M. F. Greaves, *Hybridoma 1*:109-123 (1982).
53. M. J. Embleton, G. F. Rowland, R. G. Simmonds, E. Jacobs, C. H. Marsden, and R. W. Baldwin, *Br. J. Cancer 47*:49-49 (1983).
54. J. M. Varga, N. Asato, S. Lande, and A. V. Sterner, *Nature 267*:56-58 (1977).
55. E. Hurwitz, *Cancer Res. 35*:1175-1181 (1975).
56. J. M. Varga, A. DiPasquale, J. Pawelek, J. S. McGuire, and A. B. Lerner, *Proc. Natl. Acad. Sci. USA 71*:1590-1593 (1974).
57. A. Klausner, *Biotechnology 4*:185-194 (1986).
58. R. W. Baldwin, M. J. Embleton, and M. R. Price, *Mol. Aspects Med. 4*:329-368 (1981).
59. H. E. Blythman, P. Casellas, O. Gros, P. Gros, F. K. Jansen, F. Paolucci, B. Pau, and H. Vidal, *Nature 290*:145-146 (1981).
60. K. A. Krolick, C. Villemez, P. Isakson, J. W. Uhr, and E. S. Vitetta, *Proc. Natl. Acad. Sci. USA 77*:5419-5423 (1980).
61. I. S. Trowbridge and D. L. Domingo, *Nature 294*:171-173 (1981).
62. P. E. Thorpe and W. C. J. Ross, *Immunol. Rev. 62*:119-158 (1982).
63. A. C. Allison and G. Gregoriadis, *Nature 252*:252 (1974).
64. G. Gregoriadis, in *Enzyme Therapy of Lysomal Storage Diseases* (J. M. Tayloir, G. J. M. Hooghwinkel, and W. Thdaems, eds.), North-Holland, Amsterdam, Oxford, 1974, pp. 131-148.
65. I. R. McDougal, J. K. Dunnick, M. C. McNamee, and J. B. Kriss, *Proc. Natl. Acad. Sci. USA 71*:3487-3491 (1974).
66. R. J. Juliano and D. Stamp, *Biochem. Biophys. Res. Commun. 63*:651-658 (1975).
67. Y. E. Rahman and B. J. Wright, *J. Cell Biol. 65*:112-122 (1975).
68. A. L. Werner, J. B. Cannon, and P. Tyle, in *Topics in Controlled Relese Science* (M. Rosoff, ed.), VCH, Deerfield Beach, Fla., 1989, pp. 217-253.
69. G. Gregoriadis and E. D. Neerjunjun, *Biochem. Biophys. Res. Commun., 65*:537-544 (1975).
70. E. J. Eylar and A. Hagopian, in *Methods in Enzymology* (W. B. Jacoby, ed.), Academic Press, New York, 1971, pp. 123-130.
71. G. Gregoriadis, in *Drug Carriers in Biology and Medicine*, Academic Press, London, 1979.
72. G. M. Barratt, B. E. Ryman, K. A. Chester, and R. H. Begent, *Biochem. Soc. Trans. 12*:348-349 (1984).
73. H. Koprowski, Z. Steplewski, K. F. Mitchell, M. Herlyn, D. Herlyn, and J. P. Fuhrer, *Somat. Cell. Genet. 5*:957-972 (1979).

74. M. Herlyn, Z. Steplewski, D. Herlyn, and H. Koprowski, *Proc. Natl. Acad. Sci. USA 76*:1439-1442 (1979).
75. F. E. Zwaan and J. Jansen, *Semin. Hematol. 31*:36-42 (1984).
76. F. R. Appelbaum and E. D. Thomas, *J. Clin. Oncol. 1*:440-447 (1983).
77. G. W. Santos and H. Kaizer, *Semin. Hematol. 19*:227-239 (1982).
78. J. Ritz, R. C. Bast, L. A. Clavell, T. Hercend, S. E. Sallan, J. M. Lipton, M. Feeney, D. G. Nathan, and S. F. Schlossman, *Lancet 2*:60-63 (1982).
79. M. Feeney, R. C. Knapp, J. S. Greenberger, and R. C. Bast, *Cancer Res. 41*: 3331-3335 (1981).
80. R. Taetle, D. To, A. Caviles, S. W. Norby, and J. Mendelsohn, *Blood 61*: 548-555 (1983).
81. K. A. Krolick, J. Uhr, and E. Vitetta, *Nature 294*:604-605 (1982).
82. T. Ghose and S. P. Nigam, *Cancer 29*:1398-1400 (1972).
83. I. Flechner, *Eur. J. Cancer 9*:741-745 (1973).
84. M. Szekerke, R. Wade, and M. E. Whisson, *Neoplasm 19*:199-209 (1972).
85. G. F. Rowland, G. J. O'Neill, and I. D. A. L. Davies, *Nature 255*:487-488 (1975).
86. D. Herlyn and H. Koprowski, *Proc. Natl. Acad. Sci. USA 79*:4761-4765 (1982).
87. R. K. Oldman, *J. Clin. Oncol. 1*:582-590 (1983).
88. R. K. Oldham, *Cancer Treat. Rep. 68*:221-232 (1984).
89. R. Levy and R. A. Miller, *Fed. Proc. 42*:2650-2656 (1983).
90. J. Ghrayeb, *World Biotech. Rep. 2*:15-19 (1986).
91. S. L. Morrison, *Science 229*:1202-1207 (1985).
92. M. S. Neuberger, G. T. Williams, and R. O. Fox, *Nature 312*:604-608 (1984).
93. D. M. Goldberg and F. H. DeLand, *J. Biol. Resp. Modif. 1*:121-136 (1982).
94. S. M. Larson, J. P. Brown, and D. W. Wright, *J. Nucl. Med. 24*:123-129 (1983).
95. J. W. Goding, in *Monoclonal Antibodies: Principles and Practice*, Academic Press, London, 1983, p. 36.
96. Pharmacia Fine Chemicals, SPDP Heterobifunctional Reagent Brochure, Pharmacia, Sweden, 1978.
97. D. C. Edwards and D. P. McIntosh, in *Methods of Drug Delivery* (G. M. Ihler, ed.), Pergamon Press, Oxford, 1986.
98. K. E. Hellstrom, I. Hellstrom, and G. E. Goodman, in *Controlled Drug Delivery* (J. R. Robinson and V. H. L. Lee, eds.), Marcel Dekker, New York, 1987, pp. 623-653.
99. J. D. Rodwell, *Antibody-Mediated Delivery Systems*, Marcel Dekker, New York, 1988.

part two
MONOCLONAL ANTIBODIES

2

Evaluation of Hollow-Fiber Bioreactor Systems for Large-Scale Production of Murine Monoclonal Antibodies

THOMAS L. EVANS and RICHARD A. MILLER
IDEC Pharmaceuticals Corporation, Mountain View, California

I. INTRODUCTION

Recently, considerable emphasis has been placed on the development of large-scale cell culture technology for the manufacturing of products derived from mammalian cells [1,2]. Driving this technology has been the need to produce monoclonal antibodies and other therapeutic proteins in genetically modified mammalian cells. Production of these proteins by bacterial or yeast systems is desirable because mammalian cells require complex and expensive media, are fragile and thus shear-sensitive, and have population doubling times that are often in excess of 24 hours. Moreover, although a few of these proteins (e.g., insulin) can be produced by fermentation of microorganisms, the majority require the complex protein folding and other post-translational processing available only through mammalian cell production methods.

Historically, mammalian cell bioreactors have been adapted from microbial fermentors. Generally these aerobic fermentations have certain basic minimum requirements: a method of heat and mass transfer, usually by aeration or agitation; sufficient nutrient supply and waste removal by batch, fed-batch or continuous-feed systems; and the ability to sample and monitor on line without the risk of contamination [3]. Because mammalian cell lines exhibit a wide heterogeneity of characteristics (i.e., growth rate, secretion rate, optimal pH, carbon and energy substrate utilization rates) a single, optimal technology for all cultures is not feasible. Since mammalian cells also have a prefered mode of growth, either in suspension or attached to surfaces, a deep-tank bioreactor would be appropriate for suspension cultures such as hybridomas and a packed-bed design more correct for anchorage-dependent cell lines such as CHO cells [4]. Cell culture methods that are currently employed include but are not limited to: suspension [5], perfusion reactors [6,7], encapsulation in semipermeable vessicles [8], solid-phase cell immobilization [9,10], and hollow-fiber systems [11,12]. While advantages and disadvantages have been found for each system, no technology has been found to be superior. In fact, the system of choice is highly dependent on the cell line and both the desired characteristics and amounts of the final product.

In developing a strategy for the manufacture of therapeutic products from mammalian cells, one must consider the amount and composition of media, cost of media and serum, concentration and purity of harvested crude product, nature of contaminants, product recovery and final purity, labor and cost of production method, and quality assurance/quality control issues.

We have produced murine monoclonal antibodies for human therapeutic use using hollow-fiber bioreactors (HFBRs). Patients with B-cell lymphomas have been treated with hybridoma-derived monoclonal antibodies reactive with the immunoglobulin idiotype (id) expressed on the surface of their malignant cells. Since the id expressed by each tumor is unique, a custom-made hybridoma was developed that produced an antibody (Ab) specific for each patient. These antibodies, called anti-idiotypes, are essentially tumor-specific and have been reported to produce significant antitumor responses [13].

The production of large quantities of different monoclonal antibodies has presented us a unique cell culture scale-up problem. A system was needed that was highly adaptable to the requirements of each hybridoma cell line developed, efficient, and of sufficient capacity. The HFBR system described in this chapter has enabled us to produce up to 15 g of anti-idiotype antibody from each of 47 different hybridomas. Several advantages of HFBR production of antibodies include high cell densities ($>10^8$ cells/ml), highly concentrated Ab in the harvested supernatants (>1 mg/ml), high purity of secreted products in the harvest medium (up to 85%), and a simple and relatively inexpensive operation. The development of the processes necessary for producing patient-specific antibodies required optimization of the HFBR system. The results of our experience preparing antibodies using two different HFBR systems are presented.

II. MATERIALS AND METHODS

A. Cell Lines

Mouse X mouse and mouse X rat hybridoma cell lines were developed in our laboratories. These lines were derived by fusing $SP^{2/0}$ mouse myeloma cells with splenic lymphocytes of either mouse or rat origin [14]. All cell lines were grown in seed cultures of Iscoves' Modified Dulbeccos Medium (IMDM, J. R. Scientific, Woodland, Calif.) containing 4.5 mg/ml glucose with 4% fetal calf serum (FCS; J. R. Scientific) and 3% horse serum (HS: J. R. Scientific). Prior to cell inoculation into bioreactors, all lines were screened for and found free of mycoplasma, ecotropic and xenotropic murine leukemia viruses and were negative in the Mouse or Rat Antibody Production Test.

B. Bioreactor Setup

The bulk of our investigation has been done using a semibatch-feed system. More recently we have begun producing our Ab in a process-controlled/continuous-feed system and in a novel "2x" HFBR.

Batch-Feed System

Two different models of HFBRs were used (CD Medical, Inc., Miami Lakes, Fla.): the series 1900 (1x), which has a surface area of 19 ft^2 and an extracapillary space (ECS) of 125 ml; or the series 3500 (2x), which has a surface area

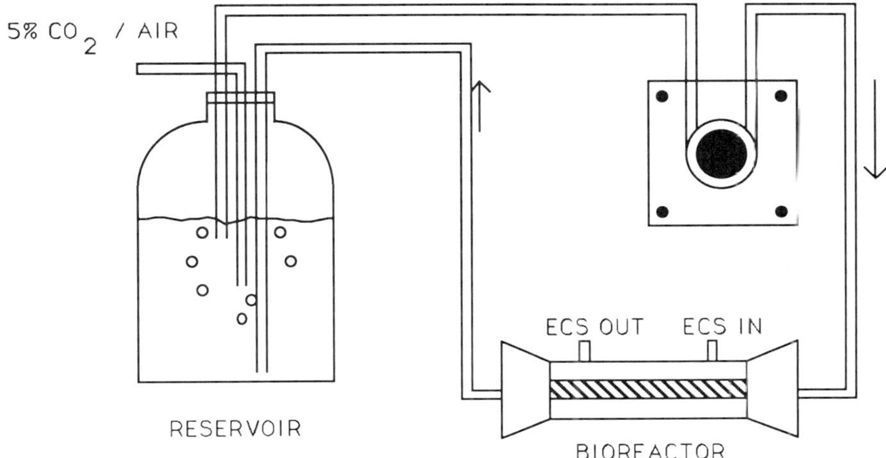

Figure 1 Schematic diagram of hollow-fiber bioreactor setup in the semi-batch-feed system.

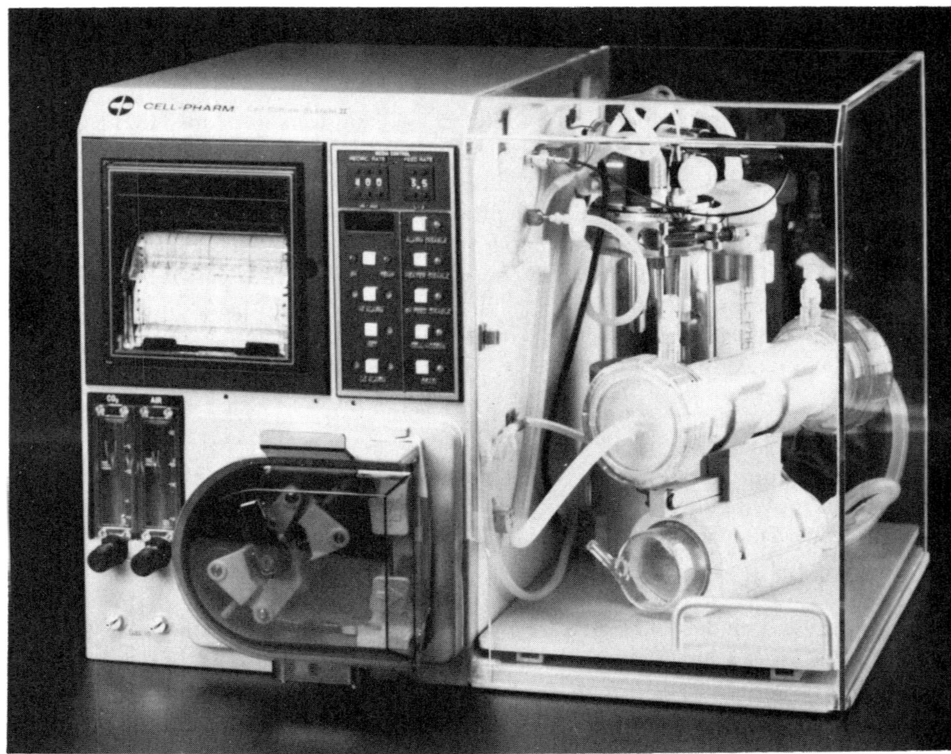

Figure 2 Continuous-feed Cell-Pharm cell culture system II with series 3500 HFBR. (Courtesy of G. Fleig, CD Medical, Inc., Miami Lakes, Fla.)

35 ft^2 and an ECS volume of 250 ml. Cell-Pharm hollow-fiber bioreactors (CD Medical, Inc.) with capillary membrane molecular weight cutoffs (MWCO) of 10,000 (10K), 30,000 (30K), and 70,000 (70K) daltons were assembled in a sterile manner in a laminar-flow hood. Silicone tubing (0.25-in. I.D.) was attached to the inlet and outlet lines of a 10-liter media reservoir flask (Bellco Glass, Vineland, N.J.) containing IMDM plus HS (Figure 1). The assembled unit was placed in a 37°C warm room and medium was recirculated through the hollow fibers at a rate of 200–400 ml/min using a peristaltic pump (Cole Parmer, Chicago). A humidified 5% CO_2-in-air mixture was passed through a 0.2-μm filter and bubbled into each reservoir. The extra capillary space was filled with approximately 125 ml (series 1900) or 250 ml (series 3500) of IMDM containing FCS. Cells in 20-30 ml of ECS medium were inoculated into the ECS and flushed down the inlet port by an additional 6–10 ml of ECS fluid. In general,

the entire 10-liter reservoir media flask was changed every 3–4 days during the production phase. Bulk Ab in spent ECS fluid was collected in sterile conical vials while instilling 60–100 ml of fresh medium into the ECS. The harvested Ab was separated from the cellular components of the supernatant by low-speed centrifugation (200 rpm) and further clarified by a high-speed centrifugation (4500 rpm) and then stored at -20°C until purified.

Continuous-Feed System

The series 3500 HFBRs were interfaced with the Cell-Pharm System II (CD Medical, Inc.) cell culture apparatus (Figure 2). Reservoir medium was recirculated at a rate of 400–600 ml/min. Humidified and filtered CO_2 and air were passed through a Cell-Pharm Oxygenator (OXY 10; CD Medical, Inc.) in line with the HFBR. Temperature was maintained at 37°C and the monitored pH was adjusted by altering the CO_2/air mixture. Feed rates were altered both to meet cellular demands as bioreactors became confluent with cells and to maintain glucose levels above 2 mg/ml. Ab was harvested as described above.

C. Assays

Cells were counted in a hemocytometer using Trypan blue dye exclusion for evaluation of viability and density. Glucose, ammonia, and lactate levels were measured by standard enzymatic assays (Sigma Chemical, St. Louis) or by a Glucose/L-lactate analyzer (YSI Instruments, Yellow Springs, Ohio).

ECS or reservoir media samples were assayed for Ab concentration using an ELISA technique. In brief, 96 well microtiter plates were coated with either purified goat antiserum to mouse, rat, or horse IgG. After washing the wells three times, purified isotype-specific IgG standards, HFBR supernatant samples, or spent reservoir medium (100 μl each) were added to the wells and serially diluted. After incubation at room temperature for 45–60 min, the plates were washed as above and 50–60 μl of horseradish peroxidase-conjugated goal anti-species-specific IgG (1:1000 dilution) was added. After an additional incubation period of 45–60 min, the plates were washed and 150 μl of ABTS substrate solution was added. Optical density at 480 nm was measured using an automatic ELISA reader.

D. Ab Purity in Harvested Supernatants

The purity of harvested Ab was determined by SDS-PAGE analysis or by fast-protein liquid chromatography (FPLC) separation.

SDS-PAGE

In general, 2.5 μg total protein was run on a 12.5% SDS polyacrylamide gel at a steady-state current of 40 mA for 3.5 h. The gel was stained with Coomassie

Blue, then destained in a mixture of acetic acid, ethanol, and water (7.5:5.0: 87.5). Protein bands were analyzed by scanning densitometry (Hoeffer Instruments, San Francisco) to obtain antibody purity levels.

FPLC Separation

0.5 ml of bulk ECS supernatant from harvested HFBRs was loaded onto an ion-exchange column (Pharmacia, Uppsala, Sweden) at a flow rate of 2 ml/min. Protein separation was performed using a step salt gradient at constant pH. Detection of protein elution was done by absorbance at 280 nm and recorded at 0.3 cm/min.

E. Suspension Culture

5×10^7 viable cells were added to an 11-spinner flask (Bellco Glass) in IMDM plus 10% FCS. The flask was stirred using a magnetic stirrer at low rpm, sparged with 5% CO_2 in air, and maintained at 37°C. For static culture studies, 2.5×10^5 viable cells were inoculated into each of three 25-ml tissue culture flasks containing 5 ml of IMDM plus 10% FCS. All flasks were maintained in a humidified, 37°C, 5% CO_2-in-air incubator. Cell viability, density, and antibody concentration were assayed daily for 6 days as described above.

III. RESULTS

A. Monoclonal Antibody Production Using HFBRs

Thirty mouse × mouse and three rat × mouse hybridoma monoclonal antibodies (MoAbs) have been produced using the series 1900 HFBRs in the batch system (Table 1). In every case but one, clones produced Ab in bioreactors for up to 6 months without evidence of hybridoma instability. Ab productivity in the bioreactors correlated with both the secretion of MoAb by the clone in a static tissue culture flask and the hollow fiber membrane MWCO (Figure 3). The greatest correlation between Ab production in HFBRs and that of static cultures was found in HFBRs with MWCO of 30K and 70K. HFBRs with 10K MWCO produced less MAb than the 30K or the 70K bioreactors.

The rate of MoAb production for the HFBRs is shown in Table 1. Between 0.80 and 2.20 g of Ab was harvested from the 10K bioreactors per month, compared to 0.92 to 3.51 g of Ab per month for the 70K bioreactors. As expected, hybridomas with high secretion rates produced more Ab for each group of HFBR studied. Moreover, within each secretion group, bioreactor productivity was increased with increasing MWCO as a result of enhanced mass transfer.

Table 1 Summary of Antibody Production Using 1x HFBRs in a Semibatch-Feed System

Clone species[a]	No. of clones	Static culture production (μg/ml)	MWCO[b]	gAb/BR/mo[c]
Ms X Ms	12	≤30	10K	0.80
			30K	0.81
			70K	0.92
	11	31-60	10K	0.99
			30K	1.21
			70K	NT
	4	61-100	10K	0.94
			30K	1.09
			70K	2.29
	3	>100	10K	2.20
			30K	3.72
			70K	3.02
Rat X Ms	3	≥70	10K	1.39
			30K	1.57
			70K	3.51

[a]Ms = mouse.
[b]Molecular weight cutoff of capillary membrane (K = 1000).
[c]BR = bioreactor.

B. Effect of Cell Inoculum

The kinetics of MoAb production in each of two series 1900 HFBRs inoculated with 1×10^8 or 1×10^9 cells is shown in Figure 4. The HFBR that received the larger cell inoculum had a shorter lag phase (approximately 7 days) and had more MoAb productivity at each time point studied. The HFBR receiving the lower cell inoculum needed a longer amount of time to reach maturity, but eventually produced levels of Ab comparable to the reactor that received the larger cell inoculum. These experiments have been repeated with similar results of longer lag periods and productivity using inoculums as small as 1×10^6 and 1×10^7 cells.

C. Consumption of Reservoir Media

The average consumption of reservoir medium was evaluated for each type of series 1900 HFBR. As shown in Table 2, 95.3 liters of medium per gram of MoAb produced (liters/g) was consumed by the 10K bioreactors compared to 26.4 liters/g for the 70K bioreactors. The concentration of MoAb in the extra-

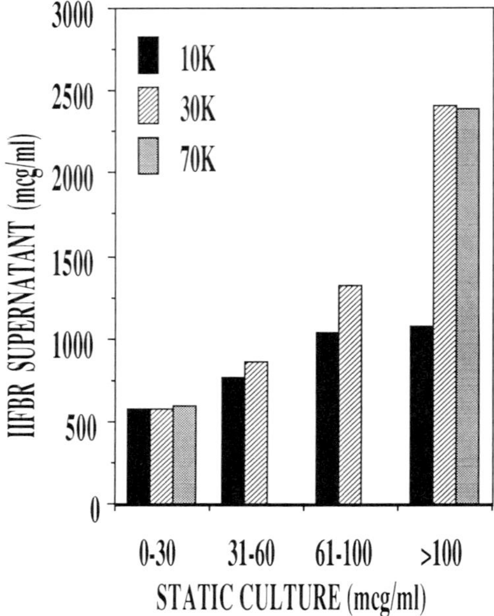

Figure 3 Correlation of Ab production in static culture for 33 different hybridomas with productivity in HFBRs of 10K, 30K, and 70K MWCO.

capillary fluid ranged from 150 μg/ml to 5000 μg/ml. Based on total media usage, the MoAb productivity was 10.5 μg/ml to 37.9 μg/ml.

Figure 4 shows a representative example of glucose utilization in 10K and 30K HFBRs inoculated with 1×10^8 or 1×10^9 hybridoma cells. In both cases, glucose consumption rapidly increased to a peak level of 6–7 g/day by days 20–30. A plateau in glucose use is reached at the time of bioreactor maturity and is maintained during MoAb production.

Ammonia and lactic acid production have been measured in the reservoir medium recirculating through the HFBRs. Fresh media contained an average ammonia concentration of 0.24 ± 0.01 mM/liter. Spent medium was found to have ammonia concentrations of 1.73 ± 0.16 mM/liter. The molar ratios of lactic acid production to glucose consumption have ranged from 0.98 to 1.23. The pH of spent reservoir medium has ranged from 6.77 to 7.23 during the production phase. Neither ammonia, lactic acid, nor pH was found to correlate with MoAb production or inhibition of secretion.

D. Effects of Serum Concentration on MoAb Production

The effect of serum source and concentration on MoAb production in the series 1900 HFBRs was evaluated (Table 3). The content of serum in the reservoir

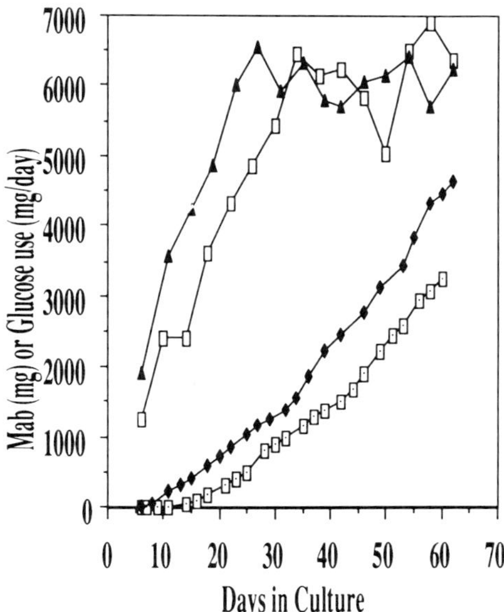

Figure 4 Kinetics of MoAb production and glucose consumption for HFBRs inoculated with 10^8 or 10^9 cells. Cumulative Ab (mg) production for 10^8 (- □ -) and 10^9 (- ◆ -) cells. Glucose consumption (mg/day) for 10^3 (- □ -) and 10^9 (- ▲ -) cells.

Table 2 Amount of Medium Required to Produce 1 g Antibody in 1x HFBRs

HFBR MWCO[a]	Number of clones	Number of bioreactors	liter/g bulk Ab
10K	34	109	95.3
10K[b]	3	4	42.4
30K	34	98	89.9
40K	2	5	43.4
55K	2	2	42.2
70K	3	6	26.4

[a]Molecular weight cutoff HFBR capillary membrane (K = 1000).
[b]Cell-Pharm bioreactor model 3510, which has surface area two times standard 10K reactor (model 1910).

Table 3 Effect of Decreasing Serum Concentrations on Antibody Production in 10K MWCO HFBRs

Clone[a]	Percent HS	Percent FCS	μg Ab/ml[b]
1	10	10	383
	7.5	10	530
	5	5	1241
	2.5	5	1915
	2.5	4	2293
2	10	10	231
	7.5	10	412
	5	10	627
	5	7.5	740
	5	5	1103
3	5	4	1064
	2.5	4	992
4	5	4	1181
	2.5	4	1289

[a]Five different hybridomas were evaluated.
[b]Average concentration of Ab in ECS.

medium could be decreased to 2% from 10% without substantial adverse effects on MoAb productivity. Reduction of FCS content in the ECS medium to 4% was tolerated well. Similar experiments were performed with 30K HFBRs with identical results (data not shown).

E. Ig Leakage Across HFBR Membranes

As shown in Table 1 and Figure 3, Ab levels in harvested ECS fluids are higher in the larger MWCO bioreactors. Although secretion rates are increased, there exists the possibility of Ig leakage across the fiber membrane. Both the concentrations of murine MoAb in the reservoir medium and horse immunoglobulin (Ig) in the ECS medium were measured in order to evaluate hollow-fiber membrane leakage. As shown in Figure 5, HFBRs, with capillary membrane MWCO of 10K showed no evidence of Ig leakage. Ig leakage was barely detectable in the 30K MWCO HFBRs. With the 30K HFBRs, there was less than 3 μg/ml of horse Ig in the ECS and 0.28 μg/ml of murine MoAb in the reservoir medium. HFBRs with capillary membrane MWCO of 50K and 70K were both found to contain substantial amounts of contaminating horse IgG in the ECS which had leaked across from the reservoir and murine Ab leakage into the reservoir. By Western

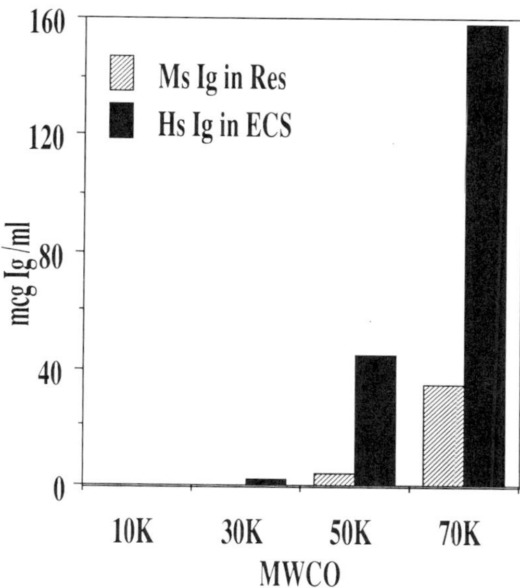

Figure 5 Assay for murine and horse IgG in reservoir and ECS medium for 10K, 30K, and 70K HFBR. Contamination of horse Ig in ECS and murine Ig in reservoir (Res) indicates protein leakage across capillary membrane.

Table 4 Comparison of MoAb Production in HFBR Versus Spinner Flask and Static Suspension Culture

Culture system	μg Ab/ml[a]	g Ab/liter medium[b]
HFBR 10K (1x)	568	0.008
10K (2x)	1189	0.024
30K (1x)	655	0.013
Spinner flask[c]	13	0.013
Static culture[d]	21	0.021

[a]Concentration of Ab in harvested fluid.
[b]Calculated concentration of Ab in total media consumed.
[c]1 liter culture volume.
[d]Average of three T-25 flasks.

blotting of nonreduced gels we have identified these potentially immunogenic proteins as whole molecules and not fragments and have therefore produced all our subsequent Ab in the 10K and 30K MWCO bioreactors. For the 70K HFBRs, more than 30 μg/ml of murine MoAb was found in the reservoir medium. This corresponded to a loss of 300 mg of Ab for each 10-liter reservoir change. The level of contaminating horse IgG in the 70K ECS amounted to about 10% of the total IgG present in the ECS compartment for the low (<30 μg/ml) to medium (31-60 μg/ml) Ab secretors. In the higher-producing cell lines, contaminated by horse Ig was less than 5% in the harvested ECS.

F. MoAb Production in HFBRs Compared to Suspension Culture

A single hybridoma cell line was grown in HFBRs, in suspension, and in static culture. The Ab productivity was compared for the different culture systems. Fully spent medium from the suspension culture contained MoAb levels of 13-21 μg/ml compared to MoAb concentrations up to 1189 μg/ml in the ECS of the HFBR (Table 4). Including the reservoir medium used to support the HFBR, the yield of MoAb for the total medium consumed was up to 24 μg/ml. Although total medium consumption per gram of Ab produced was similar for each system evaluated, the HFBR system produced a highly concentrated (up to 90-fold) product as compared to the static or spinner-flask systems. Final purification volumes calculated for 1 g of bulk Ab ranged from a low of 0.84 liter (10K-2x HFBR) to 77 liters (spinner flask).

G. Antibody Purity

An advantage of the HFBR is that high concentrations of MoAb are maintained in the ECS compared to suspension culture (Table 4). The MoAb accounts for up to 40% of the total protein in ECS, compared to a much lower percentage (<1%) in suspension cultures. Using standard Ig purification techniques, we have routinely obtained final product Ab purity in excess of 95%.

H. Novel 2x HFBRs and Continuous-Feed Systems

As mammalian cell culture processes are passed on from the research laboratory to the large-scale production facility, the need to optimize these scale-up processes becomes critical [15]. For processes that are not time-limited, such as MoAb production, continuous-feed systems are more desirable than the batch mode of operation. As the biomass increases during the course of a bioreactor run, the demand for oxygen, nutrients, and pH buffering capacity increases [16]. Thus there is a need for increased capabilities to monitor and maintain reactor conditions. By employing a continuous-feed system, a constant environment is maintained in the HFBR by adding fresh nutrient medium and removing waste

Table 5 MoAb Productivity Comparison in 2x HFBRs: Semibatch/Versus Continuous-Feed Systems

	System type[a]					
	Semibatch			Continuous		
Cell Line	g Ab	Conc. Ab in ECS (mg/ml)	l_{ECS}/g Ab	g Ab	Conc. Ab in ECS (mg/ml)	l_{ECS}/g Ab
1	2.00	1.12	0.89	5.76	1.66	0.50
2	2.28	1.02	0.99	7.39	4.13	0.24
3	2.05	0.44	2.28	2.74	0.55	1.82
4	0.80	0.50	2.01	5.80	1.58	0.63

[a]As described in Section II.

products. We have recently compared batch to continuous-feed systems in the 2x HFBR with a capillary MWCO of 10K.

Productivity Comparison: Batch Versus Continuous Feed

Table 5 summarizes the comparison data of four cell lines tested in both systems. In three of four cases, Ab productivity was increased dramatically (2.9 to 7.3-fold) when placed in the continuous-feed system. Purification volumes for bulk Ab contained in the ECS fluids were reduced from 20% to 76%. Total medium consumption (reservoir and ECS) was similarly reduced (data not shown).

Table 6 presents the performance characteristics of six hybrids in the 2x HFBR during the MoAb production phase under continuous-feed conditions. As previously shown in Table 1 for the 1x HFBRs, Ab productivity in the 2x HFBRs correlated with the secretion of MoAb by the clone in static culture. We have not evaluated the correlation between MWCO and static culture secretion in the larger HFBRs. At all secretion rates tested, as shown in Table 1 for the batch-fed 1x HFBRs, the rates of MoAb production in the 2x HFBR were greater than two- and threefold higher for either the batch or continuous-feed systems, respectively. Total medium consumption for the 10K MWCO bioreactors was reduced from 95.3 liters/g bulk Ab in the 1x batch-fed system (Table 2) to 67.7 liters/g for 2x batch-fed and 31.6 liters/g bulk Ab for the 2x continuous-feed systems.

In cell inoculum studies, the kinetics of MoAb production in the continuous-feed system were similar to those presented earlier (Figure 4). HFBRs receiving 1×10^8 cells lagged behind again by approximately 7 days to those HFBRs

Table 6 Performance Characteristics of 2x HFBRs During MoAb Production Phase in the Continuous-Feed System

Cell line	T_{25} μg/ml[a]	Total g	Total liters/g[b]	l_{ECS}/g[c]	Product concentration (mg/ml)	Production rate[e]
1	20	5.27	68.5	1.11	0.90	1.89
2	21	2.79	43.0	0.75	1.34	2.47
3	60	5.51	28.5	0.50	1.99	4.04
4	82	3.79	28.5	0.67	1.49	3.23
5	130	7.23	5.6	0.19	5.16	11.67
6	133	7.45	15.3	0.31	3.24	5.67

[a]Static culture.
[b]Reservoir plus ECS medium.
[c]Purification volume for 1 g Ab.
[d]Ab concentration in ECS fluid.
[e]grams Ab/BR/month.

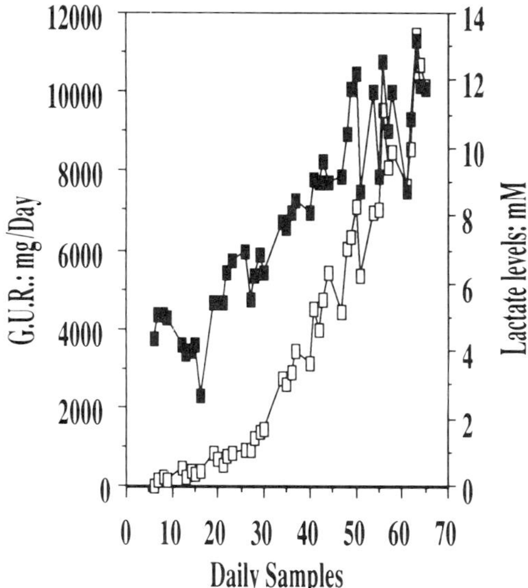

Figure 6 Metabolic profile for glucose utilization rate (GUR – □ –) and lactate levels (mM – ■ –).

receiving 1 × 10^9 cells, yet produced comparable Ab levels during the production phase. Glucose consumption in reactors with these cell inocula were similar once bioreactor maturity was reached and averaged 11,000 mg/day for this cell line and as high as 15,000 mg/day in other cell lines. The molar ratios of lactic acid production to glucose consumption in the continuous-feed system (Figure 6) were not different from those reported earlier in the batch-feed system (Figure 4), nor were ammonia levels found to be inhibitory to MoAb secretion.

Figure 7 shows a comparison of Ab production in the batch-feed model 1910 HFBR system verses the continuous-feed model 3510 HFBR system using the same cell line. Ab production rates were increased more than threefold in the larger HFBR using continuous-feed (3.44 versus 11.67 g/BR/month). Under continuous-feed conditions, production times in culture were generally decreased by 20 days when compared to the 1x HFBR batch-feed system.

No evidence of Ig leakage across the 10K MWCO membrane in the 2x HFBR could be shown in either the batch- or continuous-feed systems. Initial Ab purity in bulk ECS fluid ranged from 45% (low producers) to greater than 85% (high producer) as compared to 40% or less in the 1x HFBR batch-feed system.

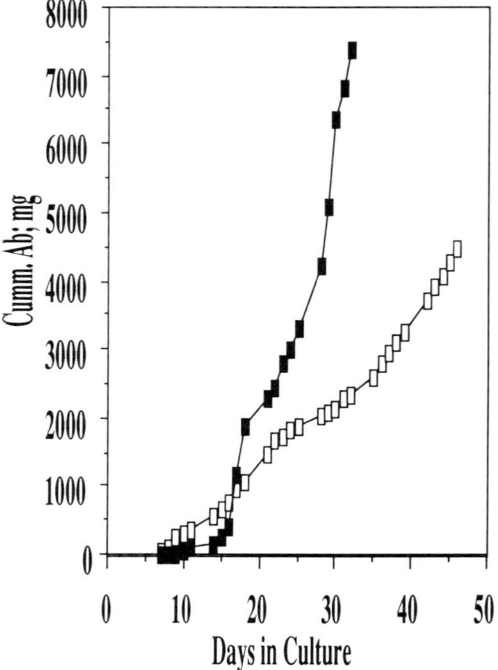

Figure 7 Comparison of MoAb production 1x HFBR batch-feed (– □ –) and 2x HFBR continuous-feed systems (– ■ –).

IV. DISCUSSION

We have examined the production of murine MoAbs from 47 different hybridomas in various HFBRs under batch-feed or continuous-feed conditions. Individual cell lines frequently need to be adapted for particular cell culture systems. Reports of MoAb productivity are frequently given for at most a few different cell lines used in each system [17,18]. Our experiences indicate that hybridomas can vary greatly in their antibody secretory capacity. As shown in Tables 1 and 6, antibody production in static culture is a predictor of productivity in HFBRs. This will likely be the case for other cell culture systems. It should be noted that the majority of hybridomas have secretion levels less than 60 μg/ml by our production study methods, and many (40%) secrete less than 30 μg/ml. These factors should be taken into consideration when evaluating other cell culture methods. The same cell line should be used when comparing different cell culture methods as was done in many of the experiments described here.

We have examined the effect of cell inoculum on HFBR productivity in both the batch-feed (Figure 4) and continuous-feed systems. As expected, bioreactors seeded with a greater number of cells reach maximum antibody output more rapidly. If possible, inoculation of a large number of cells is desirable, since it results in greater MoAb production in a shorter period of time. With lower cell inocula, high levels of MoAb production will be achieved over a longer time period. Once the bioreactor is mature, the output of antibody is the same regardless of cell inoculum.

Total medium requirements using HFBRs are similar to that found with other cell culture systems (Table 4). Even in our process-controlled/continuous-feed system, the overall grams of Ab per liter of medium was low (average = 0.059, range 0.015–0.179) in five of six clones tested. Although no less total medium is required using the HFBR, the composition of the medium may be different than that required in other systems. We have found that the composition of the reservoir medium may be less stringent than that used in the extracapillary medium. For example, reservoir medium contained horse serum instead of the more expensive fetal bovine serum. Media costs were reduced considerably in our HFBRs because greater than 95% of the total media used had the less stringent serum requirements.

MoAb concentration in the ECS usually exceeds 1 mg/ml in either system employed. This simplifies downstream processing and should result in a greater final product purity. Moreover, in a typical batch-fermentor production, product containing medium must be reduced in volume 50- to greater than 1000-fold and may possibly alter the biological activity and yield of the product. In our system, the average volume reductions are on the order of 20- to 30-fold depending on cell secretion rates and the system design. The MWCO of the hollow-fiber capillary membrane has a considerable effect on MoAb productivity. The increase in antibody production seen with larger MWCO is counterbalanced by membrane leakage, which results in both antibody loss and decreased purity. Future studies will be aimed at using medium with lower protein content.

The ability to optimally control bioreactors is limited by the lack of quantitative descriptions of mammalian cell culture kinetics [18]. Oxygen and a reduced carbon source such as glucose or glutamine are the chief metabolic nutrients required by mammalian cells [19]. Both glucose and glutamine are utilized to provide the energy and carbon source in cellular protein synthesis. We have found glucose measurements to be important in evaluating HFBR performance. The kinetics of glucose consumption, as shown in Figures 4 and 7, are a useful predictor of maximum bioreactor maturity. Measurements of lactate, ammonia, and pH were not found to be useful in the systems described here. Other metabolic parameters, such as dissolved oxygen, glutamine, and amino acid content, are being examined to determine the limiting nutritional factors. An important parameter that is difficult to monitor in the HFBR system is cell mass and viability. Methods are needed to calculate cell mass in the HFBR in order to optimize productivity.

The HFBR is an appealing method for large-scale MoAb production. It is both a simple system to operate and relatively inexpensive. Its modularity and versatility offer particular advantages in our application. The high concentration of MoAb in the harvested product is desirable and facilitates downstream processing. We have used this system to produce therapeutic-grade MoAb in lot sizes up to 15 g. For larger lot sizes, larger hollow-fiber bioreactors or other cell culture systems may have advantages.

V. ACKNOWLEDGMENTS

The authors wish to thank Gordon Fleig, C. D. Medical, Inc., for his expert advice and guidance. We also thank Gary Thelen, Susan Elrod, Karen Kondo, Cristine Coulter, Khanh Pham, Xochi Geiger, and Loriann Blomquist for their expert technical support.

Parts of this chapter are reprinted with permission of BioTechniques.

VI. REFERENCES

1. W. R. Arathoon and J. R. Birch, *Science 232*:1390 (1986).
2. M. Ratafia, *Pharm. Technol.* November 1987, p. 48.
3. P. Knight, *Bio/Technology 6*:506 (1988).
4. Y. Fouron, *BioPharm 1*(6):34 (1988).
5. D. F. Parson, *Am. Biotech Lab. 9*(3):15 (1988).
6. L. Rainen, *Am. Biotech Lab. 6*(3):20 (1988).
7. C. P. Prior, G. M. Prior, and J. A. Hope, *Am. Biotech Lab. 6*(3):25 (1988).
8. E. G. Posillico, M. S. Kallelis, and J. M. George, in *Commercial Production of Monoclonal Antibodies* (S. S. Seaver, ed.), Marcel Dekker, New York and Basel, 1987, p. 139.
9. J. E. Putnam, in *Commercial Production of Monoclonal Antibodies* (S. S. Seaver, ed.), Marcel Dekker, New York and Basal,1987, p. 119.
10. P. C. Familletti, *BioPharm. 1*:48 (1987).
11. D. H. Randerson, *J. Biotechnol. 2*:241 (1985).
12. R. Swartz, *Genet. Eng. News 5*(7):16 (1985).
13. T. C. Meeker, J. Lowder, D. G. Maloney, R. A. Miller, K. Thielemans, R. Warnke, and R. Levy, *Blood 65*:1349 (1985).
14. R. H. Kennett, in *Methods in Enzymology* (W. G. Jackoby and I. H. Pastan, eds.), Academic Press, New York, 1979, p. 345.
15. G. P. Cortessis and C. M. Proby, *BioPharm. 1*:30 (1987).
16. P. Wobber, M. Dosmar, and J. Banks, *BioPharm. 1*(6):38 (1988).
17. W. R. Tolbert and W. R. Srigley, *BioPharm. Manuf. 1*:42 (1987).

18. M. W. Glacken, E. Adema, and A. J. Sinskey, *Biotechnol. Bioeng. 32*:491 (1988).
19. D. D. Drury, B. E. Dale, and R. J. Gillies, *Biotechnol. Bioeng. 32*:966 (1988).

3

Large-Scale Purification of Monoclonal Antibodies for Therapeutic Use

VIPIN K. GARG
Bio Response, Inc., Hayward, California

I. INTRODUCTION

Isolating a pure protein from its biological environment in large quantities remains a major concern in the biotechnology industry today. Over the last few years, many challenges have arisen in protein purification as a consequence of advances in molecular biology, protein engineering, recombinant DNA techniques, and cell culture methodologies. Central to these developments has been a growing demand for the large-scale purification of pharmaceutical-grade immunoglobulins. The commercial interest in monoclonal antibodies is rapidly increasing, especially in the area of diagnostics, cancer therapy, clinical imaging, and immunoaffinity purification. Recently, it has been predicted that the market for monoclonal antibodies will reach $3.3 billion by the 1990s [1].

The potential for production of monoclonal antibodies (MoAbs) in a uniform state and in unlimited quantities demands that the ability to harvest, isolate, and purify these proteins be well understood and documented. To date, however, the information available on downstream processing of MoAbs is very sparse and by no means constitutes the whole story. Many chromatographic purification procedures have been suggested and successfully applied, such as anion exchange [2-4], protein A affinity [5,6], hydroxyapatite [4,7], hydrophobic interaction [4], and cation exchange [8-10]. However, none of these methods can truly be applied generally, because of the heterogeneity in isoelectric point, hydrophobicity, and size, as well as biological activity of different MoAbs. The medium in which the MoAb is present is also of importance. The choice of method(s) must therefore be optimized for the individual antibody to be purified and also for the amount of material purified at each time. High-performance liquid chromatography (HPLC), for instance, is well suited for up to gram-scale amounts, while low-pressure production-scale columns are used when hundreds of grams of material must be processed.

This chapter discusses the current status of purification processes available for MoAbs. Purification of milligram to multigram quantities of MoAbs using modern chromatographic techniques is reviewed. Special emphasis is given to the production of therapeutic-grade material.

II. PRODUCTION TECHNOLOGY VERSUS PURIFICATION

Since a number of production technologies can be applied to monoclonal antibodies, it is important to consider the impact of these techniques on the purification strategies. Before the advent of monoclonal antibodies, the only source of antibodies was heterogeneous serum or plasma. Immunoglobulins and particularly IgG were present as a major protein component. The immunoglobulin fraction of serum or plasma could be greatly enriched by selective salt precipitation, followed by some form of crude ion-exchange separation [11]. Final purification often was not essential for the end use of antibody, which could be in vitro diagnostic assay or as a research tool. Usually the major concern was the elimination of nonspecific antibody cross-reaction rather than absolute protein purity. In contrast to the serum- or plasma-derived antibodies, monoclonal antibodies must now be purified to a much greater level of purity, ranging from 70% to 90% pure for in vitro diagnostics to near homogeneity for in vivo diagnostics and therapeutic applications. Furthermore, the purification must be performed on product from such diverse sources as ascites fluid or cell culture media. In the former, the antibody is usually present in large amounts but is contaminated by host antibody. In the latter, the antibody is present in very low concentrations and may have to be purified in the presence of varying amounts of serum added to the growth medium. While the ultimate choice of production method may depend on such factors as the amount of monoclonal antibody required, desired end use for the purified antibody, and even the availability of necessary technology, it usually has a profound effect on the purification strategy.

A. Production in Ascites Fluid

The simplest approach to growing cells that produce monoclonal antibodies is to inject the cells into the mouse peritoneal cavity, where they proliferate and secrete antibodies into the ascites fluid, which is then collected. Ascites fluid is a very concentrated starting material, ranging from 2 to 15 mg of antibody per milliliter of fluid. However, since a given mouse has only a limited amount of ascites fluid, producing large quantities (gram to kilogram) of antibodies using this technique would require a large number of mice and numerous manipulations.

From a protein purification point of view, ascites fluid represents a potential advantage in having a higher starting antibody concentration. This eliminates the need for any prepurification concentration, as the product can be processed directly on a chromatography column. However, since the ascites fluid is an exudate of plasma, contamination from host proteins often causes problems during purification. The main contaminants of ascites fluid are proteins such as albumin, transferrin, and nonhybridoma immunoglobulins. While albumin and transferrin can be removed by conventional chromatography techniques, removal of nonhybridoma immunoglobulins can represent a significant challenge. This becomes particularly important when the antibody is used for in vivo application.

B. In Vitro Cell Culture

While production in ascites fluid is appropriate for making small volumes of relatively concentrated antibody, for large-scale production, in vitro cell culture is usually the method of choice. Technology for the in vitro culture of mammalian cells requires a high degree of sophistication because of the complex metabolic and growth requirements of animal cells and their sensitivity to mechanical stress. Recently, several in vitro methods, such as stirred fermentors [12], air-lift fermentors [13], hollow-fiber cartridges [14,15], ceramic cartridges [16], porous beads or capsules [17], tubular reactors [18], and even glass-bead bioreactors (unpublished data, Process Development group, BioResponse, Inc., Hayward, Calif.) have been used to produce MoAbs. Clearly, cell culture systems take many forms and include a number of diverse approaches, and it would be beyond the scope of this chapter to evaluate all these approaches. However, for the consideraion of purification-related aspects, a basic distinction can be made between cell culture systems based on whether they utilize batch or perfusion methods.

A batch culture, for industrial-scale production, typically utilizes deep-tank reactors having a volume of 100 to 1000 liters (e.g., stirred fermentors). These batch reactors are generally charged at the beginning of the production cycle with seed culture and sufficient medium to sustain the anticipated cell mass during the entire production cycle. As the culture grows, the cells utilize the nutrients and produce desired MoAbs and waste products, both of which remain in the vessel. At the completion of production, the contents of the reactors must be processed to eliminate the cells (and the cell debris), waste products, and remaining nutrients. The desired MoAb that remains can then be purified. In contrast, in a perfusion culture (e.g., hollow fibers), medium is perfused through the reactor at a rate proportional to the number of cells and their metabolic characteristics. Once the cell concentration reaches a maximum, the growth rate becomes almost zero and the reactor attains a steady state. In perfusion culture, cell densities from 10 to 100 times those of batch culture can be achieved.

For antibody purification, continuous perfusion is more desirable from a number of standpoints. First, the starting material from batch culture is relatively dilute, but if hybridomas are produced in perfusion systems, the monoclonal antibody may be more concentrated [15,19]. This has an obvious advantage in purification. A second consideration is the purity of the starting material. Perfusion systems have the advantage that secreted proteins are removed throughout the production run and residence time within the bioreactor is minimized. Thus the purity of the resulting product is generally much greater than the product in batch culture. Furthermore, in perfusion culture, after an initial phase of cellular proliferation, the serum requirements of the cells can be lowered dramatically [19,20], and in some cases serum can be completely eliminated from the culture medium [20]. This has a significant advantage, as the number of purification steps required can be reduced. The purification yield goes up, while the overall production cost is lowered. In some cases (unpublished data, Process Development group, Bio-Response, Inc.), hybridoma cell lines in perfusion cultures can even be adapted in protein-free, chemically defined media, further simplifying purification.

Based on these considerations, it is clear that an in vitro perfusion culture system is currently the most desirable production method for generating large quantities of highly pure monoclonal antibodies.

III. TRADITIONAL PRECIPITATION TECHNIQUES

Traditional methods for antibody purification from serum usually involve salt precipitation or organic precipitation followed by dialysis and ion-exchange chromatography [11]. Salting-out procedures are highly effective for preliminary concentration and partial purification of immunoglobulins, and can be used effectively with monoclonal antibodies. The methods are very inexpensive and simple to perform and avoid direct effects on the proteins that could lead to denaturation. The salt concentration at which precipitation occurs is specific for each protein and depends on factors such as pH, temperature, and protein concentration. For most antibodies the pH of precipitation should be at their pI (generally between 6.5 and 8.0), but some MoAbs may precipitate at a pH away from their pI. In general, it is preferable to work with antibodies at 4°C and with preservatives to inhibit enzyme activity and microbial growth. Whatever the temperature during the precipitation, however, the salt concentration and the protein concentration determine the speed and the degree of precipitation.

Ammonium sulfate and sodium sulfate are the most commonly used salts for the precipitation of immunoglobulins. It has been customary to express ammonium sulfate concentrations in percentage of saturation of the solution, but since this indication will vary with temperature, it is preferable to cite molar concentration. Saturated $(NH_4)_2SO_4$ at 0–4°C is about 5.35 M, or 700 g/liter. Concentration of Na_2SO_4, on the other hand, is usually given in percent w/v; thus differences in saturation due to temperature effects do not come about. At 5°C, however, 18% Na_2SO_4 is a supersaturated solution and is very unstable. It is therefore advisable to work at room temperature with this salt. The precipitated antibody after the salting-out procedure is pelleted out by centrifugation, washed, and then dissolved in a desired buffer. The high concentrations of salt in reconstituted precipitates may be rapidly removed by gel filtration or dialysis. While salting-out procedures have a clear advantage in achieving very high antibody concentration, it is usually not possible to achieve high product yields. Conditions for salting-out often make a compromise between yield and the degree of contamination. Furthermore, the technique is not well suited for large-scale sterile operation.

A second approach that has been traditionally used for the precipitation of antibodies is the use of organic solvents (e.g., ethanol, acetone). Addition of an organic solvent to an aqueous extract containing proteins has a variety of effects that, combined, lead to protein precipitation. This method has been especially important on the industrial scale in the plasma fractionation industry. For monoclonal antibodies, however, this method is not very useful. A major concern with MoAbs is the high probability of denaturation and loss of activity due to organic solvents.

Another precipitation method that has been employed in some cases is the use of organic complexing precipitants. Organic compounds such as Ethodine, Rivanol, or caprylic acid [11] can be used to selectively fractionate immunoglobulins from complex solutions. Caprylic acid, in particular, has been used for fractionation of monoclonal antibodies [21].

Finally, while the above-described precipitation methods may be suitable for certain crude applications, they have only a limited use in modern therapeutic MoAb purification. Based on our experience, precipitation methods are not recommended for large-scale applications [22].

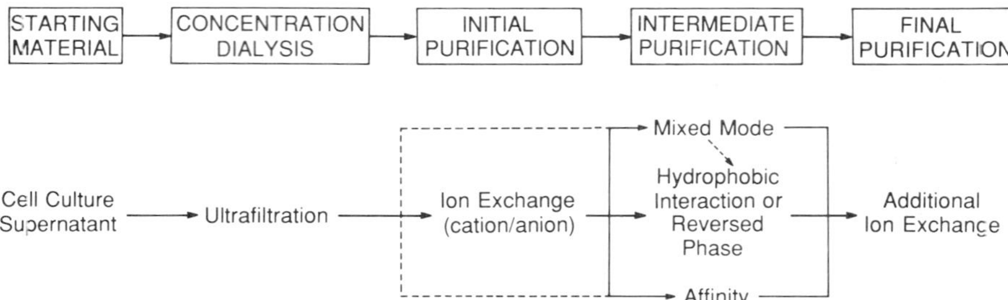

Figure 1 Potential purification schemes for monoclonal antibodies produced in mammalian cell culture.

IV. ION-EXCHANGE CHROMATOGRAPHY

Ion-exchange chromatography has become one of the most widely used methods of protein purification since it was first reported by Peterson and Sober in 1956 [23]. The basic principle of ion exchange is that the separation is achieved on the basis of the charge density of a molecule. Ion-exchange chromatography is capable of separating molecules differing by very small differences in charge, and is therefore a technique of very high resolving power. Several review articles [24-26] have previously discussed methods of protein separation by ion exchange in detail.

In the past, the most frequently used ion-exchange adsorbents for the purification of antibodies were weak anion exchangers (diethylaminoethyl, DEAE) and strong anion exchangers (quaternary aminoethyl, QAE). More recently, however, cation exchangers such as carboxymethyl (CM) and sulphopropyl (SP) have been shown to be more suitable for large-scale work.

In our laboratory we have found both the anion and the cation exchangers to be useful at different steps during MoAb purification (Figure 1). The final choice of the type of ion exchanger to be used at a given purification step should be based on the isoelectric point of the individual MoAb, the degree and type of impurities in the antibody preparation, and the desired form of final product For instance, as described later in this section, anion-exchange chromatography can be used as a final cleanup step to lower endotoxins and cellular DNA from therapeutic-grade antibody preparations.

In addition to the type of ion exchanger, the choice of the matrix should also be considered carefully. The nature of the matrix determines several important parameters, such as mechanical strength, flow characteristics, adsorbing capacity, and nonspecific binding. Similarly, the pore size of the matrix determines accessibility and protein penetration, thereby affecting loading capacity and efficiency. Widely used matrices for preparative protein chromatography include either the insoluble soft gels based on polysaccharides and other polymers (e.g., cellulose, dextrans, agarose, and polyacrylamides), or rigid supports based on inorganic substances (e.g., silica and controlled-pore glass). The exact choice of the matrix at different purification steps is also very important for scale-up considerations. A complete discussion of this aspect is presented in Section X.

A. Anion Exchangers

Historically, anion exchange has been used at research scale and has been adapted for small-scale processing of monoclonal antibodies. Most studies in the literature have utilized the weak anion exchanger DEAE [27-30]. However, functional moieties such as quaternary amine groups (QAE, QMA, etc.) that are quite strongly basic have also been employed [3,30]. The mechanism of separation seems to be very similar regardless of functional group. For instance, the chromatographic profiles of ascites fluid generally show binding of three major proteins—transferrin, immunoglobulin, and albumin, in that order—with maximum resolution being achieved by optimizing gradients of ionic strength, pH, or both. In small-scale processing (10-50 mg IgG), recoveries of up to 95% have been shown from ascites fluide [27,28]. However, such high recoveries were possible only with high-performance matrices using HPLC. Furthermore, high-purity product (>90% pure) using anion-exchange columns was possible only when the ion-exchange step was preceded by either an ammonium sulfate precipitation [31] or a dye-column affinity step [4].

Since most immunoglobulins are only weakly acidic, with isoelectric points above 6, complete binding of antibodies to anion-exchange columns is usually achieved between pH 6.5 and 8.5 [30]. The column is equilibrated with a low-ionic-strength buffer, and the starting antibody solution is also dialyzed against this buffer prior to loading. The elution is performed with an ascending ionic strength up to 300 mM salt. This protocol can be modified to accommodate double-gradient (descending pH, ascending ionic strength) [28,30] or step elution conditions [22]. The use of double-gradient elution conditions has been shown by several investigators [28,30] to maximize IgG resolution. Step elution conditions are useful for scale-up to production-scale chromatography (20-100 liters column volume), where complex gradient elutions can become very time-consuming and are often not reproducible [22].

As mentioned earlier, anion-exchange chromatography is used only at research scale and for small-scale processing of MoAbs. In fact, anion exchange has several limitations for large-scale purification of antibodies. Since the majority of serum proteins are strongly acidic, with isoelectric points below 5.5, under the binding conditions described above (pH 6.5-8.0), all the proteins in the starting material would bind to the column, thereby compromising the binding capacity. Furthermore, since the contaminating proteins are bound to the column along with the antibody, use of a gradient or a multistep elution becomes essential in order to achieve an acceptable level of purity. This produces a very dilute elution fraction and also reduces antibody recovery from the purification step. There is a further disadvantage when processing cell culture supernatants. A pH indicator dye (phenol red) is often used in the culture medium. Unless removed by prior desalting/dialysis, this too will bind to the matrix, further reducing the capacity of the anion exchangers to bind antibodies. The binding of these dyes can also decrease resolution and change the elution characteristics of the immunoglobulin peaks by affecting the mechanism of separation. In our production system (Bio-Response, Inc.), however, phenol red has now been completely eliminated from cell culture media.

Recent work in our laboratories has shown that the anion exchangers are better suited for large-scale chromatography of MoAbs under nonbinding conditions [10,32]. In the past, very few investigators have explored this possibility. A method has been described by Levy and Sober [33] in which a rabbit IgG was purified by passing it through a preequilibrate DE-52 or a DEAE-Sephacel column in 17.5 mM phosphate buffer, pH 6.5. All contaminants were retained on the column, while most of the IgG passed unretained. A similar application has also been reported using QAE-Sephadex A-50 [34]. We have developed similar applications for anion exchangers as a final cleanup step in antibody purification (Figure 1). This cleanup step is particularly useful for therapeutic-grade antibodies, where extremely high purities (>95%) and very low levels of DNA (<10 pg/mg of protein) and endotoxins (<2 EU/mg protein) must be attained in the final product [10,21]. In applications where affinity chromatography (e.g., protein A) is used for purification, a final anion-exchange cleanup step can be designed for removal of any extraneous protein leached from the affinity column [32]. Most of these applications are performed at around neutral pH and 20–150 mM buffer ionic strength.

B. Cation Exchangers

In contrast to the anion exchangers, cation-exchange supports are becoming increasingly popular for large-scale purification of monoclonal antibodies. Several process-scale antibody purifications (using cation exchange) from cell culture supernatants have recently been described [9,22,35]. The advantage with cation exchangers is that the conditions can be adjusted such that most of the product and only a few of the contaminating proteins are bound initially, thus eliminating the need for high-resolution gradient elutions. This is possible due to relatively low pIs of albumin and transferrin. Furthermore, under these conditions, bound antibody can be eluted in very high concentration in a relatively small volume. This is very important when using large volumes of cell culture supernatants. Clearly, cation-exchange chromatography is more suitable than anion exchange for process-scale purification.

Binding of antibodies to cation-exchange supports is generally achieved between pH 5.0 and 6.0 [9,22]. The actual binding pH depends on the ionic strength of the starting material. At low ionic strenghts (<25 mM salt), complete binding of some MoAbs can be achieved at pH 6.5; however, a pH below 6.0 is more suitable for most MoAbs [22]. Many investigators have suggested the use of a desalting/buffer-exchange step prior to cation-exchange chromatography in order to achieve a low ionic strength in the starting material. A procedure has been described by the process separation group of Pharmacia AB (Uppsala, Sweden) in which the first step is a buffer exchange on Sephadex G-25 [9]. This is followed by cation-exchange chromatography on S Sepharose Fast Flow (Pharmacia). The cation-exchange column was equilibrated and loaded in 20-mM sodium citrate buffer at pH 5.3. The elution buffer was 20 mM sodium citrate + 140 mM NaCl, pH 5.0. Under these conditions, the antibody fraction eluted in two or three bed volumes, resulting in a 50- to 75-fold concentration and a fivefold purification. It is also possible to elute the bound Ig at a higher pH and salt concentration. Elution with 50 mM Tris-Hcl, 1 M Na Cl, pH 7.8, reduces the elution volume to approximately one bed volume. The antibody eluted with this buffer, however, will be less pure than if eluted with the low-salt buffer [9].

The cation-exchange step can be used in two different ways depending on the requirements of the process. In the first instance, very general conditions can be chosen (e.g., low pH binding and high salt, high pH elution) for a very rapid volume reduction. However, in this case optimal purity is compromised. In the second instance, a range of individually optimized conditions can be employed for each different MoAb. These may compromise optimal process time but obtain maximum purity.

In our experience, cation-exchange supports are ideally suited for intermediate purification steps that allow large-volume throughput with 50–70% purity and yields of > 80% [22]. In many cases, hybridoma cell culture supernatants can be processed directly (without a desalting step) under low-pH (4.5–5.0) binding conditions. Since the column can be run under sterile conditions, hundreds of liters of supernatant are easily processed per column run. The bound antibody is eluted in a relatively small volume (20–30 liters), and can be further purified by affinity, hydrophobic interaction, mixed-modal, or additional ion-exchange chromatography (Figure 1).

V. HYDROPHOBIC INTERACTION AND REVERSED PHASE

Recently, hydrophobic-interaction chromatography (HIC) has shown great promise as a versatile protein-purification tool [36,37]. Hydrophobic regions on the surface of proteins cause them to adsorb to the hydrophobic ligands of the stationary phase when high-molarity salts are used. These adsorbed proteins are selectively desorbed during descending salt gradients. The high ionic strength enhances hydrophobic interactions and the decreasing ionic strength weakens this interaction, allowing the protein to elute from the column. This process is referred to as HIC.

Since ammonium sulfate fractionation has been widely used in antibody purification, it is possible that HIC could accomplish this step with greater resolution. Indeed, recent work in our laboratory [10] and those of others [4,38] indicates that HIC is a useful method for the purification of MoAbs.

Two of the most commonly used HIC media are Phenyl-Sepharose CL-4B and Octyl-Sepharose CL-4B, made by Pharmacia. However, since most antibodies are relatively strongly hydrophobic proteins, Phenyl-Sepharose is usually the best choice for their purification. Octyl-Sepharose is used for weakly hydrophobic proteins, since strongly hydrophobic proteins will not be easily eluted from this gel. A one-step purification of IgG from the serum using Phenyl-Sepharose has been described by Goudswaard et al. [39]. The IgG was bound in the presence of 1.0 M $(NH_4)_2SO_4$ and eluted in 0.6 M $(NH_4)_2SO_4$. Another matrix that has been used for MoAb purification is Alkyl-Superose (Pharmacia). This chemistry is even less hydrophobic than phenyl chemistry, and thus very well suited for antibody purification. With alkyl chemistry, the running conditions can be optimized such that many contaminants including albumin elute in the void without binding to the column. This means that there are more hydrophobic ligands available on the column for the binding of the antibody. Another mildly hydrophobic matrix that has recently been developed specifically for large-scale MoAb purification is Alkoxy-Superflow from Sterogene Biochemicals (Arcadia, Calif.).

In addition to the conventional agarose gels described above, HIC chemistries are also available on polymeric and silica-based matrices. For instance, a butyl group (C_4) chemistry is available on a semirigid polyvinyl gel (Fractogel TSK) from Toyo Soda (Japan). Phenyl group is also available on a Toyo Soda polymeric gel (Phenyl-5PW) for small-scale, high-performance applications. Among silica-based matrices, a propyl (C_3) chemistry is available from J. T. Baker Chemical Co. (Phillipsburgh, N.J.) on 5-, 15-, and 40-μm particles.

While HIC has not been used extensively for antibody purification, it offers an important purification alternative for MoAbs. HIC provides a selectivity very different from ion-exchange chromatography, thereby providing another dimension of fractionation and simplifying many otherwise difficult separations. Furthermore, since proteins are bound to HIC supports in the presence of high (rather than low) ionic strength, HIC is a convenient next step for fractions that are eluted from ion exchangers in relatively high-ionic-strength buffers. Similarly, HIC is also appropriate following salt precipitation, or affinity when the sample may be eluted in a high salt concentration. In our experience, HIC is best suited after ion exchange in antibody-purification schemes (Figure 1). Finally, HIC can also be used as a cleanup step to remove minor contaminants and to remove nucleic acid (DNA) and pyrogens in some applications.

In contrast to HIC, traditional reversed-phase chromatography is not suitable for antibody purification [40,41]. Due to their relatively strong hydrophobic nature, immunoglobulins tend to bind very tightly to reversed-phase supports and require strong organic solvents for elution. This can result in a loss of functional activity and denaturation of the antibody molecule. As discussed above, only borderline reverse-phase chemistries with short carbon chains (C_1-C_4) can be used for antibody purification. Longer chains, C_8 (octyl) and C_{18} (octadecyl), are not recommended for antibody purification.

VI. AFFINITY CHROMATOGRAPHY

Nearly all proteins can be separated by affinity chromatography on the basis of their biological function. The most important prerequisite for this technique is the availability of a ligand that can be covalently immobilized to a porous support via spacers. Certain conditions bring about a specific and reversible binding of the biomolecule to be separated to this ligand, while contaminating substances are washed out. The desorption of the bound molecules of interest can subsequently be carried out by competitive displacement with an effective solution or by influencing the structure by changing the salt or pH gradient. This method is ideal for isolating a small number of molecules from a large excess of accompanying substances or from very dilute solutions. Since the key to the method's attractiveness is its unequaled degree of selectivity, the target of affinity chromatography is generally the direct isolation of a given protein in purified form. A considerable amount of processing time can therefore be saved with affinity chromatography. Furthermore, affinity chromatography can lead to notable improvements in the yields of purified final product.

In view of the above considerations, affinity separation is becoming an increasingly preferred method for purifying monoclonal antibodies. In fact, after ion exchange, bioaffinity chromatography is the second major method currently used in the purification of MoAbs. There are two general methods for purifying a MoAb using affinity chromatography. The first method involves the use of protein A affinity, whereas the second method exploits the antigen-antibody interactions.

Protein A is a component of the cell wall of *Staphylococcus aureus* that is able to bind various mammalian immunoglobulins [5]. Its reactivity with mouse immunoglobulin subclasses has been described, and it is now widely used for the purification of MoAbs from both ascites fluid and from cell culture. Several methods for the purification of MoAbs using protein A have been described in the literature [5,6,42,43]. In general, antibodies can be bound to protein A at pH 8.0 (150-mM sodium phosphate buffer), and elution can be performed over the pH range of 6.0 to 3.0. The exact binding and elution conditions, however, are dependent on the binding specificity of protein A to different immunoglobulin types and subclasses from various species [5]. In the human, for instance, IgG1, IgG2, and IgG4 bind strongly, while IgG3 does not bind. In the mouse, initial studies indicated that IgG2a, IgG2b, and IgG3 bind to protein A, while IgG1 and IgM do not [5,6]. More recently, however, several mouse IgG1 MoAbs have been shown to bind to protein A under high-salt and high-pH binding conditions (0.5-1.5 M glycine + 3 M NaCl, pH 8.9). The bound IgG1 could be eluted with 0.1 M sodium citrate buffer, pH 6.0. This ability to purify mouse IgG1 MoAbs by protein A is of considerable importance, since the majority of commercially important murine MoAbs belong to the IgG1 subclass. Recently, protein A gene has been cloned, and recombinant protein A is now available at a fraction of the cost of natural protein A [44].

In the second affinity method, where antigen-antibody interaction is utilized, the pure antigen (Ag) is first coupled to a suitable matrix. The MoAb solution to be purified is then passed through the gel. Ideally, the MoAb will bind to the Ag and all the contaminants can be washed away. The MoAb is then eluted from the column in a relatively small volume. One important consideration in selecting this affinity approach as the method of choice is the avidity with which the MoAb binds to the Ag. If the MoAb binds only weakly, then it may be eluted prematurely when the column is washed to remove contaminants. On the other hand, if the binding is very strong, it may be very difficult to elute active antibody from the column [45]. Another consideration in selecting an antigen affinity method is the availability and cost of the antigen required. In many cases the specific antigen is either very expensive or is not always readily available.

Recently, we have compared the purification of a commercially important MoAb using both protein A and immobilized antigen [43]. The antibody purified was a murine IgG1, anti-human tissue plasminogen activator (t-PA), from cell culture supernatant concentrates. The purification was performed on a high-performance preparative column (21.4 mm × 10 cm), using a new affinity matrix, Hydropore-EP (Rainin Instrument Co., Berkeley, Calif.). The comparison data is summarized in Table 1. The binding and elution conditions for both the antigen (immobilized t-PA) and the protein A column were identical. Binding/wash buffer: 0.1 M potassium phosphate, pH 7.5, elution buffer: 0.2 M potassium phosphate, pH 4.0. While the antigen column provided greater loading capacity and loading speed than the protein A column, the capacity of the antigen column decreased with extended use (greater than 30 cycles) whereas the protein A column maintained a constant loading capacity for up to 100 cycles. This is not surprising, since the protein A molecule is very stable to denaturing conditions [5].

Table 1 Preparative Purification of a Commercially Important MoAb Using Affinity Chromatography on Immobilized Protein A and Antigen Columns

Experimental parameter	Protein A	Antigen
Anti-tPA capacity	120 mg	325 mg
Load/cycle	100 mg	238 mg
No. of auto-cycles	25	25
Anti-tPA purified	2.5 g	5.95 g
Purification time	8.3 h	8.3 h
Anti-tPA throughput	300 mg/h	714 mg/h
Product concentration	3 mg/ml	5 mg/ml
Product purity	> 98%	> 98%
Total no. of cycles without significant loss of capacity	100	30

As mentioned earlier, affinity chromatography is suitable for the purification of MoAbs from both ascites fluid and cell culture supernatants. However, since antibody concentrations in cell culture supernatants are very low, it is recommended that a concentration step be used prior to affinity chromatography. In our experience, an antibody concentration of 2–5 mg/ml is appropriate for the affinity step. As shown in Figure 1, required product concentrations can easily be achieved by ultrafiltration and (or) ion-exchange chromatography. Introduction of an ion-exchange step, particularly cation exchange, has an added advantage in that the sample can be cleaned prior to affinity purification, thus greatly reducing the total protein load per affinity column cycle. This lower protein load leads to a higher antibody binding efficiency and a longer column lifetime.

In addition to the protein A and antigen-antibody interaction methods described above, other affinity methods using ligands such as protein G [45], Jacalin (for IgA) [47], and anti-immunoglobulin antibody [48] have also recently been shown to be effective in MoAb purification.

VII. ADSORPTION CHROMATOGRAPHY ON HYDROXYAPATITE

Adsorption chromatography on hydroxyapatite has been used for the separation of proteins for over 20 years. Substances are bound to the surface of the hydroxyapatite crystals by electrostatic interaction, and are desorbed by a concentration gradient of phosphate buffer. For MoAbs, it has been suggested that separation is based on variation in the light-chain composition [49]. The advantage of hydroxyapatite over conventional, nonaffinity methods is the fact that antibodies are often bound more strongly than other protein contaminants [50]. However, hydroxyapatite has several limitations for routine use. Hydroxyapatite is rather fragile and comes in many different sizes of crystals. This causes many practical problems, such as elimination of fines, low and unpredictable flow rates, difficulty of using large columns, and often short column lifetime. Furthermore, long chromatographic run times are required in order to resolve the MoAb from numerous other bound proteins, resulting into an antibody preparation that is extremely dilute [51].

A recent study comparing ion exchange, hydroxyapatite, and protein A affinity chromatography for purification of three different MoAbs found that hydroxyapatite was the least efficient in terms of yield and purity [29]. A similar result was reported in another study where HIC and ion exchange were found to be superior to hydroxyapatite [4].

VIII. MIXED-MODAL CHROMATOGRAPHY

The concept of mixed-mode chromatography is not new to protein purification chemists. For several years, it has been suggested that ion-exchange chromatography usually accompanies some mixed-modal effects. The most obvious of such effects is a hydrophobic contribution to ion-exchange protein chromatography. Alternatively, a purely electrostatic interaction could also be responsible for mixed-modal effects. For instance, the magnitude of ionic interaction can be altered by modification of surface chemistries. Recently, investigators have taken synthetic approaches to construct chromatographic matrices that show precise mixed-mode effects [52]. Mixed-modal matrices are currently available that bind one class of proteins more selectively than others, thereby providing specific purification applications.

For monoclonal antibodies, a mixed-mode chemistry has been recently developed by J. T. Baker. The matrix is known as Bakerbond ABx. The surface chemistry of the ABx matrix has been optimized such that it behaves like an ion exchanger in that immunoglobulins are resolved via the manipulation of buffer pH and ionic strength, but which exhibits an affinity-like sensitivity toward all immunoglobulins [53,54]. ABx is a mixed-mode ion exchanger containing both weakly anionic and weakly cationic functional groups.

The ABx material has been used to purify antibodies from ascites fluid, cell culture supernatants, and plasma [53,54]. It is available in a variety of particle sizes ranging from HPLC-grade 5 μm to preparative-grade 40 μm for batch extraction and low-pressure open column chromatography. This introduces a significant advantage, in that method development can be performed on analytical HPLC (3- to 5-μm particle size) and scaled up to either preparative HPLC (15-25 μm) or low-pressure production-scale chromatography (40-80 μm).

Researchers from J. T. Baker have published several articles regarding the use of ABx matrix for antibody purification [30,53-55]. These investigators have described detailed methods for the purification of MoAbs of different types, subclasses, and from different species sources [53,54]. In some cases, it was reported that a one-step purification of highly pure (>95%) antibody could be achieved with ABx matrix [53,54]. Recently, Ross et al. [56] have shown that mouse MoAbs produced in ascites could be purified to 90-95% homogeneity in a single ABx step, and the column was also useful for detecting and separating multiple immunoglobulins present in the same ascites. However, most of this work was conducted using HPLC columns, and the considerations associated with the large-scale purification were not really taken into account.

We have evaluated the use of ABx matrix in large-scale, low-pressure chromatography mode [22]. Several mouse monoclonal IgG1 antibodies produced in cell culture were purified on 40-μm ABx matrix. In general, methods described for HPLC columns [53,54] could also be used for large-scale purification. However, it was noteworthy that different IgG1s behaved very differently in their elution profiles, indicating the need for a careful method development for each MoAb [22]. In terms of the overall column performance, in large-scale chromatography it was not possible to achieve therapeutic-grade purity (>95% pure) by a single ABx step, even when the antibody was produced in a serum-free medium. Combined with other chromatographic modes, however, ABx provided a powerful alternative for MoAb purification (Figure 1). In several applications, we were able to achieve >95% MoAb by combining a conventional ion exchanger with Abx chromatography [22]. A combined purification effect of ion-exchange and ABx chromatography columns in a large-scale therapeutic application is shown in Figure 2. The purified product (lane C) not only meets the protein purity requirements, it is also within specifications in terms of DNA and endotoxin requirements for therapeutic application. A major advantage of the ABx column was the high product recovery (usually 80-90%) under mild buffer conditions [22].

Since the ABx matrix is relatively new in the market, it has not been used extensively for large-scale work. However, as the demand for large quantities of highly pure MoAbs grows, ABx chromatography could prove to be a major method for their purification.

In addition to ABx, other matrices involving mixed interactions have been used for MoAb purification. One such matrix that has already been described in this chapter is hydroxyapatite. However, the exact nature of the multiple interactions on hydroxyapatite is not totally understood. Other mixed-function chromatography matrices used for the purification of MoAbs are DEAE and CM

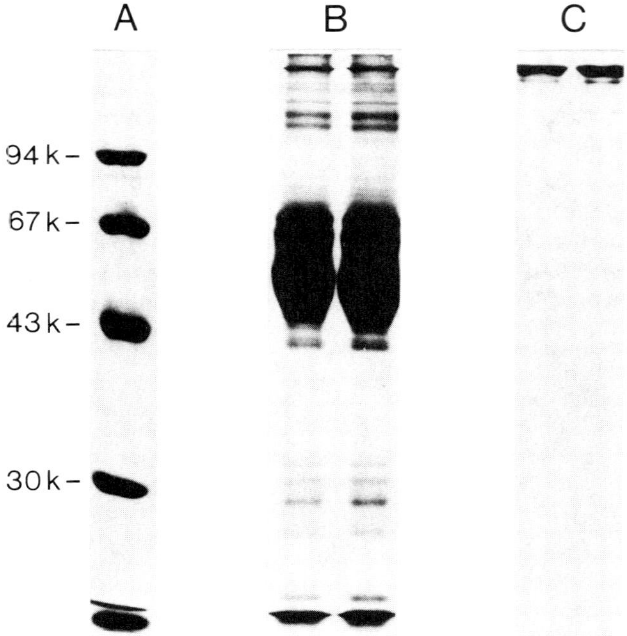

Figure 2 SDS-PAGE analysis of an IgG1 monoclonal antibody purified by a three-step protocol using ion-exchange and mixed-mode chromatography. A, MW standards; B, starting material after concentration; C, purified therapeutic-grade antibody (>98% pure).

Affi-Gel Blue (both produced by Bio-Rad Laboratories, Richmond, Calif.). These two gels combine ion-exchange and Cibachrone Blue functionalities for antibody purification. DEAE Affi-Gel Blue has been used for the purification of MoAbs from mouse ascites fluid [57]. The purified MoAb preparation was free of all contaminants except transferrin, and was also free of any detectable proteolytic activity [57]. Similarly, CM Affi-Gel Blue was reported to provide a partially purified protease-free antibody preparation with a minimum of work and expense [58]. Neither of these gels, however, has been used for large-scale purification.

IX. IMPACT OF HIGH-PERFORMANCE LIQUID CHROMATOGRAPHY

In the late 1970s, the methods of protein purification were revolutionized by the advent of high-performance liquid chromatography (HPLC) techniques per-

Table 2 Comparison of Classical LC with High-Performance Liquid Chromatography (HPLC) with Respect to MoAb Purification

Classical LC	HPLC
1. Provides moderate resolution and high recovery of bioactive MoAb	Provides higher resolution and recoveries
2. Column bed compression limits flow velocity	Rigid supports permit use of very high flow velocity
3. Low flow velocity permits use of large particle supports	High flow velocity requires use of small particle supports
4. Low operating back pressure	High operating back pressure
5. Low chromatographic efficiency: long separation times dilute elution bands lower resolution	High chromatographic efficiency: fast separation (up to 200-fold) concentrated elution band (up to 100X) higher resolution
6. High-capacity scale-up (for 50 g of product: 50 g/run, 1 run/day)	Lower-capacity scale-up (for 50 g of product: 1 g/run, 50 runs/day)
7. Automation difficult	Ease of automation: Automatic sample injection, fraction collection, data acquisition

formed on bonded-phase inorganic supports [59,60] and rigid organic gels [61]. Initially, these techniques were used mainly for method development and analytical-scale work. More recently, however, there has been growing interest in the use of HPLC for semipreparative to preparative-scale (milligram to gram quantities) protein purification. This is particularly true for monoclonal antibodies. Virtually every technique discussed in this chapter for the purification of MoAbs is currently available in the preparative-scale HPLC mode.

A comparison of classical liquid chromatography (LC) with HPLC is shown in Table 2. While HPLC seems to have several advantages over classical liquid chromatography, the major limitation with HPLC is its lower capacity for scale-up. A list of considerations for preparative-scale HPLC is shown in Table 3.

Table 3 Considerations for Preparative Protein HPLC

1. Matrix rigidity and mechanical strength, allowing high flow rates and long-term use for large-scale chromatography.
2. Relatively small particle size (15–40 μm), with a narrow size distribution for high resolution and efficient chromatography. However, particle size of <10 μm is not compatible with preparative work due to high back pressure.
3. Uniform particle shapes (preferably spherical) for close packing and efficiency.
4. Presence of large pores (>150 Å) for greater binding capacity.
5. Column design and biocompatibility.
6. Cost effectiveness.

Currently, the outside limit for preparative HPLC is approximately 1 g MoAb/run. However, only a few laboratories are actually using HPLC purification at this scale. In most cases, an HPLC-based purification is limited to milligram quantities of antibodies. In our laboratory, we have routinely purified 0.5- to 1-g lots of MoAbs by preparative HPLCs using ion exchange [10], hydrophobic interaction [10], and affinity column chemistries [43]. These separations were achieved on either a 21.4-mm-I.D. × 25-cm-long or a 41.4-mm-I.D. × 25-cm-long column. However, in one case an 80-mm-I.D. × 20-cm-long column custom packed by the manufacturer was also tested. One consideration with these preparative columns is that as the column size increases, some of the advantages associated with HPLC begin to diminish due to increasing particle size of the packing materials. Clearly, the performance and resolution of a 3- to 5-μm particle size analytical column cannot be expected from a 25- to 30-μm particle size preparative column. Based on our experience, a best compromise between the particle size and the scale-up is somewhere between 15 and 20 μm. However, in some cases (e.g., affinity chromatography), packing materials of up to 40-μm particle size can be used with significant high-performance advantage. In addition to the particle size, particle shapes and pore sizes are also important considerations for preparative-scale HPLC columns (Table 3). As with analytical scale, a uniform spherical particle shape is most desirable for preparative HPLC. This will ensure close column packing and thereby improve efficiency. In terms of the pore sizes, large pore (>150 Å) column packings are necessary for efficient mass transfer of protein molecules. Narrow pores (50–100 Å) impede the mass transfer of proteins between the matrix and the mobile-phase buffer, resulting in broad peaks, poor resolution, low recovery, and long separation times. Several manufacturers currently make 300-Å or greater pore size matrices, which are ideal for antibody-purification work.

A second limitation of HPLC-based purification systems is the high cost of packing materials. Typically, for an equivalent amount of purification capacity, HPLC matrices are 10- to 20-fold (in some cases even up to 100-fold) more expensive than the classical gels. This becomes even worse if one considers the cost of column hardware and equipment required for HPLC. One way of dealing with the high cost of column matrices is to develop matrices that allow multiple retention mechanisms, separately, from the same column. This concept, referred to as "multimodal" [62] or "polytypic" [63] chromatography, differs greatly from mixed-mode chromatography described earlier in this chapter. In multimodal chromatography, a particular retention mechanism is induced by the mobile-phase system used while other potential interactions are suppressed. A multimodel preparative HPLC column for MoAb purification has recently been introduced by Rainin. The column (Dynamax 300A AX) is silica-based and contains a polyethyleneimine (PEI) monolayer as the ligand. The separation mode of the column is governed by the ionic strength of the binding buffer. In low-salt buffers the column functions as an anion exchanger; whereas in the presence of high salt concentration the charges on the column are neutralized, making the packing hydrophobic. We have recently used this column for antibody purification both in anion-exchange and HIC mode [10]. Similarly, other multimodal columns are now being developed [64] for preparative HPLC.

In spite of the above limitations, HPLC is becoming an increasingly popular tool for the purification of MoAbs. Several HPLC companies have introduced biocompatible systems for protein-purification applications. The pumps and the flow paths in these systems are made with corrosion-resistant materials such as titanium. Similarly, new column designs and packing materials have been developed for improved handling of protein solutions and biocompatibility. Preparative-size columns are now available in medium-pressure glass and polymeric materials that are inert to salts and buffers. While such inert materials are not essential for antibody purification, they do provide a safeguard against antibody denaturation and loss of activity.

HPLC finds its greatest use toward the end of a purification scheme where many of the easily separated proteins, which would reduce the capacity of the HPLC column, have been removed [10,43,65]. Furthermore, by this time the sample volume has been considerably reduced in comparison to that which existed at early purification steps. This eliminates long sample loading times or the use of larger, expensive HPLC columns to reduce sample loading time. In our experience, HPLC is particularly suited for situations where multiple lots of a highly pure MoAb are required in 1- to 2-g lot sizes [10,43]. For instance, HPLC could be ideal in the early phase of a new product development, when multiple lots of a therapeutic-grade product are required for clinical trials, product characterization, and other experiments associated with the new drug application (NDA).

X. ASPECTS OF SCALE-UP

The degree of scale-up required for a given product is dependent on the amount of final product needed for a particular application. Some applications require only milligram quantities of monoclonal antibody, whereas others—such as cancer therapy—use much greater quantities: Doses may be in grams per patient per year. This range causes confusion as to what is meant by scale-up. For the majority of antibodies, however, a 1- to 100-g final product lot size would fall under large-scale production. Although potential for up to 1-kg lot sizes exists for some MoAbs, to our knowledge, no MoAb is currently required in such large quantities.

From a purification point of view, once the final product lot size is determined, the first step is to define the size of the starting material (total volume and amount of MoAb present). The initial amount of MoAb required would depend on the overall product yield from the purification process. Typically, for therapeutic-grade MoAbs, an overall yield of 40-50% can be expected from the purification process [22]. In other words, to generate 10 g of a therapeutic-grade MoAb, approximately 20-25 g of crude product would be required. In some cases (e.g., in the presence of high serum), however, the yield can be as low as 20%; while in other cases (serum free), up to 75% of the product can be recovered. An important point to remember here is that the yield usually goes up when the process is scaled up from method development to intermediate-scale and finally to large-scale operation.

Once the size of the starting material is determined, the next step is a determination of the sequence of purification steps. Since the volume of the starting material to be processed can be in the region of several thousand liters, it is necessary to start with steps that are capable of handling large volumes of liquids and have a concentrating effect. As discussed earlier, a suitable starting method is cation-exchange chromatography (see Section IV). In some cases, an initial concentration and dialysis can be achieved by ultrafiltration prior to cation-exchange chromatography (Figure 1). Following these initial concentration and purification steps, the rest of the steps for intermediate and final purification are determined individually for each MoAb. As discussed in this chapter, the various techniques used, i.e., affinity, HIC, mixed-mode, or additional ion exchange, each have their advantages and disadvantages. These should be carefully considered before deciding a final purification sequence. One other technique that is not discussed in this chapter is gel filtration chromatography. Several investigators have reported the use of this technique in large-scale purification of MoAbs [9,35,66]. In our experience, however, this is not a suitable purification method for large-scale work [22]. In comparison to other techniques, gel filtration is very slow and can handle only limited volumes of sample. We recommend the use of gel filtration only as a desalting or buffer exchange step, either in between other purification steps or as a final product formulation step.

Once the sequence of purification steps required for a particular process is determined, the next stage is to consider the choice of gel matrix and the choice of column. As pointed out in Section IV, the choice of matrix at different purification steps is very important for efficient scale-up. For instance, one of the problems of large-scale chromatography using conventional soft gels is the tendency of the gel to compress at high flow rates. The current generation of highly cross-linked gels, such as Sepharose CL-4B (Pharmacia) and Trisacryl (IBF biotechnics, Savage, Md.), are considerably more rigid than uncross-linked materials, but are not as rigid as an inorganic support. Inorganic supports, such as silica, are incompressible, but may suffer from high back pressures and nonspecific adsorption of proteins. Another disadvantage of silica-based matrices is their inability to withstand chemical disinfection treatment with 0.5 M NaOH. This process is usually required to prevent buildup of biological residues, including pyrogens. As an alternative, however, a variety of other mobile phases are available for cleaning silica-based matrices [67]. Results in our laboratories have also shown that silica-based matrices (e.g., ABx matrix) can be kept pyrogen-free if the chromatography is performed using sterile buffers and the column is periodically cleaned with organic solvents.

In terms of resolution and capacity, silica-based matrices show a considerably superior performance compared to soft gels. This, combined with the ability to use higher flow rates, allow for substantially higher throughput on silica-based matrices. Clearly, both types of matrices (soft gels versus rigid inorganic supports) have their advantages and disadvantages. In our experience, highly cross-linked sepharose-type gels are best suited for initial purification steps, in which large volumes and highly impure feed materials are usually processed. As the product becomes concentrated and the purity levels are improved, high-resolution silica supports become more appropriate.

As an alternative to the gel matrices, several bioprocess companies have recently developed membrane and composite structure matrices in cartridge form [68-70]. Composite matrix cartridges for large-scale chromatography have been developed by Cuno, Inc. (Meriden, Conn.) [68], and are available both in ion-exchange and in affinity mode. These cartridges have several advantages over gel matrices [68], and have been successfully used for MoAb purification [71]. Similarly, membrane-based purification devices have also been developed both in ion-exchange and affinity modes [69,70]. A hollow-fiber affinity membrane device, recently developed by Sepracore, Inc. (Marlborough, Mass.), has shown exceptional promise for antibody purification [70].

A final consideration for scale-up is column design. Scaling up in chromatography essentially means increasing the column diameter. In this way, the same linear flow rate can be obtained in production columns as in laboratory columns, but with a much greater throughput. Keeping the bed height small permits good flow rates without having to use high pressure and expensive pumping systems. The actual column design is determined by the specific application and the conditions to be used, which in turn are determined by the gel and the flow rates that can be used. The materials used in the construction of large-scale chromatographic columns should be compatible with biological molecules, offer good chemical resistance and mechanical strength, and if possible should be transparent so that the packing material can be observed.

Several bioprocess equipment companies currently supply columns that meet the above criteria. Three different types of process-scale columns are available from Pharmacia. These are the stack columns (KS 370/15), the bioprocess columns (BP 113 and BP 252), and the Sephamatic stainless settl columns. Similarly, Whatman Biosystems (Clifton, N.J.) supplies Prep-25 and Prep-100 process columns, and Amicon (Danvers, Mass.) supplies adjustable-bed-height industrial columns made from acrylic, glass, and stainless steel. In addition to the above conventional (axial-flow) columns, a novel radial-flow column has recently been developed by Sepragen Corporation (San Leandro, Calif.). These columns allow significantly higher flow rates at low pressures [72], which could have a significant advantage with some of the high-resolution packing materials.

XI. CONSIDERATIONS FOR THERAPEUTIC-GRADE PURITY

The production of therapeutic-grade proteins requires strict adherence to established pharmaceutical manufacturing principles and practices. Both the Food and Drug Administration (FDA) in the United States and the National Institute for Biological Standards and Control in Great Britain have produced guidelines for the therapeutic use of monoclonal antibodies. In the United States, all operations for the production and purification of therapeutic-grade MoAbs must be carried out in compliance with the FDA's current Good Manufacturing Practices (cGMPs). A detailed description of FDA regulations and cGMPs for MoAb production can be found elsewhere in this book (Chapter 4).

In terms of purification, the actual methods themselves are not as important as the final result of the purification process. The purification process must remove any components that may generate undesirable reactions of either protein or virus or nucleic acid origin. In the specific case of MoAbs, the risk of carrying contaminant proteins and viruses along with nucleic acids must be considered when the producer cells are of malignant origin. The importance of bacterial toxins such as pyrogens is well recognized; these must also be removed during the purification process.

A general summary of the specifications required for therapeutic-grade MoAbs is given in Table 4. To meet these specifications, it is important that the purifica-

Table 4 Purity Requirements for a Therapeutic-Grade Monoclonal Antibody

Specification	Testing method	Acceptability criteria
Protein purity	SDS-PAGE, HPLC	>95% pure
Endotoxin[a]	Rabbit pyrogen test	Negative
	LAL[b]	<2 EU/mg protein
DNA	DNA hybridization	<10 pg/patient dose
Microbiological	Sterility test	Negative

[a]Test dose and specifications depend on patient dose.
[b]Limulus amebocyte lysate.

tion process be designed not only to maximize the protein purity, but also to remove DNA and pyrogens (endotoxins) at each step. Too often, purification chemists concentrate only on the protein purity, and ignore the nonproteinaceous contaminants until the final purification steps. In our view, contaminants such as DNA and pyrogens should be removed gradually as the relative purity of the MoAb increases from one purification step to the next. With regard to pyrogens, it is important that the product be kept free from any microbial contamination during the purification process. When possible, the chromatographic steps should be performed sterily; otherwise the eluted MoAb should be sterily filtered immediately after elution. Regeneration and sanitization of the gel matrix is also very important. The matrix should be stripped of any bound proteins, and depyrogenated, before re-use. With many matrices (e.g., Sepharose, cellulose), regeneration and depyrogenation can be achieved by washing the matrix with 0.5-1.0 M NaOH. However, if the matrix is unstable at high pH (e.g., silica), the depyrogenation and sanitization can be achieved with mobile phases such as 70% ethanol, 10% acetic acid, 1% SDS, or 8 M urea.

If affinity chromatography is used as one of the purification steps, removal of any extraneous protein leached from the affinity column should also be taken into consideration. For instance, leaching of protein A from affinity matrices is a major concern when purifying therapeutic-grade MoAbs [73]. As described earlier in this chapter, additional purification steps should be included to ensure the removal of any leached protein A. Similarly, additional cleanup steps are required when using other affinity purification methods.

XII. CONCLUSION

As described in this chapter, a whole array of chromatographic techniques are available for the large-scale purification of monoclonal antibodies. Ion exchange and affinity are currently the most widely used modes of chromatography for MoAb purification. However, as the need for novel purification protocols increases, other modes of chromatography (e.g., HIC, mixed-mode) are also becoming popular.

Innovative bioprocess companies are constantly developing new technologies to meet the growing demand for large-scale processing. For instance, improvement of standard chromatographic supports have led to the availability of more rigid matrices with better flow properties suitable for industrial-scale processing. Similarly, membrane-based purification systems, currently under development, could prove to be of major significance in the future. Another area that is generating a lot of interest is the development of automated purification systems. Several companies (e.g., Biorad, Sepracore, Oros Systems) are currently in the process of developing automated chromatography systems for monoclonal antibody purification.

XIII. ACKNOWLEDGMENTS

The author wishes to acknowledge the technical support of Ivan Crane, Barbara Czuba, and Maureen Costello. Word processing support by Diane Franklin is also gratefully acknowledged.

XIV. REFERENCES

1. L. Baskin, Mammalian cell culture techniques allow for new opportunities in commercial scale-up of biotech drugs. *Med. Business J.*, July 15, 1988, pp. 198-202.
2. J. W. Goding, Antibody production by hybridomas. *J. Immunol. Methods 39*:285-308 (1980).
3. S. W. Burchiel, J. R. Billman, and T. R. Alber, Rapid and efficient purification of mouse monoclonal antibodies from ascites fluid using high performance liquid chromatography. *J. Immunol. Methods 69*:33-42 (1984).
4. B. Pavlu, U. Johansson, C. Nyhlen, and A. Wichman, Rapid purification of monoclonal antibodies by high performance liquid chromatography. *J. Chromatogr. 359*:449-460 (1986).
5. J. W. Goding, Use of Staphylococcal protein A as an immunological reagent. *J. Immunol. Methods 20*:241-253 (1978).
6. P. L. Ey, S. J. Prowse, and C. R. Jenkin, Isolation of pure IgG1, IgG2a and IgG2b immunoglobulins from mouse serum using protein A sepharose. *Immunochemistry 15*:429-436 (1978).
7. L. H. Stanker, M. Vanderlaan, and H. Juarez-Salinas, One-step purification of mouse monoclonal antibodies from ascites fluid by hydroxylapatite chromatography. *J. Immunol. Methods 76*:157-169 (1985).

8. M. Carlsson, A. Hedin, M. Inganas, B. Harfast, and F. Blomberg, Purification of in vitro produced mouse monoclonal antibodies. A two-step procedure using cation exchange chromatography and gel filtration. *J. Immunol. Methods 79*:89–98 (1985).
9. B. Malm, A rapid method for isolation of monoclonal antibodies from large volumes of hybridoma cell culture supernatants. Presented at the Biotech '85 USA Exhibition, Washington, D.C., October 21–23, 1985.
10. V. K. Garg, Use of preparative HPLC in large scale purification of therapeutic grade proteins from mammalian cell culture. Paper 911, presented at the Seventh International Symposium on HPLC of Proteins, Peptides, and Polynucleotides, Washington, D.C., November 2–4, 1987.
11. H. F. Deutsch and J. L. Fahey, Purification of antibody. In *Methods in Immunology and Immunochemistry, Vol. 1* (C. A. Williams and M. W. Chase, eds.), Academic Press, New York, 1967, pp. 315–332.
12. W. R. Lebherz, III, Batch production of monoclonal antibody by large-scale suspension cultures. In *Commercial Production of Monoclonal Antibodies* (S. S. Seaver, ed.), Marcel Dekker, New York, 1987, pp. 93–118.
13. W. R. Arathoon and J. R. Birch, Large scale cell culture in biotechnology. *Science 232*:1390–1395 (1986).
14. J. Hopkinson, Hollow fiber cell culture systems for economical cell-product manufacturing. *Bio/Technology 3*:225–230 (1985).
15. P. C. Brown, M. A. C. Costello, R. Oakley, and J. L. Lewis, Applications of the mass culturing technique (MCT) in the large-scale growth of mammalian cells. In *Large-Scale Mammalian Cell Culture* (J. Feder and W. R. Tolbert, eds.), Academic Press, New York, 1985, pp. 59–71.
16. J. E. Putnam, Monoclonal antibody production in a ceramic matrix. In *Commercial Production of Monoclonal Antibodies* (S. S. Seaver, ed.), Marcel Dekker, New York, 1987, pp. 119–138.
17. E. Posillico, Microencapsulation technology for large scale antibody production. *Bio/Technology 4*:114–117 (1986).
18. H. Katinger, A tubular biological film reactor concept for the cultivation and treatment of mammalian cells. In *Animal Cell Biotechnology, Vol. 3* (R. E. Spier and J. B. Griffiths, eds.), Academic Press, London, 1988, pp. 240–250.
19. S. S. Seaver, Culture method affects antibody secretion of hybridoma cells. In *Commercial Production of Monoclonal Antibodies* (S. S. Seaver, ed.), Marcel Dekker, New York, 1987, pp. 49–71.
20. B. L. Brown, Reducing costs up front: A method for adapting myeloma and hybridoma cells to an inexpensive chemically defined serum-free medium. In *Commercial Production of Monoclonal Antibodies* (S. S. Seaver, ed.), Marcel Dekker, New York, 1987, pp. 35–48.
21. M. Steinbuch and R. Audran, The isolation of IgG from mammalian sera with the aid of caprylic acid. *Arch. Biochem. Biophys. 134*:279–284 (1969).
22. V. K. Garg and R. E. Branson, Applications of mixed and multi-modal chromatography in protein purification. Paper S-66, presented at the

Society for Industrial Microbiology Annual Meeting, Baltimore, Md., August 9-14, 1987.
23. E. A. Peterson and H. A. Sober, Chromatography of proteins. I. Cellulose ion exchange adsorbents. *J. Am. Chem. Soc. 78*:751-755 (1956).
24. S. R. Himmelhock, Chromatography of proteins on ion exchange absorbents. In *Methods in Enzymology, Vol. 22* (W. B. Jacoby, ed.), Academic Press, New York, 1971, pp. 273-286.
25. Pharmacia Fine Chemicals, *Ion Exchange Chromatography: Principles and Methods*, Pharmacia, Uppsala, Sweden, 1982.
26. R. Scopes, *Protein Purification Principles and Practice*, Springer Verlag, New York, 1982, pp. 75-101.
27. J. R. Deschamps, J. E. K. Kildreth, D. Derr, and J. T. August, A high performance liquid chromatographic procedure for the purification of mouse monoclonal antibodies. *Anal. Biochem. 147*:451–454 (1985).
28. M. J. Gemski, B. P. Doctor, J. K. Gentry, M. J. Pluskal, and M. P. Strickler, Single step purification of monoclonal antibody from murine ascites and tissue culture fluids by anion exchange high performance liquid chromatography. *Bio/Techniques 3*:378-384 (1985).
29. L. Manil, P. Motte, P. Pernas, F. Troalen, C. Bohuon, and D. Bellet, Evaluation of protocols for purification of mouse monoclonal antibodies. Yield and purity in two-dimensional gel electrophoresis. *J. Immunol. Methods 90*:25-37 (1986).
30. L. J. Crane, Purification of monoclonal antibodies by high performance ion-exchange chromatography. In *Monoclonal Antibodies: Production, Techniques and Applications* (L. B. Schook, ed.), Marcel Dekker, New York, 1987, pp. 139-171.
31. M. J. F. Schmerr, K. R. Goodwin, H. D. Lehmkuhl, and R. C. Cutlip, Preparation of sheep and cattle immunoglobulins with antibody activity by high performance liquid chromatography. *J. Chromatogr. 326*:225-233 (1985).
32. M. A. Costello, E. Reed, R. Dias, R. Krishman, B. Czuba, and V. Garg, Purification of therapeutic proteins produced from mass culture of mammalian cells in gram quantities: Validation of protocol with respect to DNA, endotoxin, and extraneous protein removal. *Miami Bio/Technology Winter Symposium*, Miami, Fla., February 8-12, 1988.
33. H. B. Levy and H. A. Sober, A simple chromatographic method for preparation of gamma globulin. *Proc. Soc. Exp. Biol. Med. 103*:250-252 (1960).
34. P. Tijssen, *Practice and Theory of Enzyme Immunoassays*, Elsevier, Amsterdam, 1985, pp. 101-102.
35. C. Ostlund, Large-scale purification of monoclonal antibodies. *Tibtech 4*: 288-293 (1986).
36. D. L. Gooding, M. N. Schmuck, and K. M. Gooding, Analysis of proteins with new, mildly hydrophobic HPLC packing materials. *J. Chromatogr. 296*:107-114 (1984).
37. Y. T. Kato, T. Kitamura, and T. Hashimoto, Operational variables in high performance hydrophobic interaction chromagraphy of proteins on TSK gel Phenyl-5 PW. *J. Chromatogr. 298*:407–418 (1984).

38. S. C. Goheen and R. S. Matson, Purification of human serum gamma globulins by hydrophobic interaction high-performance liquid chromatography. *J. Chromatogr. 326*:235-241 (1985).
39. J. Goudswaard, G. Virella, A. Noordzij, and J. Pol, Isolation of equine IgG(T) by hydrophobic interaction chromatography. *Immunochemistry 14*:717-719 (1977).
40. K. M. Gooding, High performance liquid chromatography of proteins—A current look at the state of the technique. *Biochromatography 1*:34-40 (1986).
41. D. R. Nau, S. A. Berkowitz, M. P. Henry, and L. J. Crane, A multidimensional approach to monoclonal antibody purification. Presented at the Fifth Annual Congress of Hybridoma Research, Baltimore, Md., January 26-29, 1986.
42. J. J. Langone, Applications of immobilized protein A in immunochemical techniques. *J. Immunol. Methods 55*:277-296 (1982).
43. R. A. Kagel, G. W. Kagel, and V. K. Garg, High performance affinity chromatography for commercial protein purification. Presented at the HPLC '88, Twelfth International Symposium on Column Liquid Chromatography, Washington, D.C., June 19-24, 1988.
44. D. Colbert, A. Anilionis, P. Gelep, J. Farley, and R. Breyer, Molecular organization of the protein A gene and its expression in recombinant host organisms. *J. Biol. Resp. Modif. 3*:255-259 (1984).
45. Pharmacia Fine Chemicals, Monoclonal antibody purification. *Pharmacia Sep. News 13*(4):1-5 (1986).
46. S. R. Fahnestock, Cloned streptococcal protein G genes. *Tibtech 5*:79-83 (1987).
47. D. R. Nau, Chromatographic analysis and purification of antibodies. In *Microbial Detection* (B. Swaminathan and G. Prakash, eds.), Marcel Dekker, New York, 1988, pp. 383-430.
48. H. Bazin, F. Cormont, and L. De Clercq, Purification of rat monoclonal antibodies. In *Methods in Enzymology, Vol. 121* (J. J. Langone and H. Van Vunakis, eds.), Academic Press, New York, 1986, pp. 638-652.
49. H. Juarez-Salinas, G. S. Ott, J-C. Chen, T. L. Brooks, and L. H. Stanker, Separation of IgG idiotypes by high-performance hydroxylapatite chromatography. In *Methods in Enzymology*, Vol. 121 (J. J. Langone and H. Van Vunakis, eds.), Academic Press, New York, 1986, pp. 615-622.
50. T. L. Brooks and A. Stevens, Techniques for purifying monoclonal antibodies. *Am. Lab.*, October 1985, pp. 54-64.
51. D. R. Nau, M. P. Henry, S. A. Berkowitz, L. J. Crane, H. E. Ramsden, J. G. Guenther, and M. Zief, Purification of monoclonal antibodies by solid phase extraction. Presented at Bio Expo '86, Boston, Mass., May 1, 1986.
52. W. Kopaciewicz, M. A. Rounds, and F. E. Regnier, Stationary phase contributions to retention in high-performance anion-exchange protein chromatography: Ligand density and mixed mode effects. *J. Chromatogr. 318*: 157-172 (1985).

53. D. R. Nau, A unique chromatographic matrix for rapid antibody purification. *Biochromatography 1*:82–94 (1986).
54. D. R. Nau, ABx: A novel chromatographic matrix for the purification of antibodies. In *Commercial Production of Monoclonal Antibodies* (S. S. Seaver, ed.), Marcel Dekker, New York, 1987, pp. 247–313.
55. D. R. Nau, Rapid antibody purification. Bakerbond ABx. *J. T. Baker Product Information Bull. No. 8506*, J. T. Baker Chemical Company, Phillipsburg, N.J., 1986.
56. A. H. Ross, D. Herlyn, and H. Koprowski, Purification of monoclonal antibodies from ascites using ABx liquid chromatography column. *J. Immunol. Methods 102*:227–231 (1987).
57. C. Bruck, D. Portetelle, C. Glineur, and A. Bollen, One-step purification of mouse monoclonal antibodies from ascites fluid by DEAE affi-gel blue chromatography. *J. Immunol. Methods 53*:313–319 (1982).
58. Bio-Rad Laboratories Information Letter, Choosing an antibody purification method. *Bio-Radiations No. 56*, 1985, pp. 4–5.
59. S. H. Chang, K. M. Gooding, and F. E. Regnier, High-performance liquid chromatography of proteins. *J. Chromatogr. 125*:103–114 (1976).
60. A. J. Alpert and F. E. Regnier, Preparation of a porous microparticulate anion-exchange chromatography support for proteins. *J. Chromatogr. 185*: 375–392 (1979).
61. O. Mikes, P. Strop, and J. Sedlachova, Rapid chromatographic separation of technical enzymes on Spheron ion exchangers. *J. Chromatogr. 148*:237–245 (1978).
62. L. A. Kennedy, W. Kopaciewicz, and F. E. Regnier, Multimodal liquid chromatography columns for the separation of proteins in either the anion-exchange or hydrophobic-interaction mode. *J. Chromatogr. 359*:73–84 (1986).
63. Z. E. Rassi and C. Horvath, Metal chelate-interaction chromatography of proteins with iminodiacetic acid-bonded stationary phases on silica support. *J. Chromatogr. 359*:241–253 (1986).
64. Rainin Instrument Company HPLC Information Letter, New dual-mode support for biopolymers. *Rainin Dynamax Review 1*(2):7–8 (1987).
65. S. A. Berkowitz, M. P. Henry, D. R. Nau, and L. J. Crane, A strategic approach to the use of silica-based chromatography media (bonded phases) for protein purification. *Am. Lab.* May 1987, pp. 33–42.
66. C. Ostlund, P. Borwell, and B. Malm, Process scale purification from cell culture supernatants: Monoclonal antibodies. Presented at 7th General Meeting of European Society for Animal Cell Technology, Vienna, Austria, September 30–October 4, 1985.
67. C. T. Wehr, Sample preparation and column regeneration in biopolymer separations. *J. Chromatogr. 418*:27–50 (1987).
68. K. C. Hou and R. M. Mandaro, Bioseparation by ion exchange cartridge chromatography. *Biotechniques 4*:358–367 (1986).
69. Kontes Life Sciences Products Information Letter, Kontes fastchrom membrane chromatography. *Bio/Separations No. 8*, March 1988, pp. 2–9.

70. S. Brandt, R. A. Goffe, S. B. Kessler, J. L. O'Conner, and S. E. Zale, Membrane-based affinity technology for commercial scale purifications. *Bio/Technology 6*:779–782 (1988).
71. A. Jungbauer, F. Unterluggauer, F. Steindl, E. Wenisch, and H. Katinger, Purification of mouse anti-human urokinase monoclonal antibody. Presented at Recovery of Bioproducts, Uppsala, Sweden, May 11–15, 1986.
72. V. Saxena and A. E. Weil, Radial flow columns: A new approach to scaling-up biopurifications. *Biochromatography 2*:90–97 (1987).
73. A. Rosevear and C. Lambe, Downstream processing of animal cell culture products—Recent developments. In *Animal Cell Biotechnology, Vol. 3* (R. E. Spier and J. B. Griffiths, eds.), Academic Press, London, 1988, pp. 379–440.

4

Regulatory Agency Concerns in the Manufacturing and Testing of Monoclonal Antibodies for Therapeutic Use

DONALD A. BAKER
Bio Response, Inc., Hayward, California

W. SCOTT HARKONEN
Becton Dickinson Monoclonal Center, San Jose, California

I. INTRODUCTION

Although the technology necessary for the development of monoclonal-based therapeutics has arisen relatively recently, products derived from this aspect of biotechnology face the same regulatory review procedures and submission requirements as pharmaceutical products manufactured through more traditional techniques. Thus the regulatory agency's prime concern remains the safety and efficacy of a new product in its proposed use. However, the technological issues concerning the manufacture of monoclonal-based therapeutics are not ignored. An example to illustrate these points can be found in the U.S. government's 1986 Coordinated Framework for Regulation of Biotechnology [1], where the Food and Drug Administration (FDA) stated that "Although there are no statutory provisions or regulations that address biotechnology specifically, . . . [t]he sponsor's process techniques are also considered in FDA reviews and communications for the development of appropriate information."

This policy statement indicates that in addition to the usual characterization of manufacturing process, the FDA is asking for specific and detailed information on the unique aspects of the manufacture, characterization, and testing of biotechnology products. Since the development of traditional pharmaceuticals has already been estimated to average 4 to 8 years and $100 million in research and development to obtain the information necessary for produce approval, developing an effective strategy for identifying and addressing regulatory issues unique to monoclonal therapeutics has become even more critical. This chapter reviews important regulatory concerns that specifically impact the development of monoclonal antibody-based therapeutic agents.

II. REGULATORY OVERVIEW

A. Current Regulations

Most developed countries have evolved complex and demanding regulatory processes that are designed to ensure that unsafe or ineffective drugs do not reach their domestic markets. In the United States, the Food and Drug Administration (FDA), through the administration of Title 21 of the Code of Federal Regulations, regulates the manufacture, licensing, distribution, and sale of all drugs, and monoclonal-based therapeutics are subject to the relevant portions of these statutes. Similar situations prevail in other countries in that antibody-based drugs are subject to existing statutes regulating other drug substances. Thus, for example, the Pharmaceutical Affairs Law covers these products in Japan, and the Medicines Act serves a similar function in Britain. In the discussion that follows, the regulatory agency model used will be, by and large, the FDA, although mention will be made, where appropriate, of other regulatory agency concerns.

The FDA has identified a number of the issues that specifically impact monoclonal-based therapeutics, and has addressed these concerns in a series of informal "Points to Consider" documents. The first of these [2] issued in 1983 from the Office of Biologics and concerned the manufacturing and testing of murine monoclonal antibody products for human use. The original document was revised and expanded in 1987 [3] to include recommendations for human monoclonal antibody products and monoclonal antibodies coupled to radionuclides, toxins, or drugs. These "Points to Consider" documents address the FDA's expectations regarding the development of biological products for human use and the submission of Notices of Claimed Investigational Exemption (INDs) and Product License Applications (PLAs). It should be noted that these FDA guidelines are broadly relevant to regulatory concerns in other countries, as evidenced by a new publication from the commission of the European Communities that addresses many of the same issues [4].

B. The Regulatory Process

Although the FDA can have input in the pre-IND testing of new products, the regulatory process officially begins when the sponsor of the new therapeutic submits an IND application to the agency. When complete, this document will consolidate what is known of the origin, manufacture, characterization, and animal pharmacology of the new drug as well as the scientific basis of its expected therapeutic activity and the plan for at least the initial phase of the proposed clinical investigation. The preparation of the IND and licensing applications are discussed at the end of this chapter.

However, prior to the IND submission, the FDA also plays several key roles in the testing of monoclonal-based therapeutic products. The agency reviews data on the characterization of new cell lines, production processes, and quality control procedures, and determines when preclinical tests for safety are sufficient to show that the product is safe for initial testing in humans. If asked, the agency will often provide drug sponsors with advice on the adequacy of preclinical testing programs developed before animal studies are initiated or the IND is filed. The agency also sets minimum standards for laboratories conducting preclinical testing programs through good laboratory practice (GLP) regulations under 21 CFR Part 58.

Once the IND is submitted, the FDA becomes particularly involved during the phase of development when the sponsor is conducting the clinical studies necessary to demonstrate the safety and efficacy of a new drug in the treatment of a specific disease. During this stage, the FDA is concerned primarily with the accuracy of the clinical data and the safety of human subjects involved in the study. In parallel with the clinical trials, additional long-term animal toxicity studies are often conducted to further substantiate the safety of a new agent. Large-scale manufacturing processes are implemented and validated, and site inspections are conducted. Throughout this phase of product development, the sponsor generally maintains close communication with the FDA via periodic update reports and meetings.

The Product Licensing Application (PLA) for biologics or New Drug Application (NDA) for drugs is a particularly important element in the drug approval process. In the application, the sponsor proposes that the drug be approved for marketing and uses test data to show that the drug is safe and effective in its proposed use or indication. The FDA bases its approval/disapproval decision on the data presented in the application. With an approved PLA, the manufacturer of a monoclonal antibody product is licensed to market and sell the product in the United States for the approved indication. The typical time frames involved in this development process are illustrated in Figure 1.

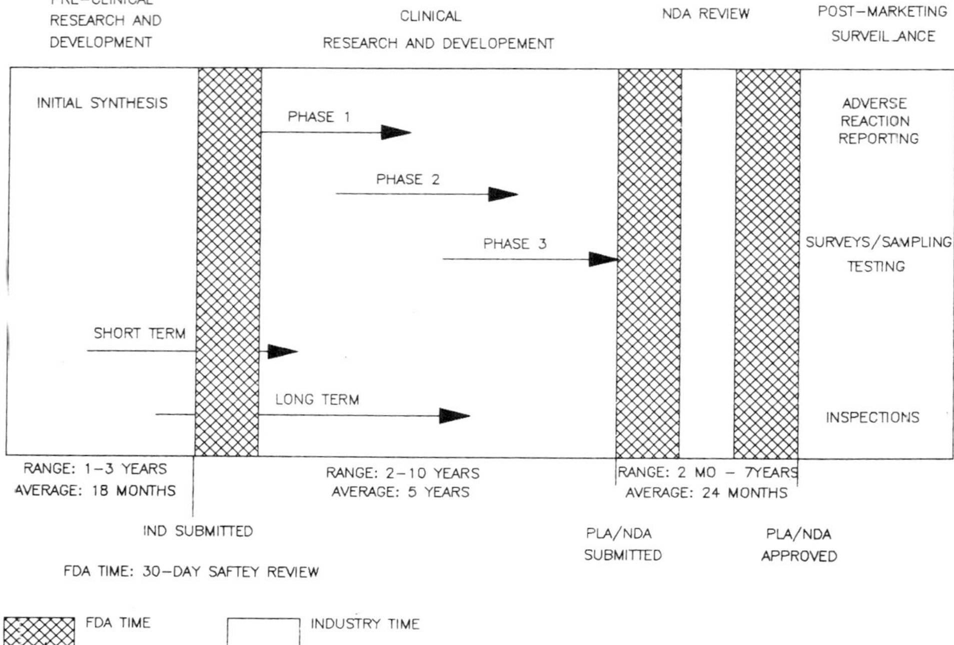

Figure 1 Time frames for new drug development. (Adapted from *New Drug Development in the United States*, FDA Consumer Special Report, January, 1988.)

III. THE CELL LINE

A. Acquisition of the Cell Line

A key element in the development of a monoclonal-based therapeutic is the cell line used to produce the antibody. Should it be necessary to import the cell seed stock, this will be a regulated process in most countries. In the case of the United States it is necessary to obtain a permit from the U.S. Department of Agriculture (USDA) in advance of the planned shipment. This entails completing an application (VS Form 16-3) and submitting it to the Import-Export Staff, VS, APHIS, USDA, 6505 Belcrest Road, Hyattsville, MD 20782. Depending on the country of origin, the conditions under which the cell line has been cultured, and the proposed use of the cell line, the USDA may require that the cell line be tested for the presence of animal pathogens. A memo detailing the applicable procedures can be obtained by writing the USDA.

Further to the process of cell line acquisition, it should be noted in passing that unequivocal permission for the commercial use of any biological material should be obtained from the donor before any extensive commitment to a particular project is made.

B. Origins of the Cell Line

In the review of a monoclonal-based therapeutic product, the initial area of FDA scrutiny will be the origin of the cell line to be used to produce the antibody. To address the agency's concerns, documentation of the origin and characterization of the cell stocks used to prepare the Master and Working cell banks (the cell banks used to generate the production cell stocks) is very important.

Although the documentation of the isolation and derivation of the cell line might seem a rather simple mechanical process, it has been the authors' experience that the timely recording of the origin of the cell line is crucial, as this information tends to be somewhat ephemeral and what data is available often contains a considerable body of anecdotal material. With the passage of time, accurate reconstruction of the cell line origin can become very difficult. The message here is to formally establish this record as soon as possible.

The information typically required to document the provenance of a cell line can be found in Ref. 3 and is supplemented by a recently revised publication, "Points to Consider in the Characterization of Cell Lines Used to Produce Biologicals," [5] from the Office of Biologics Research and Review. In addition to the information requested in these documents, in the specific case of human cell lines, it is very useful to have an extensive individual donor and family medical history. In addition, although not required, retained (–70°F storage) donor serum or plasma samples, if available, could also prove useful, should questions later arise regarding possible exposure of the donor to various infectious agents.

As well as describing the origins of the cell line, the procedures used in the derivation of the Master and Working cell banks must be clearly outlined. Also, regulatory agencies will normally ask, and common sense would insist, that the manufacturer of the therapeutic antibody be able to demonstrate that sufficient cell stocks are available to allow for extended production runs. Redundant storage, in liquid nitrogen, in at least two physically segregated areas is suggested. Archival storage of the cell line in an appropriate center should also be considered.

C. Characterization of the Cell Line

The parameters to be addressed in the characterization of the cell bank cell stocks are listed in Table 1. Two items, cell line viral contaminants and cell line

Table 1 Points to Address in the Characterization of a Monoclonal Antibody-Producing Cell Line

A.	Morphology
B.	Karyology
C.	Growth characteristics
D.	Freedom from adventitious agents
E.	Stability of immunoglobulin secretion
F.	Freedom from irrelevant immunoglobulin or light-chain secretion

stability, should receive particular attention during the development process. Both of these issues can affect the suitability of a cell line for the production of therapeutic material in serious but not necessarily immediately obvious ways.

The theoretical and real hazards associated with the presence of a pathogenic virus in a cell line used for the production of a biological therapeutic are readily apparent. Unfortunately, as there is no single test or combination of tests that will absolutely rule out the presence of all viruses within a cell line, it becomes a judgment call as to what testing is appropriate at each stage of the development process.

During the preclinical investigational stage, the cell line or, as appropriate, conditioned media from the cell line, should be subjected into laboratory animals (mice, Guinea pigs, embryonated chicken eggs), murine antibody production (MAP test, murine lines only), cytomegalovirus, hepatitis, Epstein-Barr and human immunodeficiency virus screening (human and human-derived lines only), cytopathic effect testing using at least three indicator lines, and reverse transcriptase testing.

Very frequently, electron microscopy examination of murine hybridomas will show the presence of endogenous murine retroviruses [6-8]. The presence of these viral particles does not necessarily bar the utilization of a cell line.

Negative results obtained with this level of screening will probably be considered sufficient by the FDA for initiation of Phase 1 clinical studies. Factors that reviewers will weigh in this risk/benefit assessment include the cell line history, production and purification procedures, and the proposed patient population. Sponsors should, however, be prepared to deal with additional viral concerns either at a pre-IND meeting or during the IND review process. Depending on the specifics of a given product, expanded viral testing may be required concurrent with the Phase 1 clinical investigation or in later phases of the clinical program. It is likely that studies such as repeat viral testing at a passage

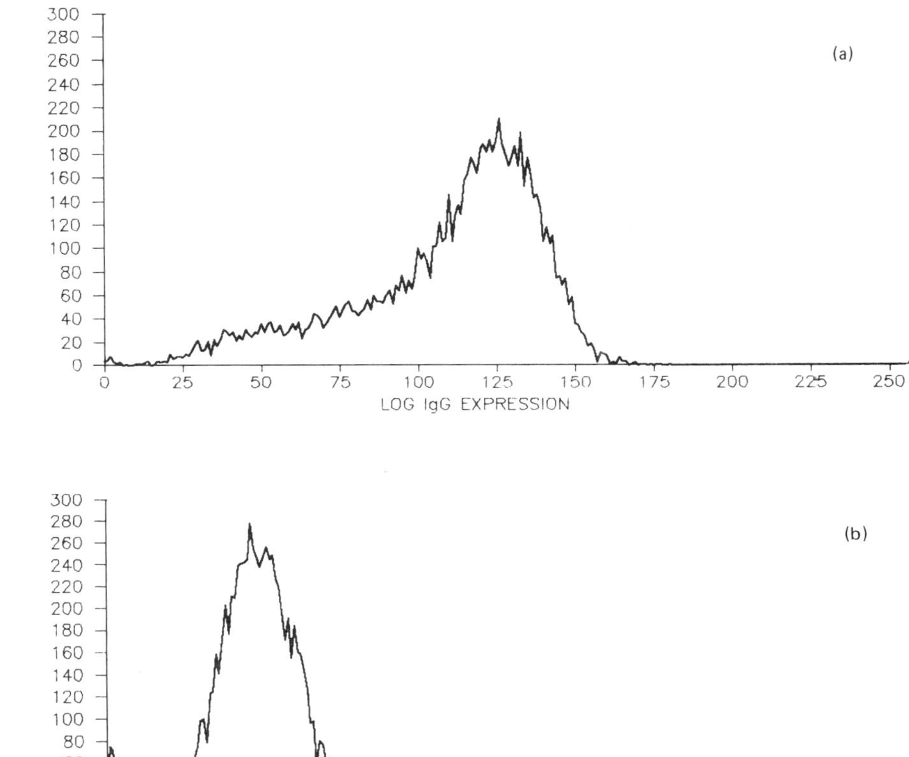

Figure 2 Cell line heterogeneity in immunoglobulin secretion for a murine hybridoma. At early passage (a), the majority of the cells are high secretors of IgG; with extended passage (b), the population shifts so that the majority of the population consists of low or nonsecretors.

number in excess of that reached during the production process, or testing for the presence of pathogenic viral subgenomes, will be requested. Human-derived lines in particular can be expected to be subjected to extremely rigorous scrutiny for vital contaminants. Investigators using such lines are advised to include, if possible, validated viral inactivation procedures in their manufacturing protocols. Since the experimental effort involved in some of these viral studies can be very costly in terms of both time and resources, it is prudent to seek expert consultation in these proposed studies in order to ensure that appropriate investigations are undertaken and that the concerns addressed are real and reasonable.

A second area of major concern is the stability of the cell line, an area that is frequently given too little attention in the early development process. An unstable cell line can result in significant wasted time and effort should the cell line prove to be unsatisfactory, and new Master and Working cell banks have to be qualified. In a recent review of over 20 hybridomas proposed for large-scale antibody production at Bio-Response, one-third demonstrated significant problems with long-term antibody secretion. In the problem cell lines, it was typically observed that with increased passage number, either nonsecreting subpopulations or subpopulations secreting nonfunctioning antibody would arise in the production cell stock. An example of this phenomenon is the spontaneous generation of a mixture of functional and nonfunctional hybridoma populations, as shown in Figure 2. In this case, a contaminating subpopulation incapable of assembling a functional antibody reproducibly arose in culture. This type of defect, which manifests itself by decreased production, can be readily detected through appropriate flow cytometry analysis of the cell line. Given the inherent genetic instability of hybridoma or virally transformed cell lines, the FDA will expect to see evidence of the stability of the production cell line. In our experience this is best achieved through multiple reclonings or other selection processes during the derivation of the cell line, combined with a critical examination of the growth and secretion characteristics of the proposed production cell line.

IV. THE PRODUCTION PROCESS

The nature of the regulatory agency concern with the specific method used to produce the antibody will emerge during the IND process. At the initiation of the Phase 1 trials, the pilot-scale production process will typically be rather loosely characterized. However, to avoid extensive recharacterization of the product during the clinical program, the sponsor would be well advised to have at least made the choice between in vivo or in vitro production systems. While either system is acceptable in principle, the comment [3] of the European Ad Hoc Working Party on Biotechnology/Pharmacy, "in vitro production of monoclonal antibodies at finite passage level offers prospects of a higher degree of control and standardization than in vivo production and is the preferred method," probably reflects current opinion on this matter. Since production processes vary widely, and agency concerns will be specific to a given production process and product, only items of general importance will be discussed here.

A. In Vitro Production

For in vitro production, two major points are the culture medium and the production passage number. Initial antibody production for limited clinical use may take place in serum-supplemented medium, but regulatory concern over product contamination with medium-derived proteins or adventitious agents will favor the development of a production process utilizing a defined serum-free medium. To further avoid regulatory concerns, other additives in the culture medium, such as nonhuman proteins, antibiotics (particularly the penicillins), steroids, endotoxins, and indicator dyes should be avoided or minimized. Since the culture medium markedly impacts the final product in terms of quality, price, and composition, a significant effort to develop a defined medium, early in the development process, will be amply repaid.

A limitation on passage number of production cell lines results from the knowledge that hybridomas and virally transformed cells are genetically unstable. This inherent instability has prompted regulatory agencies to limit the passage number (more properly the average number of cell divisions) a cell line can undergo in the production process. This is usually a somewhat arbitrary determination, and we recommend that sponsors generate data demonstrating cell line stability at passage numbers well in excess of that contemplated to be used. The small additional effort this requires can prove very valuable should modifications to the production system requiring extended culturing be desired.

B. In Vivo Production

If an in vivo production system is utilized, particular attention must be paid to the maintenance of the animal colony, the design of the documentation system, and the control of microbial contaminants. Animals should be obtained from a closed, specific pathogen-free colony. Documentation recording the freedom from viral contamination, and frequency of testing for viral contamination in the colony, is strongly recommended.

Each animal utilized is in effect an individual production run. To keep the records manageable when large numbers of animals are involved, extreme care must be taken in the design of the documentation system used to record the production process. In addition, given the inherent risk of adventitious agent contamination of the bulk ascites during collection, particular attention should also be paid to the methods given to control microbial, fungal, and viral contamination during ascites processing.

V. PRODUCT PURIFICATION AND CHARACTERIZATION

Regardless of the production system used, elaborate purification and characterization of the antibody will be required. Careful product examination should not be delayed in the race to the clinic, since without an extensive characterization it will be difficult to demonstrate that a newly devised production process yields product identical to that originally produced and tested. In addition, an adverse reaction observed in clinical studies, arising from an impurity that should have been detected and removed, could delay a development program for years. Of special note in the characterization of immunotoxins is the determination of the degree of heterogeneity in the product, as the biological properties of immunotoxins with different antibody/toxin ratios will be different. Since these materials will unavoidably have some degree of heterogeneity, careful product examination to determine the relative proportion of the various subspecies present and their antibody/toxin ratio is recommended early in the preclinical investigation process. It is, of course, very desirable to develop a manufacturing process that minimizes the degree of heterogeneity.

The design and validation of a purification process that will withstand regulatory scrutiny will be product-specific and is beyond the scope of this chapter. However, concerns most often addressed by regulatory agencies are addressed below.

1. What degree of purity is sufficient for a therapeutic antibody?

Based on our own experience and that of our colleagues, we consider it unlikely that parenteral-grade antibody, produced by a currently commercially feasible production process, can exceed a demonstrable protein purity in excess of 95%, when examined by all available analytical techniques.

Although it is not reasonable to expect a demonstration of purity >95%, it is reasonable to expect that no single, nonimmunoglobulin protein impurity should exceed 1% of the total protein, and that the identification of all contaminants has been attempted. Contaminating substances in this context include nonimmunoglobulin materials originating from the manufacturing process, immunoglobulin degradation products, chemically modified species, and aggregates. Degradation products are of particular concern for antibody conjugates, since fragmentation of the molecule can destroy the drug's targeting mechanism and yield free toxin.

2. How is DNA contamination measured and controlled?

Despite a lack of clear evidence demonstrating an associated hazard [9-11], DNA contamination in biotechnology-produced products for human use, such as monoclonal-based therapeutics, is currently limited to 10 pg/patient dose. In the early stages of the IND process, it is possible that clinical trials will be allowed to proceed with contaminating DNA levels in excess of that level. Nevertheless, it should a major goal to drive DNA contamination to as low a level as possible early in the development process.

Normally DNA levels will be determined by hybridization with a labeled probe; however, for products in which relatively large doses (>5 mg/kg) are anticipated, the accurate determination of 10 pg contaminating DNA in a patient dose may present considerable technical difficulties. Here, the theoretical DNA clearance value determined for the purification process (typically a six-to-eight log reduction) can be used to support a claim of acceptable DNA levels in the final product.

3. What is the permissible level of pyrogenic material in the final product?

Pyrogens, like DNA, can present significant difficulties both in control and measurement and, as with DNA, this may be particularly difficult to control when large doses of antibody products are contemplated. Determination of an acceptable level of pyrogens in the final product is product-specific and related to both total dose and rate of administration. Levels of pyrogens acceptable to regulatory agencies will likely decrease during the development process as the patient exposure expands.

For determination of endotoxin the Limulus Amebocyte Lysate (LAL) endotoxin test procedure, the LAL, has many advantages over the older, rabbit pyrogen test. However, at the present time, the rabbit procedure is still the "gold standard" for pyrogen determination. Thus, during the initial stages of the IND process, a rabbit pyrogen test is likely to be necessary for end-product testing. However, the speed, convenience, and inherent economies of the LAL test will motivate the eventual move away from rabbit testing. A guideline for the use of the LAL test as an end-product endotoxin test is available from the FDA [12].

4. How stable is the product?

At a minimum, sponsors need to have in hand sufficient stability data to demonstrate product stability throughout the planned duration of the initial clinical trial phase. However, since product licensure is very difficult without appropriate, long-term, real-time stability data, the sponsor should be well advised to begin these studies as soon as possible. It is particularly important that sufficient material (at least 30 vials) from three early lots utilizing the final product formulation be set aside to permit adequate stability testing during the course of the clinical investigations.

VI. QUALITY CONTROL/QUALITY ASSURANCE

Quality control/quality assurance considerations have been implicit in much of the preceding discussion of regulatory concerns in the manufacturing process of monoclonal therapeutics. Quality considerations for monoclonal therapeutics are discussed in the relevant "Points to Consider" [3], and specific topics such as quality assurance in cell culture [13], adventitious agents and other contaminants [14], and quality control in the production system [15] for these biopharmaceuticals have recently been reviewed.

Although an in-depth discussion of regulatory concerns with regard to drug quality would be too lengthy to present here, one area in particular, cGMP compliance, merits special consideration.

In the drug development process, questions frequently arise regarding the degree of compliance to cGMPs necessary for each stage. Fortunately, the FDA has provided relatively clear guidance on this issue in a recent draft guideline on the Preparation of Investigational New Drug Products [16]. In summary, it is the determination of the FDA that while material produced for preclinical study need not be produced in accordance with cGMPs, when the drug under development reaches the clinical investigation stage, compliance with cGMPs is required.

In reality, it is obvious that manufacture of investigational materials, particularly in the early stages of the clinical study, is unlikely to conform completely to cGMP requirements. Conspicuous deviations are especially likely to be found in areas such as process control and validation. Nevertheless, the sponsor would be well advised, even at the earliest point in the manufacture of clinical material, to produce the agent under conditions that will ensure the purity and integrity of the product, to keep complete records, to maintain an adequate quality control unit, and to retain samples of each lot. While it is possible, although unlikely, that the initial IND filing will trigger an FDA inspection, at some point prior to PLA or NDA approval, an agency inspection will occur. The more time and effort the sponsor has spent in putting a cGMP compliance program in place, the greater the likelihood of a satisfactory inspection.

The early establishment of a well-characterized "in house" reference standard, derived from a production lot, is crucial to product characterization and the development of product quality standards. This reference material will serve as the control for product purity, reproducibility, stability, and potency. The evaluation of these product characteristics will rest heavily on the twin analytical techniques of high-performance liquid chromatography (HPLC) and polyacrylamide gel electrophoresis (PAGE). At present in our laboratories, we find that PAGE techniques combined with sensitive silver staining methods offer the greatest protein resolution and most sensitive contaminant detection.

VII. PRECLINICAL SAFETY STUDIES

A. General Safety Concerns

The FDA currently recommends animal toxicity testing on all monoclonal products; the design and extent of these studies will depend, in large part, on the nature of the specific product under development. Traditionally, pharmaceutical products have undergone extensive safety studies, often involving years of testing in multiple animal species prior to entering clinical trials. In comparison, biologically derived products such as monoclonal antibodies have been subjected to a relatively limited amount of animal testing, especially by established toxicologic standards. This is due largely to the fact that the in vivo activities of many biological compounds are species-specific, and nonprimate animal models contribute only limited information about the potential toxicity of these agents. Nonetheless, animal studies are often required to obtain safety information about new therapeutic agents prior to their administration to human subjects.

A unique concern regarding the safety of monoclonal-based therapeutics is that the target antigen, or an immunologically related epitope, may be present on cells other than the intended target tissue and lead to undesirable cross-reactions and tissue damage. With drug, toxin, and radionuclide immunoconjugates, this type of cross-reactive toxicity is of even more concern. Adverse effects of monoclonal-based products due to this type of cross-reactive binding could be unique to each antibody, and therefore have to be carefully evaluated for each monoclonal-based product. To minimize the risk of cross-reactions, careful antibody binding studies are generally required by the FDA before a product can enter clinical trials. Cross-reactivity testing is discussed in more detail below.

Monoclonal antibodies coupled to drugs, toxins, or radionuclides will generally require more extensive preclinical safety studies than are required for unmodified antibodies. Obviously, adverse effects due to cross-reactive binding of the immunoconjugates is of significant concern, as inappropriate targeting of potent toxins or drugs could be directed to vital normal tissues. But nonspecific reactions must also be considered as well. Degradation of the conjugates in vivo could lead to nonspecific tissue localization of free drug, toxin, or radionuclide. There is concern that accumulation of conjugate in tissues responsible for clearance of these molecules, such as the reticuloendothelial system, could result in indirect toxicity as well. For some immunoconjugates, extensive preclinical testing programs that resemble traditional toxicology programs may be required. It should be emphasized, however, that the appropriate preclinical evaluation of a new drug should be decided in consultation with the FDA.

B. Design of Toxicology Studies

Preclinical toxicology programs should be designed to obtain sufficient safety and toxicity information about a new product to warrant the initiation of clinical trials. For an unmodified monoclonal antibody, rodent studies may be sufficient. For immunoconjugates, however, safety studies should generally be conducted in at least one nonrodent species, such as dog or primate, and preferably a species that shares the relevant target antigen to which the antibody is directed. It should be emphasized again that the extent of preclinical toxicology studies necessary to demonstrate sufficient safety information to initiate clinical trials will be specific to each new product.

Toxicity testing usually begins with a series of dose-ranging studies in rodents designed to determine the maximum tolerated dose, or MTD, after a single or short series of injections. With unconjugated monoclonal antibodies, this MTD may be difficult to determine because the antibodies generally exhibit little toxicity. Nonetheless, it is often helpful to demonstrate that short-term administration of doses 10 to 100 times the proposed clinical dose are well tolerated in animals. For monoclonal antibodies coupled to toxins, drugs, or radionuclides, these dose-ranging studies can be very important in determining the range of doses that can be safely used in later studies, including human clinical trials.

Once a dose range is defined, subacute toxicity studies of 2 to 4 weeks duration are usually conducted prior to submission of the IND. Immune responses to the monoclonal-based agents in treated animals make longer-term studies difficult to interpret. These studies should be designed so that the route and frequency of antibody administration are similar to that proposed for clinical use. The range of doses selected for study should include one equivalent to the highest dose anticipated in clinical trials and one or two doses that are multiples of this dose (usually 5-25 times the anticipated maximum proposed clinical dose). In general, studies should be designed to evaluate the maximum toxic effect of the agent, and animals should be sacrificed for histopathologic evaluation accordingly. Some animals should be sacrificed 24 hours after the final dose and others should be observed for 14 days after treatment and then sacrificed. And all animals should undergo clinical, hematologic, biochemical, and histopathological evaluations. It is important to emphasize that these preclinical studies should be conducted with antibody product in its final configuration and should comply with FDA Good Laboratory Practice (GLP) regulations (21 CFR Part 58). Long-term chronic toxicity studies, and reproductive and teratogenicity studies, when indicated, are often conducted after the IND is filed, in parallel with ongoing clinical trials.

C. Cross-reactivity

Cross-reactive binding activity with normal tissues other than the targeted tissue is a major concern regarding the safety of monoclonal antibody products, especially agents containing drugs, toxins, or radionuclides. Therefore, the FDA currently recommends that an immunohistochemical survey of human organs, blood components, and intended target cells or tissues can be performed to evaluate potential cross-reactive binding activity.

Recommendations regarding immunohistochemical staining procedures and the acquisition of tissue specimens are summarized in the most recent "Points to Consider" document for monoclonal antibodies [3]. The FDA recommends that some attempt should be made to quantitatively assess binding to target tissue relative to binding of cross-reactive tissues, which can be expressed as a target-to-nontarget binding ratio. It is important to note that these cross-reactivity studies should be performed using product in its final configuration; changes in product design or formulation will require retesting for cross-reactivity.

If an antibody demonstrates cross-reactive binding activity with a normal population of cells in a tissue section, a comprehensive evaluation should be conducted to determine its physiological significance. All immunohistochemical studies should be reviewed by an experienced pathologist. Some patterns of tissue cross-reactivity are due to physiologically insignificant intracellular antigens or to an artifact in the detection system. However, any cross-reactive binding activity with normal human tissues that cannot be attributed to an artifact of the staining procedure should prompt a comprehensive investigation to determine the possible in vivo sequelae of this binding activity.

Studies to determine the in vivo significance of binding cross-reactivity should be conducted, if possible, using an animal species that shares the same cross-reactive antigen detected in normal human tissue. Prior to conducting toxicity studies in an animal model, tissue sections from the animal should be evaluated for cross-reactive binding activity with immunohistochemical techniques similar to those used for human tissue. A model that shares a similar pattern of cross-reactive ginding activity to human tissue for a given antibody may be a suitable system to determine the physiologic significance of cross-reactive binding. Alternatively, an isolated human organ perfusion system may be employed in this investigation, e.g., a perfused human kidney for the evaluation of in vitro binding to renal tubular epithelium.

Even when appropriate test systems are not available to evaluate cross-reactive binding activity, clinical studies may still be warranted, even with immunoconjugates. These clinical studies, however, should proceed very cautiously, and they should include extensive evaluations for biodistribution of the agent using radiolabeled monoclonal antibody. The proper design and evaluation of studies of cross-reactivity will be facilitated by close communication with the FDA.

D. Immunogenicity

The development of immune responses to monoclonal antibodies and their conjugates is now well recognized [17,18]. Over 50% of patients treated with murine monoclonal antibodies have developed measurable human antimurine antibody responses, which can consist of both anti-idiotypic and antiframework components [19]. These immune responses have generally not been associated with increased risk of immediate hypersensitivity or serum sickness reactions, even with continued administration of the antibody. However, the host immune response may have an impact on the clearance, biodistribution, and efficacy of exogenously administered antibodies. It is important to note that neutralizing antibodies can be anti-idiotypic as well as antiframework, implying that treatment with human and chimeric antibodies may also be affected by the development of a host immune response.

From a regulatory point of view, the host immune response to monoclonal antibodies is important for several reasons. First, in preclinical testing, the development of an immune response in the test animals may affect the pharmacodynamics of the test article and, as a result, will make studies of more than 7-14 days in duration difficult to interpret. For this reason, some have advocated that the duration of repeated-dose toxicity tests should be determined by the length of time needed for the test species to develop a strong neutralizing response. In human studies, the development of an immune response markedly reduces the half-life of monoclonal-based drugs and may also reduce if not completely abolish the efficacy of these agents. This makes repeated courses of treatment with the same antibody unfeasible, and may prevent patients from receiving similar antibody preparations in the future. Therefore, the immunogenicity of antibody products should be evaluated and well characterized in all preclinical and clinical studies.

E. Monoclonal Antibody Conjugates

In general, the preclinical evaluation of monoclonal antibodies conjugated to drugs, toxins, or radionuclides should address issues of purity and safety in a manner similar for unmodified monoclonal antibodies. There are additional regulatory concerns, however, specific to the safety of immunoconjugates. Inappropriate targeting of a potentially toxic agent is a major theoretical concern regarding the human use of these immunoconjugates, and thorough testing for cross-reactive binding with normal human tissues should be conducted using the antibody conjugate in its final configuration. Special care should be taken to ensure that the antibody preparations used are free of immunoglobulin contaminants that could lead to the production of immunoconjugates of inappropriate specificity. Because immunoconjugates may dissociate in vivo, the toxicity profile of the free drug, toxin, or radionuclide to be used in the immunoconjugate must also be fully understood.

The FDA recommends that preclinical toxicity testing of immunoconjugates be conducted in at least one nonrodent species in addition to studies in mice or rats. Because immunoconjugates could produce harmful cross-reactivities in humans that could go undetected in small animals studies, the FDA recommends preclinical primate studies designed to evaluate in vivo toxicity and biodistribution. If another species can be identified that has the relevant target antigen or a similar profile in vitro cross-reactive binding activity, then this model could be substituted in the preclinical toxicity program. The preclinical evaluation program for an immunoconjugate should be evaluated on a case-by-case basis with the FDA.

Regulatory concerns with regard to the clinical use of monoclonal antibodies coupled to radionuclides, either for diagnostic or therapeutic purposes, include information about the absorbed dose, distribution, excretion pattern, and kinetics of the radioimmunoconjugates. Federal regulations state that sufficient data from animal studies or previous human studies should be available to allow a reasonable calculation of the radiation absorbed dose following administration of the radioimmunoconjugate to human patients [referenced in 21 CFR 312.23(a), Form FDA 1571, Section (10)(ii)]. Dosimetry calculations on animal data should also be performed prior to initiating clinical studies. Detailed recommendations regarding the conduct of dosimetry studies is contained in Ref. 3.

VIII. CLINICAL STUDIES

A. General Concerns

Virtually all of the preclinical work discussed so far—the characterization, production, and safety testing of monoclonal antibody products—is undertaken to obtain regulatory permission to initiate human clinical trials. During these trials a new compound is evaluated for its safety and efficacy in the treatment of a specific disease. Because human subjects are involved, the FDA becomes particularly active during the clinical testing phase of a new drug. Clinical trials are regulated to ensure that data derived from clinical trials is accurate and to protect the rights and safety of participating human subjects.

When monoclonal antibodies were first introduced into clinical trials a number of years ago, safety concerns centered around potential adverse effects, such as anaphylaxis and serum sickness, that might be caused by the antibody protein or its contaminants. However, clinical experience with monoclonal antibodies has shown that the actual incidence of hypersensitivity and anaphylactic reactions is quite low, even after repeated courses of antibody therapy [19]. Reactions consisting of low-grade fever and myalgias have been associated with treatment with anti-T-cell antibodies, but these reactions were non-life-threatening and were probably unique to antilymphocyte antibody therapy [20]. A symposium presented by the American Association of Immunologists reviewed the clinical use of monoclonal antibodies and concluded that "there has been very little toxicity of mouse monoclonal antibody administered to humans" [21].

One question that remains unresolved is the role of skin testing prior to the administration of murine monoclonal antibody products. A number of clinical protocols have incorporated skin testing with a small amount of antibody product to identify patients at risk for immediate hypersensitivity or anaphylactoid reactions. It is known that some patients develop IgE antimouse antibodies [19], but whether skin testing identifies these patients or whether patient with IgE antimurine antibodies are at risk for anaphylaxis has not been established. Some clinical studies have discontinued the use of pretreatment skin testing, arguing that the incidence of life-threatening allergic reactions is too low to justify the practice. The issue of skin testing prior to the administration of monoclonal antibody products is currently under consideration by the FDA and should be discussed with the agency prior to initiating clinical trials.

B. The Clinical Research Plan

The purpose of the clinical research plan is to outline in the IND application the general approach the sponsor plans to take in the clinical development of a new drug. The plan should identify the proposed indications, or labeling, for which the new drug is being developed. At the time the IND is submitted, the FDA does not expect a fully developed clinical research plan outlining all of the trials ever to be conducted in the development of the drug, but it does require that the initial Phase I studies be well designed and planned carefully. The better defined are the objectives and endpoints of the study, the more likely that the course of FDA review will run smoothly. As each phase of clinical research is completed, new studies can be amended to the IND for review and approval by the FDA.

Clinical trials generally are conducted in three phases, in which each phase provides the data to advance to the next, more extensive phase. In general, the objectives of each phase of clinical testing are as follows:

Phase I: To obtain information about the safety and pharmacology of a new compound in human subjects.
Phase II: Preliminary evidence of efficacy in a specific disease condition and additional safety information.
Phase III: Adequate and well-controlled studies to obtain evidence of efficacy in the treatment of a specific disease condition.

The design of the individual trials within each phase will be highly dependent on the nature of the antibody or antibody conjugate, results of preclinical studies, proposed use or indication of the product, and the results of any other clinical studies that may have been conducted.

To facilitate the design and implementation of sound clinical trials, the FDA has developed a guideline entitled "General Considerations for the Clinical Evaluation of New Drugs." These guidelines address study planning and design, and detail acceptable approaches to meeting FDA requirements. In addition, the FDA has published a series of guidelines on the clinical evaluation of specific classes of drugs in the treatment of certain diseases, such as "Guidelines for the Clinical Evaluation of Antineoplastic Drugs." These publications discuss general considerations that are relevant to the clinical development of monoclonal antibody based therapeutic agents.

C. Clinical Monitoring

Monitoring of a clinical study can be thought of as the principal method of communication between the study sponsor and the physician-investigator. In addition, existing regulations require that sponsors monitor the progress of clinical investigations to ensure the safety of human subjects and integrity of the resulting data submitted to the FDA. The monitoring functions may be performed by the sponsor's own employees or may be delegated to a contract research organization. It is considered desirable that the sponsor employ medical doctors as principal monitors of clinical studies, but veterinarians, clinical research associates, nurses, paramedical personnel, or other trained individuals may be acceptable monitors depending on the type of product and the nature of the clinical study.

The sponsor is responsible, through the clinical monitor, for assuring that the physician-investigator clearly understands and accepts the obligations incurred in undertaking a clinical investigation, and that these obligations are being fulfilled throughout the clinical trial. In addition, the sponsor is responsible for assuring that the data submitted to the FDA in support of the safety and efficacy of a new product are accurate and complete. The clinical monitor reviews patient records for accuracy during a series of periodic visits to the investigative sites. These monitoring requirements are outlined in a recently updated FDA document entitled: "Guideline for the Monitoring Clinical Investigations; January 1988."

IX. REGULATORY SUBMISSIONS

A. FDA Review Responsibility

It is often tacitly assumed that for monoclonal products the FDA group that will have the primary responsibility for review will be the Center for Biologics Evaluation and Review (CBER). Those products clearly related to traditional biologics (vaccines, blood and blood products) will be reviewed by the CBER. However, monoclonal conjugates could well fall under the jurisdiction of the Center for Drug Evaluation and Review (CDER). The determination of which review staff will be assigned to a particular therapeutic is determined by the FDA. Criteria used for making this decision are factors such as the intended clinical use and the underlying biological and chemical science base. Neither the method of manufacture nor the sponsor's wishes are considered the primary basis for determining the regulatory path.

In addition to determining which review staff will be responsible for evaluating a drug, the FDA will also assign an internal review priority for the investigational agent. The assignment of priority is based both on the novelty of the agent (a numerical Chemical Novelty Rating) and the importance of the therapeutic gain (a letter Therapeutic Potential Rating), with new entities having important therapeutic potential having the highest priority. Since monoclonal therapeutics are most probably going to be new molecular entities and thus receive the highest priority, a rating of 1, in the chemical novelty category, the sponsor's role in affecting the priority will resolve around the indication chosen for clinical investigation.

The choice of initial indication involves many issues such as potential market, competitive advantage, existing alternative therapies, drug efficacy, patient availability, etc., and a discussion of these issues and how they relate to drug development strategy is beyond the scope of the present discussion. The sponsor's decision with regard to indication typically has a relatively minor impact in the initial phases of the IND process, but the consequences of this choice radically effect the final licensing process. As with many decisions in the drug development process, the wisdom or lack therefore of a particular choice manifests itself after a period of many years and an expenditure of millions of dollars.

B. Elements of the IND

Prior to beginning clinical trials, sponsors of new drugs must package the results of preclinical testing into a document called a "Notice of Claimed Investigational Exemption for a New Drug," also known as an IND. It is important to note that the IND is not an application for approval; rather, it is an application for exemption. Current law requires that a drug must have an approved federal license (an approved NDA or PLA—see below) before it can be transported or distributed across state lines. To enable a drug sponsor to freely ship experimental drug to various clinical sites around the country, an exemption from that legal requirement must be obtained. Thus the IND is essentially a proposal asking for exemption from current law prohibiting the distribution of nonlicensed drugs.

Consistent with the introductory discussion of this chapter, there are no provisions in the regulations governing the IND process with specific application to monoclonal therapeutics. Therefore the discussion that follows is broadly applicable to all drugs.

In an IND application, the sponsor submits information addressing four major areas:

1. Information about the origin, characterization, and manufacture of the new agent.
2. Results of acute and short-term toxicity testing indicating that the drug is reasonably safe for use in humans.
3. Detailed protocols for the clinical studies proposed in the clinical plan. Generally, at the time of the IND submission, complete protocols are provided only for the Phase I studies and additional protocols are included at a later date.
4. Substantiating documents showing that the clinical investigators are adequately qualified to conduct the clinical study.

Provided these general areas are adequately addressed, the FDA is not very concerned about the actual format of the IND document. The required IND content is outlined in 21 CFR part 312, and much of the requested information is obvious and broadly applicable to all therapeutic entities. The specific forms needed—forms 1571, 1572, and 1573—describe exactly what needs to be put into the IND package. In May 1987, the FDA approved a set of revisions to the IND regulations—also known as the IND rewrite (IND)—that addressed the format of IND submissions. New IND submissions should generally follow these guidelines.

C. Types of INDs

There are three types of INDs: commercial, research (or physician-investigator), and treatment INDs. Commercial INDs are submitted primarily by manufacturers of new drugs whose ultimate goal is to obtain a product license approval to market the drug. The second type of IND is the research IND, often called an investigator or physician IND. These INDs are usually filed by physicians investigating a new drug for academic rather than commercial purposes. Submission requirements for research INDs are often much less stringent than those for commercial applications, primarily because fewer patients are generally involved in clinical studies. Agency requirements are even less stringent for drugs to be used in the treatment of rare diseases or diseases for which no alternative therapy exists.

The third type of IND is the treatment IND, under which promising investigational new drugs made be made commercially available to desperately ill patients before general marketing begins. This provision, which became effective in 1987, applies only to patients with serious and immediately life-threatening diseases for which no comparable or satisfactory alternative drug or other therapies exist. Generally, treatment INDs will be granted after adequate and well-controlled clinical trials have been completed and the drug is awaiting final review and approval. For drugs to treat immediately life-threatening diseases (less than 6-12 months survival), treatment INDs may be granted after Phase II studies are complete. However, the sponsor must demonstrate that providing drug through a treatment IND will not interfere with adequate accrual in controlled Phase III trials.

D. Pre-IND Meetings

The FDA offers conferences called "Pre-IND Meetings" to sponsors of new drugs who are planning or preparing an IND. The meeting can provide sponsors with an opportunity to review the adequacy of their preclinical testing program, manufacturing and characterization studies, and proposed clinical protocols. These meetings have become quite common for manufacturers of biological compounds such as monoclonal antibody-based therapeutic agents. Typically, representatives from preclinical and clinical research and regulatory affairs attend these meetings. The purpose of the meeting is to discuss the proposed research plan, and one should resist trying to get the FDA to preapprove an upcoming IND submission. Rather, use the meeting to discuss the proposed research plan and to initiate a collaborative relationship with the FDA review team.

E. Preparation of a PLA/NDA

All new drugs and biologics must have an approved New Drug Application (NDA) for drugs or Product License Application (PLA) for biologicals before they can be legally marketed in the United States. To obtain this approval, thousands of pages of data describing results of preclinical and clinical studies, drug characterization, and manufacturing processes are submitted to the FDA reviewers to demonstrate that (1) the agent is both safe and effective in its proposed use; (2) drug's proposed labeling is appropriate; and (3) the methods used in manufacturing the drug and the controls used for the quality control are adequate to preserve the drug's identity, strength, quality, and purity.

X. SUMMARY

Monoclonal antibody-based therapeutic products are subject to the same regulatory review procedures and submissions requirements as more traditional pharmaceutical agents. In addition, the FDA often asks for specific information on the unique aspects of the manufacture, characterization, and testing of these products. Agency concern begins with the origin of the hybridoma cell line and continues through preclinical safety testing and clinical trials. Specific regulatory issues are often decided on a case-by-case basis with the FDA reviewers. Implementing an effective strategy for identifying and addressing these issues and concerns is a critical part of the development process.

XI. REFERENCES

1. Coordinated Framework for Regulation of Biotechnology, Part II, Office of Science and Technology Policy, 51 FR 23302–23350 (1986).
2. Points to Consider in the Manufacture of Monoclonal Antibody Products for Human Use, Office of Biologics, Food and Drug Administration, July 25, 1983.
3. Points to Consider in the Manufacture and Testing of Monoclonal Antibody Products for Human Use (1987), Office of Biologics Research and Review, Center for Drugs and Biologics, Food and Drug Administration, June 1, 1987.
4. On the Production and Quality Control of Monoclonal Antibodies of Murine Origin Intended for Use in Man, Commission of the European Communities, Directorate-General Internal Market and Industrial Affairs, June 1987.
5. Points to Consider in the Characterization of Cell Lines Used to Produce Biologicals (1987), Office of Biologics Research and Review, Center for Drugs and Biologics, Food and Drug Administration, 19xx.
6. A. H. Bartal, C. Feit, R. A. Erlandson, and Y. Hirshaut, Detection of retroviral particles in hybridomas secreting monoclonal antibodies. *Med. Microbiol. Immunol. 174*:325–332 (1986); The presence of viral particles in hybridoma clones secreting monoclonal antibodies. *N. Engl. J. Med. 306*: 1423 (1982).
7. D. Stavrou, T. Blizer, T. Tsangaris, E. Durr, M. Steinecke, and A. P. Anzil, Presence and absence of virus particles in hybridomas secreting monoclonal antibodies against gliomas. *J. Cancer Res. Clin. Oncol. 106*:77–80 (1983).
8. M. Lyon and J. Huppert, Depression of reverse transcriptase activity by hybridoma supernatants: A potential problem in screening for retroviral contamination. *Biochem. Biophys. Res. Commun. 112*:265–272 (1983).
9. J. Petricciani and P. J. Regan, Risk of neoplastic transformation from cellular DNA: Calculations using the oncogene model. *Develop. Biol. Standard. 68*:43–49 (1986).
10. M. P. Moyer and R. C. Moyer, Potential risks of tumor virus subgenomes in the production of biologicals. *Develop. Biol. Standard. 68*:51–62 (1986).
11. J. Doehmer, Residual cellular DNA as a potential transforming factor. *Develop. Biol. Standard. 68*:33–41 (1987).
12. Guideline for Validation of the Limulus Amebocyte Lysate Test as an End-Product Endotoxin Test for Human and Animal Parenteral Drugs, Biological Products, and Medical Devices, Food and Drug Administration, December 1987.
13. R. W. Johnson, Quality assurance of cell cultures. *BioPharm* 43–45 (1988).
14. D. L. Prince and R. N. Prince, Quality assurance of monoclonal products: Virologic and molecular biologic considerations. *J. Indust. Microbiol. 3*: 157–165 (1988).
15. W. R. Tolbert, W. R. Srigley, and C. P. Prior, Perfusion culture systems for large-scale pharmaceutical production. In *Animal Cell Biotechnology, Vol. 3*

(R. E. Spier and J. B. Griffiths, eds.), Academic Press, London, 1988, pp. 374–393.
16. Draft Guideline on the Preparation of Investigational New Drug Products (Human and Animal), February 1988, Center for Drug Evaluation and Research, Food and Drug Administration.
17. G. J. Jaffer, R. B. Colvin, J. V. Cosimi, G. Giorgi, T. C. Goldstein, J. T. Fuller, C. Kurnick, C. Lillehi, and P. S. Russell, The human immune response to murine OKT3 monoclonal antibody. *Transplantation Proceedings XV, Vol. 1* 1983, pp. 646–648.
18. S. Harkonen, J. Stoudemire, R. Mischak, L. E. Spitler, H. Lopez, and P. Scannon, Toxicity and immunogenicity of monoclonal antimelanoma antibody-ricin A chain immunotoxin in rats. *Cancer Res. 47*:1377–1382 (1987).
19. R. W. Schroff and H. C. Stevenson, Human immune responses to murine monoclonal antibodies. In *Monoclonal Antibody Therapy of Human Cancer* (K. A. Koon and A. C. Morgan, eds.), Martinus Nijhoff Publishing, Boston, 1985.
20. K. A. Foon, R. W. Schroff, P. A. Bunn, D. Mayer, P. G. Abrams, M. Fer, J. Ochs, G. C. Hochino, S. A. Sherwin, and D. J. Carlo, Effects of monoclonal antibody therapy in patients with chronic lymphocytic leukemia. *Blood 64*: 1085–1093 (1984).
21. J. Ritz and S. F. Schlossman. Utilization of monoclonal antibodies in the treatment of leukemia and lymphoma. *Blood 59*:1–11 (1982).

5

Therapeutic and Drug Delivery Application of Monoclonal Antibodies

KAREN KASHMANIAN OATES
George Mason University, Fairfax, Virginia

I. INTRODUCTION

In 1975, Kohler and Milstein [1] published a report that had enormous impact on the field of immunology and the applications of antibodies as clinical reagents and chemotherapeutics. They described a method for the production of virtually unlimited amounts of homogeneous monoclonal antibodies (MoAbs) against desired antigens or haptens after in vivo immunization of mice with that hapten or antigen. Myeloma-lymphocyte cell fusions result in the formation of hybrid cells. The cultures of these fused cells are referred to as hybridomas. In the production of hybridomas, cells are separated by limiting dilutions to a single (monocell) culture and are cloned as progeny of the individual hybrid cell. If these cells secrete antibody, the antibody is then called a monoclonal antibody (MoAb).

Research on MoAbs has recently reached a stage where its applications have become as diverse as scientific research itself. Current applications include protein purification by affinity chromotography, diagnostic reagents for enzyme immunoassays (EIA) and radioimmunoassays (RIA), evaluation of malignant states by whole-body imaging techniques, as well as an ever-expanding list of medically related therapeutic uses such as the identification of tumor-associated antigens and treatment of cancer. Immunoglobulin reagents produced by hybridomas are allowing immunochemists to disect out single, homogeneous antibody species with desirable and advantageous properties from among the many antibodies found in conventional heterologous antisera. From the generation of hybridoma technology a new generation of immunoassay systems and therapeutic approaches has evolved.

II. PRODUCTION OF MONOCLONAL ANTIBODIES

By fusing an antibody-producing B cell with an appropriate myeloma tumor cell, it is possible to produce an immortal cell line that secretes antibody characteristic of the B cell. This line can be maintained in tissue culture or by successive passages in vivo into the peritoneal cavity of a mouse. Mouse, rat, and human tumor lines are available and suitable for the production of hybridomas. As a result, hybridoma cell lines produce antibodies of almost any specificity and of virtually unlimited supply. The method by which monoclonal antibodies are produced today does not differ greatly from the original work of Kohler and Milstein [1].

Antibody-producing spleen cells (which have been immunized with the antigen or hapten desired) are fused by use of polyethylene glycol (PEG-1500) or Sendai virus to mouse myeloma tumor cells. The tumor cells employed have been adapted to unlimited growth in tissue culture. When the two cells are fused, the resulting hybridomas have the property of immortality, an attribute of myeloma cells, and specific antibody secretion, an attribute of the primed spleen B cells. Selection of the hybridoma in culture from nonhybrid myeloma cells is accomplished by poisoning the unfused myeloma cells and by the natural death of unfused spleen cells in tissue culture. For example, a myeloma line that is defective in the enzyme hypoxanthine guanine phosphoribosyl transferase (HGPRT) will die in tissue culture medium supplemented with hypoxanthine, aminopterin, and thymidine (HAT). Aminopterin is the poison that blocks the main pathway of DNA synthesis. Only the hybridoma cells can survive in HAT medium, because the myeloma cell genome provides the ability to grow in tissue culture and the spleen cells contribute the functional HGPRT enzyme necessary to overcome the aminopterin-induced block. After fusion, the cells are transferred to wells in a tissue culture plate and evaluated for growth. The tissue culture supernatant from growing colonies is generally evaluated for antibody production by radioimmunoassay (RIA) or enzyme-linked immunosorbant assay (ELISA) procedures. The positive wells are cloned by limiting dilution and then are reevaluated for antibody production until positive antibody-producing cells from a single clone are obtained. The resulting antibodies are monoclonal because they are produced by a colony of cells derived from a single hybridoma cell, and as a result, each contains identical hybrid genetic material. The result, being monoclonal antibodies, recognize a single antigenic determinant and have a single affinity constant for that determinant. The cells can be passaged in ascites of mice that have first been primed with pristane (2,6,10,14-tetramethyl pentadecane). The hybridoma cells grow in the peritoneal cavity and secret antibody into the ascitic fluid, which can be collected. Teh growing tumor cells can be transferred to another mouse, returned to tissue culture, or stored in liquid nitrogen (–196°C) for future use.

Each hybridoma clone represents the amplification of a single antibody-forming cell. Of the approximately 500 hybrids that could be generated from the spleen of a single animal, as many as 20% of the surviving hybrids may produce the desired antibody. Unfortunately, many positive hybrids eventually stop growing or lose the ability to synthesize antibody. It is estimated that only 10–20% of all initially positive hybridomas continue to make antibody. The entire process from the initial immunization of mice to freezing in the monoclonal antibody-producing cells take approximately 4 months and requires significantly smaller quantities of antigen than the conventional polyclonal antibody procedure.

Despite the standard use and production of MoAbs, some major problems still remain. Common problems include the uncertainty of the immunization schedule and class of antibody produced (IgG, IgM, IgA, etc.), the instability of hybrids, and the need for a very sensitive RIA or ELISA for screening. The preference for specific antibody classes is due to their different biologic activity. An example of this is the need for IgG and/or IgM antibody to obtain good precipitation reactions: IgA antibodies are not suitable. In general, a monoclonal antibody displays desirable attributes such as high affinity for a specific antigen, which allows the detection of small amounts of antigen in a mixture and selectivity for a specific antigenic determinant or epitope without the problem of cross-reactivity. A large amount of homogeneous antisera is easily obtained after the initial isolation procedure by allowing the clones to grow in tissue culture or in ascites. A portion of the clones are frozen until needed, ensuring standardization of the antibody produced. This standardization allows laboratories throughout the world to use the same reagent antisera, so that results obtained by different laboratories are comparable.

When a mouse is immunized with xenogenic cells, the mouse recognizes many different antigenic determinants on the cell surface. It has been estimated that approximately 4.0×10^6 different antibodies can be produced in a rabbit from immunization with xenogeneic cells. Under these conditions the sera obtained has four major disadvantages: (1) The antibody titer is sometimes low; (2) the antibodies are heterogeneous; (3) supply is limited; and (4) the exact combination of specific antibodies is impossible to reproduce in a different animal. A heterogeneous population of antibodies is normally produced in vivo because lymphocytes recognize the foreign antigen in different ways and each lymphocyte produces a different antibody. It is this heterogeneity that enables the host to effectively ward off infection by phagocytosis, neutralization, or complement-mediated lysis, yet reduces the usefulness of heterologous antisera as immunochemical reagents.

In the production of conventional polyclonal antisera the immunogen must be highly purified to minimize cross-reactive antibodies, whereas the unwanted cross-reactive antibodies in monoclonal antibody production are merely eliminated during the screening phase. In the production of monoclonal antibodies an impure antigen can be used successfully for immunization. Monoclonal antibodies will by no means replace polyclonal antisera, because there are certain applications best met by polyclonal antisera rather than MoAbs. The monoclonals are very specific in their reactivity and are restricted to recognition of a specific epitope on the antigen molecule. Since monoclonal antibodies recognize only a single determinant on most antigens, they cannot generate a complex lattice that permits the formation of a precipitate required in an RIA [2]. This difficulty can sometimes be overcome by combining two or three MoAbs with known epitope specificity and known composition [3]. Monoclonal antibodies may reveal new cross-reactivities that are real and cannot be selected or absorbed out. Because monoclonal antibodies are directed against short sequences, cross-reaction can occur as the result of homologous sequences on otherwise unrelated molecules. It is also difficult to make a monoclonal antibody to a weak immunogen, and when they are produced they are more sensitive to inactivation. The entire procedure is more laborious than traditional polyclonal antisera production, but once a clone is isolated it can be passed in vivo, kept in tissue culture, and/or expanded and frozen, which results in an essentially infinite amount of antibodies. This protocol requires a small amount of antigen, which becomes particularly important when the antigen is in short supply. Once such an antibody has been generated, later preparation of the antibody will require only the injection of the hybridoma cells into a mouse to form ascites tumors, which under proper conditions can yield 10–20 ml of fluid with an antibody titer from 100 to 1000 times greater than that obtained in a polyclonal antibody procedure [4].

A. Human Hybridomas

The vast majority of hybridomas produced have used mouse or rat cells, yet the most direct application of these MoAbs has been for the clinical treatment of humans. The use of rodent antibodies in humans introduces problems associated with host-immune responses against foreign proteins. Human MoAbs would be better tolerated in human immunotherapy than antibodies raised in mice or rats, and it may be more feasible to generate tumor-specific antibodies in an allogeneic system [5]. Early work in human MoAb has been done with human lymphocyte-mouse myeloma hybrids [6,7], but there are increasing reports of human myeloma cell lines being described and used successfully to generate human lymphocyte hybridomas and human MoAbs [8].

The production of monoclonal antibodies by human-human hybridization has been limited due to the lack of a suitable fusion partner, a general lack of sufficient quantities of human B cells that are immunized against the desired antigen, fusion protocols that yield only a small percentage of the number of hybrids possible, and ill-defined culture conditions both for hybrid survival and antibody production.

Considerable effort has been focused on finding a human cell line that would be suitable as a fusion partner. Such a cell line must be capable of unlimited divisions, carry a selectable marker such as 8-azaguanine or 6-thioguanine resistance, not express immunoglobulin, and have a high fusion efficiency in terms of both cellular fusion and antibody production. The production of a large number of human B cells immunized against the desired antigen is equal in importance to finding a suitable fusion partner. There are two routes of immunization: in vivo, which can be ruled out for ethical reasons (except in cases where a disease state exists or a natural encounter produces suitable numbers of "active" B cells for fusion); and in vitro, which is not problem-free. In vitro immunization requires a cell source, which is most often peripheral blood lymphocytes (PBL) but occasionally involves spleen cells obtained from human splenectomies. Pokeweek mitogen and lipopolysaccharides have been shown to stimulate PBL and increase the number of functional hybrids [9,10]. Recently, techniques have been used to remove suppressor cells that can interfere with in vitro immunization [3]. In addition, most methods of in vitro immunization result in hybrids that produce IgM, which may not be suitable for the intended purposes.

Given a suitable fusion partner and a sizable population of "active," immunized B cells, the next challenge is to fuse the cells. Although the protocol (varied as it may be) for mouse-mouse hybridization is well established and functional, it may not be ideally suited for human-human hybridization. Current methods using PEG may result in hybrids formed of more than two cells. Statistically this may reduce the number of functional hybrids obtained, and when coupled with the lower number of "active" B cells obtainable for fusion, may significantly contribute to the lower numbers of functional hybrids often observed in human-human hybridizations. Although still in its infancy, cell hybridization by electrofusion promises to provide a means of fusing one lymphocyte to one myeloma cell [11]. This should allow an efficient utilization of "active" B cells. Assuming that a successful fusion occurs, it is necessary to provide conditions that increase the likelihood of hybrid survival. Mouse macrophages have been used as feeder cells, but there are reports of extensive phagocytosis of the hybrids, and variable results have been achieved with mouse spleen cells. Human monocytes have been used with success.

Although the production of human MoAbs is far from routine at this time, the stated problems may be minor compared to the challenges envisioned for in vivo applications. One must recognize that high antibody specificity may be due to the sum of all individual specificities present [12,13]. Ethical questions about using the products of human tumor cells for treatment also cloud the picture and force one to cautious evaluation of the promised revolution in the treatment of human disease with monoclonal antibodies.

III. APPLICATIONS

Current applications of MoAbs include its use in such diverse ways as reagents for diagnostic test procedures, enumeration and identification of infectious agents, purification of antigens, evaluation of biopsy material for the presence of tumor-associated antigens, and monitoring of lymphocyte subsets. The full potential of these versatile antibodies has not yet been realized. I will attempt to describe several of the current procedures that take advantage of the selective specificity of the MoAbs and introduce several experimental uses of these antibodies.

A. Immunochemical Agents

One of the earliest applications of MoAb was in purification of antigens from complex mixtures by affinity chromatography. Biochemists have used the absorption of heterologous antibodies onto solid-phase media with the subsequent eluting of a more purified compound from these media by differential gradients for many years [13]. With the advent of MoAbs, high-capacity, solid-phase absorbent of exquisite selectivity now allows purification of the antigen from a complex system by the same technique. In this case, however, the eluted antigen is selected for only one antigenic determinant [14]. MoAbs to cell-surface molecules have selectively isolated cell populations based on antigenic surface characteristics [15] and have been applied to identification of tumor-associated antigens on malignant cells [16]. Agarose-bound monoclonal anti-HLA-DR antibodies have been used to selectively purify lymphocytes from HLA-DR negative cells [17].

One of the most immediate applications for MoAbs has been their use in replacing conventional polyclonal or heterologous antisera in several RIA [18] and enzyme-immunoassays EIA [19]. The advantages of MoAbs is clearly that one can engineer an antibody specific for any given determinant and yield several antibodies with a range of affinities. All antibodies recognize the precise determinant, whereas only a small faction of antibody present in rabbit antisera would be expected to be specific for that determinant. Hybridoma cell lines also last forever—rabbits do not. The high specificity of the MoAbs adds greatly to the accuracy of the assay and allows laboratories throughout the world to use the same antibody for their determinations. This aids in standardizing hormone, enzyme, and tumor-associated antigen assays worldwide. In the heterologous rabbit system, even after laborious immunizations and purifications of antibody, there are often interfering, nonspecific antibodies within the heterologous mixture. Assays performed with heterologous antibodies that may sterically inhibit binding are most reproducible when carried out in a two-step RIA or EIA. MoAbs often have single-step, single-incubation protocols [20]. An inherent disadvantage of MoAbs is its single epitope binding to antigen, which is not always sufficient to detect antigen; yet the heterologous antisera are not suitable because of their many disadvantages. A mixed monoclonal preparation has been produced [21]. This "cocktail" of several MoAbs of known specificity and affinity allows reactivity of all the major epitopes represented on the antigen. This selected blend of MoAbs has been used successfully for quantitative assays and retains the specificity of each other MoAb employed. MoAbs can also be too specific in that if an epitope is altered only slightly, the antibody may no longer bind.

MoAbs also have potential use in affinity electrophoresis. MoAbs to human alpha fetal protein (AFP) have been produced and used in various experimental systems, such as crossed-affinity immunoelectrophoresis, rocket-affinity immunoelectrophoresis, and zone-affinity electrophoresis in agarose [22]. Experimental data indicates that MoAb-AFP and AFP-MoAb-AFP complexes were formed during the electrophoresis procedure. Affinity electrophoresis is a convenient and sensitive method for studies on the reaction of MoAbs with their antigens [23]. The technique offers the possibility of calculating the dissociation constants of MoAb-antigen and of antigen-MoAb-antigen complexes and evaluating the reaction of MoAbs with different antigenic determinants [22].

B. Immunoscintigraphy

Immunoscintigraphic procedures have become an exciting area for the use of MoAbs. Initial attempts at treating certain tumors with radiolabeled MoAbs have used iodine-131 (^{131}I). The MoAb serves as the specific binder to a tumor cell of known specificity and delivers radioactive tracers to the tumor site, allowing effective localization of the primary tumor and its metastases. Whole-body immunoscintigraphy after a single injection of the radioactive-labeled antibody may be helpful in staging patients and determining the appropriate next step to therapy. The radiolabeled MoAbs to tumor-associated antigens have been used for scintigraphic determination of tumor invasion in a number of clinically important malignancies [24]. The low levels of MoAb that are delivered to the tumor following I.V. administration, however, has limited this approach to the systemic therapy of tumors. Initial attempts with ^{131}I have met with problems arising from the in vivo deiodination of the radiolabeled MoAb, resulting in lower-than-anticipated levels of conjugate reaching the tumor site.

Recently, antibody fragments have been used in lieu of the intact polypeptide antibody [25]. Clevage products, most notably $F(ab')_2$ fragments of immunoglobulins, have been used successfully, with an improvement in reducing nonspecific uptake of the conjugate in both spleen and bone marrow [26]. When one fragment conjugate $F(ab')_2$ of anti-CEA (carcinoembryotic antigen) was used, the technique was found to be highly specific for radioimmunolocalization with no false positive images and imaging resolution in the range of 50-70% [27,28].

Radioimaging techniques have not only experimented with intact and whole antibody molecules but also with the proper choice of radiolable. More recently, several investigators have used bifunctional chelates for labeling of MoAbs with other beta- and alpha-emitting radionuclides including ^{90}Y [20], ^{67}Cu [30], ^{199}Au [31], ^{212}Bi [32], and ^{153}Sm [33] in an attempt to produce radioconjugates with increased in vivo stability and enhanced tumor deposition. To further improve immunoscintigraphy sensitivity, a cocktail of MoAbs (intact or fragments) has been used instead of only one [34]. This technique is similar to mixed MoAbs used in RIA procedures and should increase accumulation and binding sites on the tumor yet maintain the specificity to the tumor antigen.

C. Immune Therapy

One of the most dramatic potentials for the use of MoAbs is not in reagents or diagnostic procedures but in the field of therapy. This approach, although still in its infancy, utilizes MoAbs as "magic bullets" to target therapeutic agents to the site of malignant or diseased tissue. Advantages occur when drugs are coupled to MoAbs rather than used as free drugs. The drug-antibody complexes can concentrate at the tumor site, and any unwanted side effects associated with free drug should be diminished. In addition, the binding of drugs to antibodies may protect the drug from enzymatic degradation and prevent its rapid excretion, thereby increasing the drug half-life. Selective delivery to the target site could also lead to the use of smaller doses of drugs for treatment. Thus, the advantages are clear—no side effect, smaller doses, and selective targeting. One of the most widely used therapeutics has been ricin A [35]. The ricin molecule is a highly toxic protein isolated from castor beans. The molecule consists of two subunits that have different but complementary functions linked by a labile disulfide bond. The A chain is the enzymically active subunit, which functions only after absorption into living cells by interfering with protein synthesis at the 60S ribosomal subunit. The B chain serves to bind the molecule to carbohydrate moieties on the cell surface and mediates entry into the cell. The A chain, when not attached to the B chain, is incapable of binding to cell surfaces and its toxic effects are not seen. Other potent protein toxins such as abrin and diphtheria toxin have also been shown to consist of a toxic portion (A chain) covalently bound to a portion (B chain) that can bind to surface moieties on cells and thereby facilitate entry of the toxic peptide into the cell. The internalized toxic peptide kills the cells by catalytic inhibition of protein synthesis. By substituting specific antibody for the B chain, it should be possible to direct the peptide to cells for which the antibody is specific.

Ricin A chain has been used in multiple applications when coupled to agents that bind to cell surface membrane components. This coupling combines the potent intracellular cytotoxicity (via protein inhibition) of ricin A chain with the ability to bind to the cells of choice. Ricin A chain has been bound to monoclonal antibodies, hormones, and other cell surface ligands. Conjugates containing ricin A chain exhibit selective cytotoxicity for a variety of mammalian cells bearing cognate antigens or appropriate receptors. Ricin A chain conjugates are being increasingly evaluated for their ability to selectively kill specific cell types in vivo or in vitro.

Applications of ricin A chain conjugated to other materials include uses in histologic identification of normal cells and cellular function, histopathologic identification of abnormal cells, removal of unwanted normal cells from cellular suspensions, and removal of abnormal cells from cellular suspensions with potential therapeutic applications both in vitro and in vivo.

The possibility of utilizing the exquisite specificity of antibodies to direct cytotoxic agents to tumor cells has been considered since the studies of Ehrlich [36]. Studies using drug-antibody conjugates for this purpose have been hampered by the difficulty of raising antiserum specific for tumor cells, the inability to prepare purified antibody for drug conjugation, and the problems of preserving both pharmacologic and antibody activity after production of the hybrid molecules [37]. No matter what the immunotoxin, rapid clearing by the reticuloendothelial system is a major drawback. In an effort to overcome this problem with ricin A conjugates, Blakey and co-workers have developed a method to destroy the carbohydrate on ricin A chain by treating it with a mixture of sodium metaperiodate and sodium cyanoborohydride and then linking the deglycosylated A chain to the antibody. Immunotoxin prepared with deglycosylated A chain was cleared from the bloodstream more slowly than the intact molecule, so the deglycosylated ricin A chain immunotoxin should be a more effective antitumor agent in vivo because it is cleared from the blood more slowly and has greater opportunity to localize within the tumor target.

Therapeutics other than toxins have been conjugated to polyclonal antibodies since the 1950s, when Mathe bound methotrexate (MTX) to antibodies raised against L1210 leukemia cells. The conjugate was used for therapy in mice [38]. With the advent of MoAbs, the concept of using antibodies to target drugs to tumors has been reexamined. Indeed, the prospect of using antibodies as vehicles for isotopes, drugs, and toxins became a reality only with the description of monoclonal antibodies, which had specificity for tumors. Prior to this, xenogenic antitumour antibodies had little specific reactivity with tumors, and undoubtedly, little conjugate would have reached the tumor. By contrast, with most monoclonal antibodies used in vivo today, the major reaction is with the tumor.

Several factors have to be considered when drugs are conjugated to monoclonal antibodies. Anticancer drugs are of many different types, and include antimetabolites, alkylating agents, protein inhibitors, intercalating drugs, and microtubule-inhibiting drugs, all of which are complex chemical structures that can be finely tuned to act on the target so that they interfere with a particular biological function in the cell and lead to cell death. These drugs have certain chemical groups and are not amenable to chemical modification due to their importance in binding to the target. Thus, in designing drug-antibody conjugates, careful attention must be paid to the structure-activity relationship of the anticancer drugs and to the method of coupling. For example, in the case of the alkylating agent chlorambucil, any modification to the chloroethyl group will yield an inactive drug; however, the carboxyl group is amenable to coupling, as it is remote from the cloroethyl groups [39]. With methotrexate, modification of the pteridine nucleus yields an inactive drug, while the gamma-carboxyl group of the glutamic acid moiety is available for coupling to antibodies [40]. In addition, the linkage of drug to antibody must be stable in the plasma so that the drug is not released prematurely. With this in mind, spacers between drug and MoAb, cleavable at low pH or by lysosomal enzymes, have been developed that are active inside but not outside cells [41,42]. It is clear that lysosomal degradation is also necessary for activity of methotrexate and chlorambucil conjugates [43,44], and probably for several other drugs.

One of the drugs currently used is adriamycin, a member of the anthracyline class of antitumor agents that is used for the treatment of a variety of solid tumors [45]. Another drug, methotrexate, a folic acid antagonist, has an active ester of a gamma-carboxyl acid group that has been coupled to MoAb [46]. This conjugate was active in vitro, but again the activity of the methotraxate-antibody was only one-twentieth that of the free methotrexate. It would be advantageous to couple two to three times as much drug molecule onto the MoAb or to substitute a more toxic drug to deliver the therapeutic effect. Chlorambucil is an alkylating agent used for the treatment of chronic lymphocytic leukemia, lymphoma, breast and ovarian carcinoma. Bone marrow suppression is, however, a serious side effect [47]. In the initial work with chlorambucil using polyclonal antibodies, complexes were prepared by using either low or high pH, but the exact nature of the linkage was not known. Chlorambucil is of special interest because, when a nonconjugated mixture of chlorambucil and antibody was administered, it was found to be more effective than either component alone. The mechanism of this type of drug-antibody synergy is not clear. In preparing chlorambucil-antibody conjugates, an active ester derivative of the drug is made where the carboxyl group of chlorambucil is activated with N-hydroxysuccinimide and dicyclohexylcarbodimide in dimethylformamide, and this active ester is reacted with antibody [48]. Up to 30 molecules of chlorambucil can be attached to antibody with good protein recovery and antibody activity. By extracting noncovalently bound chlorambucil with ethyl acetate, it was found that 80% of the chlorambucil was covalently linked and 85% of the alkylating activity of the drug was retained. Recently, a novel "prodrug" has been used [49,50]. Since the gamma-amino group of melphalan is necessary for transport but not for alkylating activity, melphalan was converted to a monofunctional compound by acetylating the amino group. The acetyl derivative, N-acetyl melphalan (NaM), was found to be 25 times less cytotoxic than melphalan when conjugated onto MoAbs.

When drug or toxin antibody conjugates are used for small tumors that are primarily subcutaneous, they are more effective in eradicating the tumor. A major problem for their use with large tumors is the various physiological and anatomical barriers that the conjugates must cross to gain access to the tumor. In any case, no matter what the toxin or drug, the conjugate is only as good as the antibody produced.

IV. MONOCLONAL-ANTI-IDIOTYPIC NETWORKS

Idiotypes are the antigenic determinants on the antigen receptors of recognition site of cells concerned with immune regulation. The simplest example would be a small determinant consisting of four to five sequential amino acids in a protein, which could be duplicated if the same amino acid sequence occurred randomly in one of the hyeprvariable loops [51]. The hypothesis put forward by Niels Jerne about 15 years ago that the normal immune response involves, and is regulated by, an interacting network of idiotypes and anti-idiotypes continues to provoke much experiment and discussion [52]. The hypothesis envisions anti-idiotypic antibodies, or cells bearing comparable structures, interacting with target idiotype-bearing cells to suppress or stimulate them. This concept has led to the possibility of manipulating the idiotypic network.

Monoclonal anti-idiotype antibodies may serve as an internal-image anti-idiotype antibody that may mimic the structure of antigen, but an antigen present in the antigen binding site of the first antibody can block this kind of anti-idiotype antibody from binding to the site as if it were true antigen. These anti-idiotypic antibodies can serve as surrogates for the antigen and are potentially useful as vaccines and therapeutic reagents. As an alternative to the use of tumor antigens or tumor cells, so-called internal image antigens can be used to induce specific immune responses similar to responses induced by nominal antigen. These antigens can also compete with nominal antigens in binding assays. The anti-idiotype hybridoma route of making surrogate antigens could alleviate the problem of preparing large amounts of purified antigens associated with a given tumor [53]. Also, the internal antigens that are expressed in a different molecular environment may be able to overcome the immunosuppression in the host by stimulating "silent" clones, or by allowing T-cell help to become active, making the overall immune response stronger than the nominal antigen is able to do [54,55]. Furthermore, idiotype-based tumor antigens used for immunotherapy are free of the danger of transmitting tumor viruses, which may contaminate cell-based tumor viruses. This alternative and safe approach to immunotherapy with the use of internal image antigens as vaccines has generated a great deal of interest in recent years in the treatment of cancer and infectious diseases [56, 57].

Several laboratories have reported on the use of anti-idiotopes for the production of antibody reactive to virus, parasites, and bacteria as well as tumor-associated antigens. There are several reports [58,59] of the production of internal image anti-idiotypes that mimic human tumor-associated antigens. Nepom et al. [58] described polyclonal anti-idiotypic antibodies raised in rabbits against murine monoclonal antibody 8.2, an antibody specific for a human melanoma-associated cell-surface marker called p96. Mice immunized with anti-idiotype demonstrated delayed-type hypersensitive (DTH) reaction when challenged with p9-positive melanoma cells. Herlyn et al. [59] produced anti-idiotypic antibodies in goats against mouse monoclonal antibody to human gastric carcinoma antigen GA733. Raychaudhuri et al. [60] have generated monoclonal anti-idiotypic antibodies in an experimental tumor system and demonstrated the induction of tumor-specific DTH and inhibition of tumor growth.

Besides their potential use as tumor vaccines, anti-idiotype antibodies have been used to treat lymphoid tumors directly and through passive immunization. In this work, immunoglobulins present on the surfaces of neoplastic B-lymphoid cells have been used to raise anti-idiotype antibodies that have eliminated much or all of a tumor with negligible effects on residual normal tissues [61,62]. The antibodies have been covalently coupled with toxins such as ricin to deliver the latter inside tumor cells, or as surface-binding agents on tumor cells that activate cytotoxic effector functions against these cells in vivo. Individuals have reported complete remission of a follicular lymphoma with such passive anti-idiotype treatment [63], although other efforts such as passive immunotherapy with anti-idiotype antibodies have been less successful [64].

V. ANTIGENIC MODULATION

In 1963 Boyse et al. [65] first described antibody-induced decrease of membrane antigen expression in animals treated with heterologous antisera directed against a specific antigen. One of the earliest reports identified that when TL^+ thymic leukemia cells are incubated in vitro at 37°C with anti-TL antibodies (polyclonal or monoclonal), they become resistant to the lytic action of newly added anti-TL antibodies and guinea pig complement [66]. All IgG antibodies directed against TL^+ cells induced modulation, including the noncomplement fixing IgG_1 class. Modulation was found to be specific for the TL antigen and no other antigens on the surface of these cells. Both Fab and Fab'_2 or noncomplement fixing IgG_1 antibodies can also induce modulation. Since that time, antigenic modulation has become central to our understanding of the mechanisms of tumor escape from the in vivo potential therapeutic effects of monoclonal antibodies.

Schreiner and Unanue [67] described the series of events leading to redistribution of antigen and antigenic modulation following exposure of anti-Ig sera with B-cell membranes. First, within 1-2 min, membrane Igs are irregularly clustered in small patches. This patch formation is a passive phenomenon that is not dependent on cell metabolism. In contrast, it is influenced by the ligand valency. The second stage is capping, and here Igs are coordinately concentrated at a pole of the cell (the uropod) within 3-4 min; this stage is energy-dependent. Capping is completely prevented by inhibitors of the respiratory chain (for example, sodium azide, dinitrophenol) or of glycolysis (for example, fluoride, iodoacetamide). The last stage is the endocytosis and subsequent degradation of the capped Ig. New Ig will be synthesized and reappear in a diffuse pattern on B cells cultured in anti-Ig-antibody-free medium with the disappearance of surface Ig from the B-cell surface. There is an acquisition of resistance of the cells for cytolytic effects of anti-Ig sera and complement [68]. Antibodies produced against virus-induced proteins may provoke their redistribution on the virus-infected cell surface. Concomitantly, the appearance of resistance to complement-mediated lysis may be noted [69]. This has been observed both in vitro and in vivo for a number of viruses [70].

In order for MoAbs to bind successfully to the antigen or tumor-associated antigen, the antigen must be continually expressed. After an initial period in which the quasi-immediate and drastic depletion effect of MoAb is revealed, difficulties rapidly become apparent. First, after 10-15 days of treatment most patients develop anti-mouse Ig antibodies, which induce rapid catabolism of the monoclonal antibodies and annihilate their therapeutic effect. Second, and even more rapidly, the antibody target cells become resistant after losing the expression of the antibody-defined antigenic determinants.

Several protocols can be envisioned that avoid the deleterious effects of antigenic modulation on the therapeutic action of monoclonal antibodies. In the case of tumors, an attempt should be made to destroy all accessable tumor cells with the first one or two injections of the monoclonal antibody, before the onset of modulation. To achieve this goal the monoclonal antibody could be coupled to a toxin such as ricin or to a highly radioactive isotope that would lethally affect the target cells. In the case of immunosuppression, antibodies that recognize functionally important molecules should be selected, since modulated cells will remain incompetent. The search for nonmodulating antibodies is still another possible approach.

VI. FUTURE PROSPECTIVES: CHIMERIC ANTIBODIES

With all the advantages and potential of monoclonal antibodies, limitations persist. It is not always possible to generate antibodies with the precise specificity desired or with the appropriate combination of specificity and effector function. An exciting recent advance in the field of hybridoma technology is the development of transfectomas. Transfectomas are hybridomas that have utilized a combination of recombinant DNA technology and gene transfections to create a novel, chimeric immunoglobulin. The procedure involves the use of DNA segments encoding for the variable regions derived from the heavy and light chains of a mouse monoclonal antibody that are joined to the DNA segments encoding for the human immunoglobulin constants regions. Transfection of these expression vectors containing chimeric immunoglobulin genes may then be introduced into mouse myeloma cells. Stable hybrids result in the production of a hybrid immunoglobulin with human constant region function yet maintain the specificity of the immunizing agent [71]. Chimeric antibodies have taken advantages of the major strengths of the production of the classic mouse-mouse hybrids and human-mouse hybrids. These strengths include the immunizing agent's specificity, which can easily be achieved and conserved in the mouse variable region DNA segment, in addition to the nonimmunogenicity of using a human immunoglobulin constant region, which is of special interest when the antibody is used therapeutically over a long period of time.

In addition to the type of chimeric antibodies described above, others use similar recombination techniques to produce antibodies with novel effector function. These techniques allow the stable introduction of immunoglobulin DNA into myeloma cells, making it possible to use in vitro mutagenesis and DNA transfection to produce recombinant antibodies. Chimeric proteins have been produced in which the antigen-binding portion of the immunoglobulin is fused to an enzymatic moiety that has potential application for immunoassay procedures [72]. In other cases the constant region of the immunoglobulin DNA can contain DNA segments that may code for a variety of functions, including the binding of the constant domain via receptor to other effector cells of the body in an attempt to focus cytotoxicity to bacterial or tumor cells. An example of this application would be to maintain the specificity of the antibody to a tumor-associated antigen via intact variable genes and to introduce via recombinant technologies the gene segment encoding for the constant region of the immunoglobulin molecule with addition information to code a receptor for cytotoxic T lymphocytes or natural killer cells.

Initial experiments demonstrate the feasibility of producing these novel immunoglobulins molecules by gene transfection. Chimeric Ig molecules should provide a new family of reagents and therapeutics with wide potential application. Changing the Fc portion of the Ig can alter its ability to fix complement, transverse membranes, and elicit effector cells. Chimeric molecules may provide us with a useful approach to cancer therapy and treatment of autoimmune and infectious diseases.

VII. REFERENCES

1. G. Kohler and C. Milstein, Continuous cultures of fused cells secreting antibody of predefined specificity. *Nature* (London) *256*:495–497 (1975).
2. A. Nisonoff, Monoclonal antibodies produced by hybridomas. In *Introduction to Molecular Immunology* (Chapter 10). Sinauer, Saunderland, Mass., 1982, pp. 161–173.
3. P. H. Ehrlich and W. R. Moyle, Cooperative immunoassay: Ultrasensitive assay with mixed monoclonal antibodies. *Science 221*:279–281 (1983).
4. V. T. Oi and L. A. Herzenberg. In *Select Methods in Cellular Immunology* (B. B. Mischell, ed.), W. H. Freeman, San Francisco, 1968.
5. A. Rosen, G. Clements, G. Klein, and J. Zeuther, Double immunoglobulin production in cloned somatic cell hybrids between two human lymphoid cell lines. *Cell 11*:139–147 (1977).
6. M. Smith and K. Hirschhorn, Immunoglobulin production by human-mouse somatic cell hybrids. *Trans. Assoc. Am. Physicians 90*:281–285 (1977).
7. J. Schwaber, Immunoglobulin production by a human-mouse somatic cell hybrid. *Exp. Cell Res. 93*:343–354 (1975).
8. L. Olsson and H. S. Kaplan, Human-human hybridomas producing monoclonal antibodies of predefined antigenic specificity. *Proc. Natl. Acad. Sci. USA 77*:5429–5431 (1980).
9. L. K. Olsson, H. Kronstrom, A. Cambon-DeMouzon, C. Honsik, T. Brodin, and B. Jakobsen, Antibody producing human-human hybridomas. I. Technical aspects. *J. Immunol. Methods 61*:17 (1983).

10. K. A. Denis, R. Wall, and A. Saxon, Human-human B cell hybridomas from in vitro stimulated lymphocytes of patients with common variable immuno-deficiency. *J. Immunol. 131*:2273 (1983).
11. P. H. Ehrich, W. R. Moyle, Z. A. Moustafa, and R. E. Capfield, Mixing two monoclonal antibodies yields enhanced affinity for antigen. *J. Immunol. 128*:2709 (1983).
12. W. R. Moyle, C. Lin,R. L. Corson, and P. H. Ehrich, Quantitative explanation for increased affinity shown by mixtures of monoclonal antibodies: Importance of a circular complex. *Molec. Immunol. 20*:439 (1983).
13. P. Cuatrecasas and C. B. Anfinsen, Affinity chromatography. *Ann. Rev. Biochem. 40*:259–278 (1971).
14. P. Indiveri, A. K. Ng, and C. Russo, Isolation of Ia-like antigen-bearing cells from human peripheral lymphocytes through the use of a monoclonal antibody to framework determinants of Ia-Like antigens. *J. Immunol. Methods 39*:343–354 (1980).
15. P. C. L. Beverley, D. Linch, and D. Della, Isolation of human haematopoietic progenitor cells using monoclonal antibodies. *Nature* (London) *287*: 332–333 (1980).
16. V. Raso, J. Ritz, M. Basala, and S. F. Schlossman, Monoclonal antibody-ricin A chain conjugate selectively cytotoxic cells bearing the common acute lymphoblastic leukemia antigen. *Cancer Res. 42*:457–464 (1982).
17. T. A. de Kretser, J. G. Bodmer, and W. F. Bodmer, The separation of cell populations using monoclonal antibodies attached to sepharose. *Tissue Antigens 16*:317–325 (1983).
18. V. R. Zarawski, Jr., J. Novotny, E. Haber, and M. N. Margolies, Antibodies of restricted heterogenicity directed against the cardiac geycoside digoxin. *J. Immunol. 121*:122–129 (1978).
19. K. E. Rubenstein, R. S. Schneider, and E. F. Ullman, Homogeneous enzyme immunoassay. A new immunochemical technique. *Biochem. Biophys. Res. Commun. 47*:846–851 (1972).
20. G. S. David, R. Wang, R. Bartholomew, and E. D. Sevier, The hybridoma—An immunochemical laser. *Clin. Chem. 27*:1080–1585 (1981).
21. P. H. Ehrlich and W. R. Moyle, Cooperative immunoassay: Ultrasensitive assays with mixed monoclonal antibodies. *Science 221*:279–281 (1983).
22. J. Breborowicz, J. Gan, and J. Klosin, Application of monoclonal antibodies for affinity electrophoresis. *J. Immunol. Methods 102*:101–107 (1987).
23. J. Breborowicz, K. Gryska, and J. Gan, Microheterogeneity of human AFP demonstrated by lectins and with monoclonal antibody. *Tumor Biol. 7*: 230–233 (1986).
24. D. Pressman, E. D. Day, and M. Blau, The use of pair labeling in the determination of tumor localizing antibodies. *Cancer Res. 17*:845–850 (1957).
25. E. Henze, J. Kabel, W. E. Buchler, J. Waitzinger, W. E. Adam, and H. G. Beger, Tumor scintigraphy with ^{131}I anti-CEA monoclonal antibodies and $F(ab')_2$ in colorectal cancer. *Eur. J. Nucl. Med. 13*:125–129 (1987).

26. G. Mariani, L. Callegaro, N. Mazzuca, D. Macchia, S. Menacci, S. Ferrone, U. Rosa, and R. Bianchi, In vivo distribution of anti-human melanoma monoclonal antibodies. In *Protides of the Biologycal Fluids*, Vol. 31, (H. Peters, ed.), Pergamon Press, Oxford, England, 1984, pp. 971–976.
27. J. F. Chatal, J. C. Saccarini, P. Fumoleau, J. Y. Douillard, C. Curtel, M. Kremer, B. LeMevel, and H. Koprowski, Immunoscintigraphy of colon carcinoma. *J. Nucl. Med. 25*:307–314 (1984).
28. J. P. Mach, F. Buchegger, M. Forni, Ritschari, C. Berche, J. D. Lumbroso, M. Schreyer, C. Girardet, and S. Carrel, Use of radiolabelled monoclonal anti-CEA antibodies for detection of human carcinoma by external photoscanning and tomoscintigraphy. *Immunol. Today 2*:239–249 (1981).
29. L. C. Washburn, L. D. Blair, B. L. Byrd, and T. T. Sun, Comparison of [99-mTc] sodium pertechnate in an experimental blood-brain barrier lesion. *Int. J. Nucl. Med. Biol. 12(4)*:267–269 (1985).
30. S. V. Desphande, S. J. DeNardo, C. F. Meares, M. J. McCall, G. P. Adams, M. K. Moi, and G. L. DeNardo, Cooper-67-labeled monoclonal antibody Lym-1, a potential radiopharmaceutical for cancer therapy: Labeling and biodistribution in RAJI tumored mice. *J. Nucl. Med. 29*:217–225 (1988).
31. P. Anderson, A. T. M. Vaughan, and N. R. Varley, Antibodies labeled with ^{199}Au: Potential of ^{199}Au for radioimmunotherapy. *Nucl. Med. Biol. 15*: 293–297 (1988).
32. D. A. Scheinberg and M. Strand, Leukemic cell targeting and therapy by monoclonal antibody in a mouse model system. *Cancer Res. 42*:44–49 (1982).
33. G. J. Ehrhardt, A. R. Ketring, and W. A. Volkert, Production of Sm-153 for radiotherapeutic applications. *J. Labelled Comp. 23*:1370 (1986).
34. G. N. Sfakianakis and F. H. DeLand, Radioimmunodiagnosis and radioimmunotherapy. *J. Nucl. Med. 23*:840–850 (1982).
35. N. R. Worrell, A. J. Cumber, G. D. Parnell, W. C. G. Ross, and J. A. Forrester, Fate of an antibody-ricin A conjugate administered to normal rats. *Biochem. Pharmacol. 35*:417–423 (1986).
36. P. Ehrlich, *Collected Studies on Immunity*, Vol. II. John Wiley, New York, 1906.
37. G. T. Mathe, T. B. Loc, and J. Berhard, Effet sur la leucemia 1210 de la souris d'une combinaison par diazotion d'A-methopterine et de v-globulins de hamsters porteurs de cette leucemie par heterogreffe. *C.R. Acad. Sci. Paris 246*:1626–1628 (1958).
38. J. M. Cassidy and J. D. Douros (eds.), Anti-cancer agents based on natural product models. In *Medicinal Chemistry*, Vol. 16, Academic Press, New York, 1980.
39. M. J. Smyth, G. A. Pietersz, B. J. Classon, and I. F. C. McKenzie, Specific targeting of chlorambucil to tumours with the use of monoclonal antibodies. *J. Natl. Cancer Inst. 76*:503–510 (1986).
40. J. Kanellos, G. A. Pietersz, and I. F. C. McKenzie, Studies of methotrexate-monoclonal antibody conjugates for immunotherapy. *J. Natl. Cancer Inst. 75*:319–332 (1985).

41. A. Trouet, M. Masquelier, R. Baurain, and D. Deprez de Campenette, A covalent linkage between daunorubicin and proteins that is stable in serum and reversibly by lysosomal hydrolases, as required for a lysosomotropic drug-carrier conjugate *in vitro* and *in vivo* studies. *Proc. Natl. Acad. Sci. USA 79*:626–629 (1982).
42. E. Diener, U. E. Diner, A. Sinha, S. Xie, and R. Vergidis, Specific immuno-suppression by immunotoxins containing daunomycin. *Science 231*:148–150 (1986).
43. M. J. Smyth, G. A. Pieterz, and I. F. C. McKenzie, The mode of action of methotrexate-monoclonal antibody conjugates. *Immunol. Cell Biol. 65*: 189–200 (1987).
44. M. J. Smyth, G. A. Pietersz, and I. F. McKenzie. Potentiation of the intro cytotoxicity of chlorambucil by monoclonal antibodies. *J. Immunol. 137*: 3361–3366 (1986).
45. M. C. Garnett and R. W. Baldwin, An improved synthesis of a methotrexate-albumin-792T/36 monoclonal antibody conjugate cytotoxic to human osteogenic sarcoma cell lines. *Cancer Res. 46*:2407–2412 (1986).
46. W. C. Shen, B. Baliou, H. J.-P. Ryset, and T. R. Hakala, Targeting, internalization and cytotoxicity of methotrexate-monoclonal anti-stage-specific embryonic antigen-1 antibody conjugates in cultured F-9 teratocarcinoma cells. *Cancer Res. 46*:3912–3916 (1986).
47. R. A. De Weger, J. F. J. Dullens, and W. Den Otter, Eradication of murine lymphoma and melanoma cells by chlorambucil-antibody complexes. *Immunol. Rev. 62*:29–45 (1982).
48. G. A. Pietersz, M. J. Kanellos, I. Zalcoberg, and F. C. McKenzie, The use of monoclonal antibodies conjugates for the diagnosis and treatment of cancer. *Immunol. Cell Biol. 65*:112–125 (1987).
49. M. J. Smyth, G. A. Pietersz, and I. F. C. McKenzie, Selective enhancement of antitumor activity of N-acetyl melphalan upon conjugation to monoclonal antibodies. *Cancer Res. 47*:62–69 (1987).
50. M. J. Smyth, G. A. Pietersz, and I. F. C. McKenzie, The *in vitro* and *in vivo* antitumor activity of N-AcMeEL-F(ab$'$)$_2$ conjugates. *Br. J. Cancer 55*:7–11 (1987).
51. P. P. Chen, S. Fong, R. A. Houghten, and D. A. Carson, Characterization of an epibody. An anti-idiotype that reacts with both the idiotype of rheumatoid factors (RF) and the antigen recognized by RF. *J. Exp. Med. 161*:323–331 (1985).
52. N. K. Jerne, Towards a network theory of the immune system. *Ann. Immunol.* (Paris) *125C*:373–389 (1974).
53. M. Bhattachya-Chatterjee, M. W. Pride, B. K. Seon, and A. Kohler, Idiotype vaccines against human T-cell lymphobliotic leukemia. *J. Immunol. 139*:1354–1360 (1987).
54. C. Bona, E. Herber-Katz, and W. E. Paul, Idiotype-antiidiotype regulation. Immunization with a levan-binding myeloma protein leads to the appearance of auto-anti-(anti-idiotype) antibodies and to the activation of silent clones. *J. Exp. Med. 153*:951 (1981).

55. P. A. Cazenave, Idiotypic-anti-idiotypic regulation of antibody synthesis in rabbits. *Proc. Natl. Acad. Sci. USA 74*:5122 (1977).
56. A. Hatzubai, D. G. Maloney, and R. Levy, The use of a monoclonal anti-idiotype antibody to study the biology of a human B cell lymphoma. *J. Immunol. 126*:2397 (1981).
57. R. C. Kennedy, J. kw. Eichberg, R. E. Lanford, and G. R. Dressman, Anti-idiotypic antibody vaccine for type B viral hepatitis in chimpanzees. *Science 232*:220 (1986).
58. G. T. Nepom, K. A. Nelson, S. L. Holbeck, and K. jE. Hellstrom, Induction of immunity to a human tumor marker by *in vivo* administration of anti-idiotypic antibodies in mice. *Proc. Natl. Acad. Sci. USA 81*:2864 (1984).
59. D. Herlyn, A. H. Ross, and H. Koprowski, Anti-idiotypic antibodies bear the internal image of a human tumor antigen. *Science 232*:100 (1986).
60. S. Raychaudhuri, Y. Sacki, H. Fugi, and H. Kohler, Tumor specific idiotype vaccines. I. Generation and characterization of internal image tumor antigen. *J. Immunol. 137*:1743 (1986).
61. K. jA. Krolick, P. C. Isakson, J. W. Uhr, and E. S. Vitetta, BCL, a murine model for chronic lymphocytic leukemia. Use of the surface immunoglobulin idiotype for the detection and treatment of tumor. *Immunol. Rev. 48*: 81–106 (1979).
62. G. T. Stevenson and F. K. Stevenson, Treatment of lymphoid tumors with anti-idiotype antibodies. *Springer Sem. Immunopathol. 6*:99–115 (1983).
63. R. A. Miller, D. G. Maloney, R. Warnke, and R. Levy, Treatment of B-cell lymphoma with monoclonal anti-idiotype antibody. *N. Engl. J. Med. 306*: 517–522 (1982).
64. T. J. Hamblin, A. K. Abdul-Ahad, J. Gordon, F. K. Stevenson, and G. T. Stevenson, Preliminary experience in treating lymphocytic leukemia with antibody to immunoglobulin idiotypes on cell surfaces. *Br. J. Cancer 42*: 495–502 (1980).
65. E. A. Boyse, L. J. Old, and S. Luell, Antigen properties of experimental leukemias. *J. Natl. Cancer Inst. 31*:987–992 (1963).
66. E. A. Boyse, E. Stockert, and L. J. Old, Modification of the antigenic structure of the cell membrane by Thymus-leukemia (TL) antibody. *Proc. Natl. Acad. Sci. USA 58*:954–961 (1967).
67. G. F. Schreiner and E. U. Unanue, Calcium-sensitive modulation of Ig capping: Evidence supporting a cytoplasmic control of ligand receptor complexes. *J. Exp. Med. 143(1)*:15–31 (1976).
68. L. J. Old, E. Stockert, E. A. Boyse, and J. H. Kim, Antigenic modulation: Loss of TL antigen from cells exposed to TL antibody. *J. Exp. Med. 127*: 523–539 (1968).
69. M. B. A. Oldstone and A. Tishon, Immunologic injury in measles virus infection. IV Antigenic modulation and abrogation of lymphocyte lysis of virus infected cells. *Clin. Immunol. and Pathol. 9*:55–62 (1978).
70. L. Chatenoud and J. F. Back, Antigenic modulation—A major mechanism of antibody action. *Immunol. Today 5*(1):20–25 (1984).

71. S. L. Morrison, Transfectomas provide novel chimeric antibodies. *Science 28*:1202–1207 (1986).
72. C. Girardet, A. Vacca, A. Schmidt-Kessen, M. Schreyer, S. Carrel, and J. P. Mach, Immunochemical characterization of two antigens recognized by new monoclonal antibodies against human colon carcinoma. *J. Immunol. 136(4)*:1497–1503 (1986).

part three
IMMUNOCONJUGATES

6

Immunoconjugates in Drug Delivery Systems

JOHN B. CANNON and HO WAH HUI
Abbott Laboratories, North Chicago, Illinois

I. INTRODUCTION

A. When Is Targeting Necessary?

Traditional modes of drug delivery in the treatment or prevention of disease can be hampered by their nonspecific action, often leading to side effects. In addition, many drugs are not able to arrive at target areas in the body at effective concentrations, while others are prematurely inactivated or excreted. One way of circumventing this problem is to apply the drug directly to its intended site of action. The range of conventional formulations for local administration, such as sprays, inhalers, creams, eye drops, etc., has been augmented in recent years by several more sophisticated systems. These have included the Progestasert intrauterine device for contraceptive use [1] and the Ocusert system for pilocarpine delivery to the eye [2]. Other systems still under investigation include microspheres for intraarticular injection in arthritis [3]. However, this approach to "targeting" via a locally applied controlled-release system is of limited use in treating disseminated disease states or when the target is inaccessible (e.g., in cancer chemotherapy and enzyme replacement treatment). In these cases, systemic drug administration is necessary.

Indeed, if a drug entity could destroy tumor cells selectively, reach all tumor cells, and remain there for the amount of time needed to be therapeutically effective, there would be no need for targeted drug delivery. Unfortunately, such a drug is not available. When a drug is cytotoxic or has serious side effects, as is the case for cancer chemotherapy drugs, the need for targeting is even more acute. In the next section we review different modes of targeting.

B. Modes of Drug Targeting

Targeted drug delivery systems can be classified according to the methods employed for accomplished target site localization. They are classified as passive, physical, or active targeting [4].

Passive Targeting

Passive targeting refers to the natural distribution pattern of the carrier in vivo. The disposition of the carriers will be determined largely by their particle size and shape, surface characteristics (hydrophobicity/hydrophilicity), surface charge, and particle number. Therefore, it is possible to target the lungs and reticuloendothelial system (RES) passively.

Entrapment of large particles ($> 7\ \mu m$) in the capillary bed of the lungs has been investigated in radiodiagnostic imaging and also for the delivery of antitumor drug to the lungs [5]. The potential of achieving sustained release of acidic drugs in lung capillaries using diethylaminoethylcellulose microspheres has also been studied [6]. However, this material is not biodegradable and hence is not suitable as such for human clinical trial.

RES-mediated elimination of drug targeting systems is usually a disadvantage. Sequestration by RES cells not only decreases targeting efficiency but can also result in damage to the RES, particularly if the carrier payload is a highly toxic agent [7]. This can have serious consequences in cancer patients, since mononuclear phagocytes are active in preventing the spread of metastases [7,8]. In spite of these concerns, clearance of carrier particles by the RES can be investigated to gain benefit in cases where targeting to the liver and spleen is advantageous, e.g., liver disease [9].

Physical Targeting

In physical targeting, some characteristics of the environment are used to direct the carrier to a specific location or to cause selective release of its contents there. This is usually accomplished through an external mechanism, such as induced local hyperthermia or a localized magnetic field.

In the presence of specific serum proteins, principally lipoproteins, unilamellar liposomes can be designed to release their payload efficiently at their liquid crystalline phase-transition temperature [10]. Liposomes with phase-transition temperatures of about 42°C can be made to release their drug preferentially in a capillary bed subjected to local hyperthermia. Thus, increased tumor uptake of methotrexate has been achieved, resulting in moderate effects in suppressing tumor growth [11].

The use of magnetically responsive microspheres for targeted drug delivery was first reported by Widder et al. [12]. Albumin microspheres of approximately 1 μm diameter were prepared containing small particles of magnetite (Fe_3O_4) and, following intraarterial injection in rats, were made to localize in the tail at the site of an applied magnetic field. In this study, more than 50% of the microspheres were localized to the magnetized tail segment.

Subsequent studies with magnetically targeted biological response modifiers —e.g., interferon producers and T- and B-cell mitogens—have also yielded promising results [13]. Other magnetically responsive carriers, such as nanoparticles [14], erythrocytes [15], and emulsions [16], have also been investigated.

Active Targeting

In active targeting, the natural disposition pattern of a carrier is modified to target it to specific organs, tissues, or cells. Approaches include suppression of the RES or attachment of cell-specific ligands and monoclonal antibodies.

One approach in suppressing the RES is to preblock the Kupffer cells with a placebo colloid of other RES-depressant agents such as dextran sulfate 500. Subsequent injections of drug carrier in treated animals are not fully subjected to RES clearance [17,18]. The clinical potential of the approach is doubtful, since impairment of RES function can have serious side effects—particularly in cancer patients [19]. Indeed, any drug-targeting carriers accumulating in the RES must be readily degradable so that RES overload does not occur with repeated dosing.

Cell-specific ligands can also be utilized to target carriers to specific cell types. Desialylated fetuin has been used to transport a variety of DNA synthesis inhibitors to mouse hepatocytes infected with ectromelia virus [20,21]. Similar approaches may be applied to target a number of other sites [10,22]. However, this approach is limited to a small number of tumor types [23].

Since the discovery of monoclonal antibodies (MoAbs) in 1975 [24], the concept of a monoclonal antibody-targeted cytotoxic modality has been rigorously explored. Several approaches have been studied concerning the design of such a system [25]. Direct covalent attachment of cytotoxic molecules to MoAbs is generally not very efficient, resulting in only approximately 10 drug molecules per antibody and thus diminishing the antigen-binding capacity [26]. Conjugation via intermediate carrier molecules such as dextran or poly L-glutamic acid [27,28] will increase the drug-antibody ratio and retain a higher degree of antibody activity. MoAb-mediated targeting also offers a distinct advantage in terms of payload capacity [29].

II. DRUG-ANTIBODY CONJUGATES

Although the concept of targeting cytotoxic agents was proposed by Ehrlich early in this century [30], its feasibility was not very promising until the advent of monoclonal antibody production [24] and the development of improved heterobifunctional coupling reagents [31].

The rationale for the use of conjugates is that a chemical entity with good pharmacological action is coupled to an antibody reacting against the target organ in order to obtain a conjugate that is therapeutically active and selective in action. Plant and bacterial toxins have received much attention in this field [31,32] because of their cytotoxic potential. It was proposed that a single molecule of such toxins is able to kill a cell on entry [33,34]. Although this high cytotoxicity is desirable, it requires that cross-reactivity of the antibody toward normal tissues should be very low. Theoretically, due to the number of potentially different antigens available on a cell surface and the cross-reactivity of some antibodies to similar structures, one would likely not expect absolute specificity of monoclonal antibodies, and hence the usefulness of these potent toxin antibody conjugates is limited. Alternatively, investigators have tried to couple conventional chemotherapeutic drugs that have clinically known and accepted side effects.

Conjugates prepared by coupling drugs directly to antibodies have been successfully synthesized. However, this type of conjugate has a limited potential because there are only a small number of functional groups available per antibody molecule that can be used successfully for chemical coupling without significant loss of antibody binding activity [35]. In order to obtain a higher molar substitution of drug to antibody, a carrier molecule is used to which the drug is attached and which in turn is linked to the antibody [36].

A. Selection of Antibodies/Target Antigens

For the production of monoclonal antibodies, usually a mouse is immunized with the antigen of interest, e.g., human tumor cells. Once the immune response initiates, B lymphocytes from the animal's spleen cells or lymph nodes are obtained in a single-cell suspension. These cells are then fused with myeloma cells from the same species [24,37-39]. By these techniques, antibodies have been harvested representing a number of different classes and isotypes. They are usually available to different epitopes on the same antigen. In order to accomplish effective targeting, we need affinity of 10^8 M^{-1} or better. Moreover, IgG antibodies are preferable to IgM because they are easier to handle and may concentrate better at the target due to their smaller size. For therapeutic purposes we also need hybridomas that are able to produce the large quantities of antibodies required for clinical studies.

Whether whole antibodies or antibody fragments are preferable for tumor targeting is difficult to predict. There will probably be different requirements in different situations. One advantage of the fragments is that they are less immunogenic than whole antibodies and their smaller molecular size may facilitate penetration into tumor tissue [40].

Furthermore, it is advantageous to have human rather than mouse antibodies because they are "less foreign" to patients and may also possess stronger therapeutic activity. However, human antibodies may still trigger immune responses when repeatedly injected into patients because they have both idiotypic and allotypic antigenic determinants. Although the idiotypic responses are likely to be weaker and need a longer period of exposure than those against mouse antibodies, some clinical procedure needs to be designed to tailor the dosage regimens for human patients.

While considering an antigen as a target for therapy, both the degree of antigen specificity and the number of antigenic molecules per tumor cell are important. The importance of antigen specificity is obvious because it allows the conjugates to anchor on the target selectively and accurately. However, an entirely tumor-specific antigen would be useless for targeting if its expression at the cell surface would not permit the binding of the number of conjugates needed to achieve the therapeutic response. In most cases, drugs need to enter tumor cells to exert an effect. Therefore, active drug has to be released from the conjugate at the tumor site, or the drug-antibody conjugate has to be taken up by the cells. Studies have shown that some conjugates were endocytosed [41], but the degree of endocytosis may vary among different antigens. Thus, conjugates that are specific to those antigens that lead to the maximum degree of endocytosis may be more effective for intracellular drug delivery.

Certainly, procedures facilitating endocytosis will play a role in utilizing drug conjugates for therapy. Thus, a concerted effort should be made to study what factors influence the degree of endocytosis of conjugates, the rate of endocytosis, and the kinetics of regeneration of surface antigens.

B. Selection of Drugs for Conjugate Formation

The mechanism of action and pharmacokinetic profiles of the drugs are important parameters for selection of drugs to be conjugated with antibody. Most antibodies are specific for antigens at the cell surface, so a drug that is active at the cell surface would be more effective. Howerver, only a few currently available chemical entities are effective through this mechanism; most drugs are active intracellularly. Some of them exert their action at DNA/protein synthesis, while some alter the function of the spindle apparatus [42]. Therefore, the ideal drug-antibody conjugate should be endocytosed and the conjugate should be cytotoxic even if the drug remains bound to the antibody. Unfortunately, some drugs, such as anthracyclines, are inactive unless they are cleaved from the complex. Nevertheless, some drugs such as alkylating agents may be effective even as intact conjugates.

Drug-antibody conjugates that are not endocytosed may still be therapeutically effective as long as they are able to increase extracellular concentrations of free drug at the target site. The strength of the covalent bond between antibody and drug will determine the extent of free drug to be released, hence the amount of drug uptake by cells.

To determine which drug is preferred for use in the conjugate, we also need to study the intracellular metabolism of drug-antibody conjugate. However, it is not necessary to know the details of intracellular conjugate disposition before initiating clinical trials to determine its therapeutic efficacy.

The choice of coupling method is another important parameter for optimizing the efficacy of the antibody-drug conjugate. There are many choices now available, but the most success has been achieved with heterobifunctional agents, since these reagents generally allow controlled coupling and highest yields of the desired immunoconjugate. The most widely used is N-succinimidyl 3-(2-pyridyldithio)propionate (SPDP) [61]. The product of the SPDP procedure contains a disulfide bond as the linkage between the antibody and the drug. Even under the controlled conditions that this reagent allows, however, a heterogenous mixture of products is generally obtained. Thus, the molecular weight and the ratio of drug to antibody can be quite variable, particularly if the drug is a protein with a large number of amino groups, as was observed with immunoconjugates of cobra venom factor [62]. The activity of drug and antibody portions of each component of the mixture can be variable, and separation of the mixture is usually not practical. Therefore, when interpreting the results of any experiment involving immunoconjugates, it must be remembered that one is not dealing with a single chemical entity.

Furthermore, immunoconjugates can have varying degrees of stability in vivo, depending on the nature of the linkage. This covalent linkage may be subject to enzymatic hydrolysis or other reactions. The disulfide bond of the SPDP product, for example, is probably subject to reduction via thiol-disulfide exchange with glutathione and other thiols [63]. However, this factor can be utilized in the release of free drug if the linkage is more susceptible to cleavage under intracellular conditions (e.g., glutathione, low pH, and enzymes of the lysosomes) than by plasma. In this way, the stability of the immunoconjugate can be maintained until delivery to the target cells, and active drug can be released in the target cells. Whether this condition is met for a particular immunoconjugate will depend on many factors, and must be evaluated for each individual case.

C. Pharmacodynamic Considerations in the Use of Conjugates for Drug Delivery

The therapeutic efficacy or pharmacological action of a drug correlates better with the concentration-time profile of the drug (or its active metabolite) in the plasma (or some other biophase) than with the absolute amount of drug administered. For drug-antibody conjugates or immunotoxin-antibody conjugates, several other factors must be carefully investigated or considered before such conjugates can be rationally used in therapeutic trials in humans. The in vivo stability of different types of chemical linkages must be assessed, the systemic elimination rate and metabolic fate/mechanism of these immunoconjugates have to be investigated, and the optimal dosing regimens for immunoconjugate delivery to target organs/cells must be determined. Thus, in the past decades, significant efforts were contributed in understanding the biodistribution of various immunoconjugates.

For example, the biodistribution of Ricin Toxin A (RTA) chain conjugate has been studied in a number of species [43,44,55]. Immunotoxins constructed by conjugating RTA to monoclonal antibodies that react with tumor-associated antigens are cytotoxic for tumor cells in vitro [31,46]. When administered systemically, these immunotoxins have been shown to inhibit growth of solid human tumors including osteogenic sarcoma [48] and colorectal cancer [49]. However, pharmacodynamic studies of RTA showed that the glycoprotein is rapidly eliminated from the systemic circulation with predominant localization in nonparenchymal hepatic cells [50,51]. It was also shown that the conjugate is removed from the systemic circulation much more rapidly than the antibody but less quickly than the free Ricin A chain [44]. Such a pattern of biodistribution is caused by the recognition of mannose-containing oligosaccharide chains of Ricin and Ricin A chain by specific cell surface receptors on nonparenchymal hepatic cells. Several investigators have investigated the interaction between mannose-containing oligosaccharides of RTA and the mannose receptors on Kupffer cells in the liver [51], rat bone marrow macrophages [52], and rabbit alveolar macrophages [53]; this is thought to be the mechanism by which hepatic uptake of RTA-containing immunotoxins occurs and suggests that a mannose-containing blocking agent might be used to prolong the systemic circulation time of the conjugates and hence increase targeted localization [43].

A study reported by Byers et al. [43] has shown that a number of blockers are effective, including defined species such as mannosyl-lysine. The study also shows that the only other organ in which immunotoxin uptake was inhibited by mannose-containing blocking agents is the spleen. This inhibition is probably caused by blockage of the mannose receptors on dendritic macrophages in that organ.

Besides Ricin Toxin A, biodistribution of several other conjugates has been studied [45,47,54,56-60]. For example, conjugates with methotrexate in which the drug is linked to the antibody via an albumin carrier have been reported to be eliminated rapidly from the systemic circulation [58], and particularly to the liver. Parallel studies in mice with human tumor xenografts [58] have shown blood survival, biodistribution, and tumor localization to be similar with conjugate and free antibody. Gamma camera imaging studies performed to assess the biodistribution and tumor localization of methotrexate-antibody conjugate revealed a similar biodistribution and systemic clearance to that found with unconjugated antibody [57,58].

Obviously, a significant component of the therapeutic approaches with monoclonal antibodies and their drug conjugates is a good understanding of their biodistribution and tumor localization capacity. Although there is an increasing body of data on the biodistribution of drug conjugates, the pharmacodynamics of various conjugated or complexed antibodies may be entirely different and require comprehensive investigation tailored to each individual conjugate.

III. EFFECT OF DOSAGE FORM AND ROUTE OF ADMINISTRATION

A. Route of Administration

The route of administration should be chosen carefully when using an antibody-drug conjugate for therapy. The conjugates have high molecular weight, are often unstable in vivo (due to both proteolysis of the antibody portion and to lability of the antibody-drug linkage) and are expensive. Thus it is imperative that the route of administration be chosen to ensure the maximum possible bioavailability, and to deliver the highest fraction of dose to the target organ or tissue. Most of the animal or clinical studies reported to date have been performed with this in mind. Intravenous administration is in most cases the obvious route to use, since patients currently expected to receive this kind of treatment are under specialized care and are critically ill. Most of the pharmacokinetic and biodistribution studies described above (Section II.C) were performed using the intravenous route, and many of the studies cited demonstrate the efficacy of intravenously administered immunoconjugates to target to the desired organ or tissues. Multiple intravenous injections, over a course of weeks, appear to be more effective than single injections, as indicated by, e.g., animal model experiments with gelonin conjugates [64].

Some success, however, has also been achieved with direct or local injection, so that the administered dose is concentrated initially in the target organ or tissue. Local injections of immunoconjugate, e.g., intratumorally, may have clinical advantage; this procedure showed tumor reduction in an animal model with repeated intratumoral injection of an immunotoxin [65].

Intraarterial injections have been used in some cases, whereby a drug is administered directly into the supplying artery of the target organ or tissue [56, 67]. The degree of advantage of this route over intravenous administration of immunoconjugate will depend on the pharmacokinetics of the immunoconjugate, the blood flow through the target tissue, and other factors. In general, drugs with a high systemic clearance or degradation administered intraarterially into a target organ with a low blood flow are good candidates for this technique [66]. In a recent clinical trial conducted in Japan, 39 patients with colon and rectal carcinoma, including eight patients who also had liver metastasis, received an immunoconjugate composed of neocarzinostatin (NCS) conjugated to the monoclonal antibody A7 (specific for colon and rectal carcinoma cells) [68]. The patients received the immunoconjugate via the artery proximal to the tumor, either directly or by a catheter inserted from the femoral to the tumor artery, followed by surgical resection of the carcinoma. An immunoperoxidase staining study showed that in all patients, the NCS was specifically localized in the cancer cells. The eight patients who had postoperative liver metastases were also given the A7-NCS conjugate through the hepatic artery; four of them responded favorably. However, since only one patient in this trial received the immunoconjugate intravenously, no conclusions can be drawn as to whether the intraarterial route had any advantage over the intravenous route.

Intraperitoneal injections of anticancer drugs have been shown to be more effective than intravenous administration in the treatment of certain types of cancer [69], especially those that are located in or spread to the peritoneal cavity, e.g., colorectal cancer and ovarian carcinoma. In a number of studies [70,71], tumor cells were implanted into mice intraperitoneally, and after the tumor was established, an immunoconjugate was injected intraperitoneally, resulting in reduction of the tumor and thereby demonstrating the utility of an intracavity administration. Ovarian carcinoma represents a good example of situations in which intraperitoneal injection of immunoconjugate following surgical removal would be beneficial, since spread of the tumor is frequently confined to the peritoneal activity [72]. Normally, transfer of intraperitoneally injected drugs from the peritoneal cavity to the vascular system is complete within several hours, but the presence of ascitic fluid (as would sometimes be the case for ovarian cancer patients) was shown in an animal model to retard the transfer of MoAbs from the peritoneal cavity to blood [73]. The same workers showed that modification of the antibody with galactose led to its rapid systemic clearance by receptors of the liver, and this was proposed as a method to reduce the systemic toxicity of any immunoconjugate that does enter the circulation from intraperitoneal injection. Intraperitoneal administration of a radioimmunoconjugate indicated the potential of conjugates such as these in intracavity cancer therapy; bismuth-212, an alpha emitter with a physical half-life of only 1 h, was conjugated via a diethylenetriamine pentaacetic acid linkage to a MoAb specific to antigen Thy1 to murine T cells [74]. Mice were injected intraperitoneally with EL4 lymphoma tumor cells, and subsequent intraperitoneal injection of the radioimmunoconjugate resulted in prolonged survival of the animals. It was proposed that due to the known slow extravasation of large molecules from the peritoneum, the radioimmunoconjugate was retained in the peritoneal cavity for at least 2 h, when the alpha-emitting conjugate was most potent. This avoided the side effects that would be associated with the high systemic levels from intravenous administration. In another study, in which Burkitt's lymphoma cells were transplanted intraperitoneally into mice, intraperitoneal injection 10–14 days later of a gelonin-5E9 MoAb conjugate showed twice the efficacy as intravenous administration in delivery of the immunoconjugate into the tumor cells [75]. In a similar experiment, B-cell lymphoma cells were transplanted intraperitoneally into mice, and a daunomycin-antibody conjugate (linked via a dextran bridge) was prepared [76]; intraperitoneal injection of the immunoconjugate was somewhat more effective than intravenous administration in prolonging survival, even though free drug was more effective when given intravenously, apparently due to the higher toxicity of the free drug when given intraperitoneally. This may reflect the more rapid systemic clearance of the conjugate due to its larger molecular weight. In summary, intraperitoneal administration of immunoconjugate will probably have benefit in applications such as ovarian carcinoma and colorectal cancer. Nevertheless, a certain amount of the conjugate will still enter the systemic circulation from the peritoneal cavity; thus, any resulting toxicity may be reduced but not eliminated, unless special techniques as described above are used.

Reports of intramuscular or subcutaneous (other than local or intratumoral) administration of immunoconjugates are scarce. The lower systemic levels that would result from these routes of administration would be especially detrimental for immunoconjugates, because of their high molecular weight and resulting slower transport from subcutaneous or intramuscular depots into the bloodstream. The inherent instability of the immunoconjugate due to proteolysis or lability of the antibody-drug linkage could make this delay in transport into the vascular system a severe problem for intramuscular or subcutaneous injections. In addition, these modes of administration frequently lead to drainage of the dose into the lymphatics, especially for high-molecular-weight proteins [77], lowering the systemic levels. This lymphatic drainage may, however, be exploited in certain cases where introduction of the immunoconjugate into the lymphatic system is desirable. For example, in an animal model in which hepatoma cells were transplanted intradermally into mice, subsequent subcutaneous injection (1 in. distal to the tumor) of an immunoconjugate, composed of a hepatoma-specific MoAb linked to Abrin, was significantly more effective than an intravenous injection, in terms of both regression of the primary tumor and inhibition of metastasis to draining lymph nodes [78]. Although the effect could be due partly to some degree of local injection, it is tempting to conclude that subcutaneous administration led to an enhanced delivery of the immunoconjugate to the tumor by the surrounding lymphatics, as well as to higher protection of the lymph nodes.

In conclusion, it can be stated that the choice of route of administration (intravenous, intraarterial, intratumoral, intraperitoneal, subcutaneous, or intramuscular) depends on the particular disease being treated and on the properties of the immunoconjugate (e.g., stability, specificity, toxicity, and pharmacokinetics), and the choice must be made for each individual case. With regard to other nonsystemic routes of administration (e.g., oral, transdermal, inhalation, and buccal), the effective use of these alternative routes of administration for immunoconjugates is doubtful, nor is there presently a great need for nonsystemic administration of these drugs. However, one can envision that as immunoconjugates come to the marketplace and become more widely used on an outpatient basis, the need for these additional dosage forms will grow. Hopefully, the present problems associated with nonsystemic administration of proteins (e.g., poor absorption, low bioavailability, degradation in the GI tract) will be surmounted in the decades to come, which will allow widespread nonhospital use of these "magic bullets."

B. Liposomal and Polymeric Immunoconjugate Systems

In addition to "simple" immunoconjugate systems, in which drug or toxin are linked together either directly or by a relatively small linker or carrier group, there is now a large body of work encompassing more complex systems in which the drug and antibody are coupled via a common carrier, either a liposome or a polymer. In many of these cases (certainly for liposomes), the largest contribution to the mass of the system is from the carrier. Thus, the pharmacokinetics, toxicology, and metabolism of the carrier itself cannot be ignored when considering such an immunoconjugate carrier system for therapy. These systems, however, may have significant advantages over "simple" immunoconjugates, including the following:

1. Higher "payload"—the number of drug molecules delivered by one antibody is increased.
2. More flexibility in design of systems—e.g., the same antibody-carrier system can be used for many drugs with little modifications.
3. Protection of labile drugs from degradation by serum enzymes.

4. Decreased toxicity due to lack of recognition of the drug by receptors not in the "target."
5. Preservation of activity of drug and antibody–preparation of "simple" immunoconjugate often leads to a significant reduction in activity of either the drug or the antibody, and the release of active drug from the conjugate may be slow; this problem may be avoided by polymeric or liposomal systems, especially if no covalent bond exists between the drug and the antibody.
6. Opportunity for sustained release–polymeric and liposomal systems can, under certain circumstances, provide sustained release of the active drug, either from higher circulating lifetimes or from action as a "depot."

To prepare an "immunoliposome" system, typically a cytotoxic drug is entrapped inside a liposome, either in the acqueous core or in the lipid bilayer, and the antibody is covalently linked to a lipid that will be incorporated into the outer bilayer of the liposome to provide targeting. Since this subject has been reviewed recently [79,80], and is dealt with by Sullivan and Weiner in this volume, it will not be discussed extensively here. However, a few comments are in order. A multitude of methods now exist for preparation of liposomes [81] and for incorporation of antibodies onto the surface of the liposomes [79,80,82]. Those techniques will afford a wide variety of compositions, sizes, and lamellarity (i.e., unilamellar versus multilamellar). Not all of the methods will be appropriate for preparation of immunoliposomes; e.g., some methods employ the use of organic solvents or other factors that may be detrimental to the stability and integrity of the targeting antibody or the drug. In addition, some of the preparation techniques may employ steps (e.g., sonication, rotary evaporation) that have reproducibility problems or do not lend themselves well to commercial scales. Considerable progress has been made in the last few years with regard to commercial processes for liposome manufacture, as liposomes move from the laboratory to the marketplace [81], and this need must also be addressed for immunoliposomes if they are to make this transition.

While the antibody on the surface of the liposome provides the "active" targeting of the immunoliposome, as discussed above (Section I.B), the "passive" targeting of the liposome, viz., uptake by the RES cells of the body, cannot be ignored. The degree of RES uptake will depend on such factors as liposome size, charge, and fluidity. If the targeting antibody has low specificity or if the target tissue has low blood flow, the RES uptake may greatly reduce the amount of drug reaching the target. One technique to reduce the RES uptake is to incorporate a sialic acid-containing glycolipid, e.g., ganglioside GM1, into the lipid bilayer [83]. This masks the liposome from detection by the RES because of the presence of the same marker on the surface of red blood cells, and has been shown to significantly increase the circulating half-life of liposomes [83]. Indeed, an immunoliposome-targeting antibody itself often will contain sialic acid moieties that retard the RES uptake, which may be sufficient to favor uptake by the target tissue over that by the RES.

One factor that has a large effect on the in vivo fate of a liposome is its size. Intravenously injected liposomes larger than ca. 4 μm are taken up to a large extent by the lungs, since this is the first capillary bed they would come in contact with [84]. The RES uptake is more acute for liposomes in the size range 0.1-4 μm, with the result that small liposomes have a longer half-life than large ones [85]. Of relevance to immunoliposomes is the fact that, to be taken up into target cells, they must generally bind to the surface of the cell and be endocytosed. It has been shown [86,87] that, at least for some cells, this process is more efficient for small unilamellar immunoliposomes (<100 nm) than for large liposomes. Although large unilamellar liposomes conjugated to an anti-Thy 1.2 antibody (specific for lymphomas) were shown to bind to the surface of murine lymphoma cells, they failed to be internalized into the cells, as seen by the lack of effect of the encapsulated methotrexate gamma-aspartate. In contrast, analogous small unilamellar liposomes (53 nm mean diameter) were endocytosed into the lymphoma cells, and were in fact 20–40 times more effective than free drug in inhibiting cell growth [86]. Similar results were obtained with methotrexate-containing liposomes conjugated to *Staphylococcus aureus* Protein A [87].

The route of administration is also important to the in vivo fate of liposomal targeting systems. Intravenous administration leads to clearance times of several hours, which may nevertheless be advantageous for drugs with very short half-lives. Under certain conditions, alternative routes of administration may impart a sustained-release profile to the pharmacokinetics of a liposomal formulation [81]. Intramuscular injections allows the formulation to act as a depot: It is retained at the injection site and can be released into the circulation over a period of days or even weeks [88,89]. Liposome size, however, has a large impact on the degree of sustained release, and small liposomes will be cleared faster from the intramuscular depot than larger ones [90].

Similarly, subcutaneous injection of liposomes can also lead to a depot effect. In this case, however, there has been shown to be a significant amount of drainings of the administered dose into the lymphatic system [91]. As noted above, this could be an advantage where delivery into lymph nodes is important, e.g., for certain tumors and lymphomas. For both subcutaneous and intramuscular administration used as a controlled-release system for an immunoconjugate, it is unclear how much of the administered dose reaches the circulation as intact liposomes. Release of encapsulated drug alone into the circulation would, of course, negate any targeting effect of the antibody. In addition, the targeting antibody's stability to proteases and other degradative pathways at the injection site must be considered in light of any sustained-release effect.

A method to further increase the sustained-release property of intramuscularly and subcutaneously administered liposomes is to sequester the liposomes in another matrix, such as a collagen gel [92] or similar polysaccharide matrix [93]. The potential of this for immunoliposomes has not been explored, but such a system should have a significant effect on the pharmacokinetics and efficacy of the immunoliposome.

While immunoliposomes exhibit a wide variety of properties, polymer-based immunoconjugate systems can display an even greater diversity. A low-molecular-weight polymer, e.g., dextran, may simply serve as a common linking agent for the drug and the antibody, and thus have little effect on the in vivo fate of the immunoconjugate system. Such soluble systems generally are not cleared rapidly by the RES, and would be expected to have potential in gaining access to a variety of target cell types and allow favorable properties to be built into the formulation. For example, conjugates made from N-(2-hydroxypropyl) methacrylamide (HPMA) copolymers were designed such that the drug is linked to the polymer by an oligopeptide sequence that is stable in plasma but is degraded by lysosomal enzymes [94-96]. Daunomycin was linked via a degradable oligopeptide to the HPMA polymer, to which a T-lymphocyte-specific antibody was also linked. Injection of the polymer conjugate into a mouse model for autoimmune disease led to suppression of the immune response of the spleen by 60–85%. The specific conjugate was 50–100 times more cytotoxic against T lymphocytes than either nonspecific (IgG-linked) conjugates or conjugates with noncleavable oligopeptide linkages, and was 80 times less toxic to bone marrow than free daunomycin [95,96].

At the other extreme, the polymer-drug-antibody complex could approach nanoparticle size (100–1000 nm), which would of course have a large impact on the distribution and cellular interaction of the system. Normally, particles of this size would be rapidly cleared by the RES system. Just as is the case for immunoliposomes, this effect can be partially overcome by the presence of targeting antibodies on the surface of the nanoparticles. The highest efficiency of targeting would be afforded by covalent linkage of the antibody to the surface of the nanoparticle; the drug need not be covalently bound, but can be occluded in the polymeric matrix during preparation of the nanoparticles. Release of the drug from the nanoparticle would therefore depend on erosion of the polymer or on diffusion of the drug through the polymer matrix; nanoparticle stability in serum must therefore be monitored. Noncovalent adsorption of the antibody to the nanoparticle may in some case be sufficient for targeting. For example, an anti-alpha-fetoprotein antibody was introduced into the polymerization medium for cyanoacrylic nanoparticles (120 nm) when the formation of the nanoparticles was almost complete [97]. In this way, immunoreactivity was preserved, and the antibody-nanoparticles had moderate stability in serum. Similarly, adsorption of an immunotoxin onto colloidal gold particles did not interfere with the binding and subsequent internalization of the antibody with lymphoblastic cells [98]. However, tight control of nanoparticle formation conditions would be required to optimize the efficiency of the targeting ability of the antibody. Therefore, covalent coupling of antibody to the surface of the nanoparticle is probably the method that has the highest potential in targeting.

Just as in the case for immunoliposomes, polymeric systems provide the opportunity for controlled release, either from extended circulating lifetimes or from a depot effect with intramuscular or subcutaneous injections. The potential for this has only begun for polymeric immunoconjugate systems, which remain a fruitful area for future research.

IV. CONCLUSIONS

Targeted drug delivery by means of immunoconjugates holds a significant potential in the treatment of cancer and other diseases. In spite of their promise, however, it is apparent that a number of issues must be resolved before immunoconjugates see widespread acceptance for clinical use and commercialization. When considering use of an immunoconjugate system, many factors must be examined and optimized carefully. These include: antibody purity and selectivity; efficiency of the drug-antibody coupling method used; possible heterogeneity of the product; preservation of drug and antibody activity of immunoconjugate formation; stability of the immunoconjugate in vivo; pharmacokinetics and biodistribution; effect of route of administration; and effect of carrier system used. Many promising results have been obtained using in vitro studies, but examination of the same immunoconjugate system in vivo sometimes has given disappointing results, perhaps due to lack of consideration of one of the factors listed above. The same precautions should be taken when attempting to extrapolate from promising animal studies to human clinical results. Furthermore, it has not yet been demonstrated that reliable immunoconjugate production can be achieved on commercial scales. The coming years hold promise, however, that many of these issues can be resolved. Development of immunoconjugate targeted drug delivery systems would therefore open important avenues to safer and more effective treatments for a variety of diseases.

V. REFERENCES

1. Y. W. Chien, in *Novel Drug Delivery Systems* (Y. Chien, ed.), Marcel Dekker, New York, 1982, p. 112.

2. M. F. Armaly and K. R. Rao, *Invest. Ophthalmol. 12*:491 (1973).
3. J. H. Ratcliffe, I. M. Hunneyball, A. Smith, C. G. Wilson, and S. S. Davis, *J. Pharm. Pharmacol. 36*:431 (1984).
4. G. Poste and R. Kirsh, *Biotechnology 1*:869 (1983).
5. Y. Yoshioka, M. Muranishi, and H. Sezaki, *Int. J. Pharm. 8*:131 (1981).
6. L. Illum and S. S. Davis, *Int. J. Pharm. 11*:323 (1982).
7. G. Poste, *Biol. Cell. 47*:19 (1983).
8. I. J. Fidler, *Cancer Res. 34*:1074 (1974).
9. C. R. Alving, *Pharm. Ther. 22*:407 (1983).
10. J. N. Weinstein and L. D. Leserman, *Pharm. Ther. 24*:207 (1984).
11. J. N. Weinstein, R. L. Magin, R. L. Cysyk, and D. S. Zaharko, *Cancer Res. 40*:1388 (1980).
12. K. J. Widder, A. E. Senyei, and D. G. Scarpelli, *Proc. Soc. Exp. Biol. Med. 158*:141 (1978).
13. K. J. Widder, P. A. Marino, R. M. Morris, D. P. Howard, Jr., G. A. Poore, and A. E. Senti, *Eur. J. Cancer Clin. Oncol. 19*:141 (1983).
14. A. Ibrahim, P. Couvreur, M. Roland, and P. Speiser, *J. Pharm. Pharmacol. 35*:59 (1983).
15. U. Zimmerman, in *Targeted Drugs* (E. P. Goldberg, ed.), John Wiley, New York, 1983, p. 153.
16. M. Akimoto and Y. Morimoto, *Biomaterials 4*:49 (1983).
17. R. T. Profitt, L. E. Williams, C. A. Presant, G. W. Tin, J. A. Uliana, R. C. Gamble, and J. D. Baldeschwieler, *Science 220*:502 (1983).
18. J. W. B. Bradfield, in *Microspheres and Drug* (S. S. Davis, L. Illum, J. G. McVie, and E. Tomlinson, eds.), Elsevier, Amsterdam, 1984, p. 25.
19. G. Poste and R. Kirsh, *Biotechnology 1*:869 (1983).
20. L. Fiume, C. Busi, A. Mattioli, P. G. Balboni, G. Barbanti-Brandona, and T. Wieland, in *Targeting of Drugs* (G. Gregoriadis, J. Senior, and A. Trouet, eds.), Plenum Press, New York, 1982, p. 1.
21. A. Trouet, R. Baurain, D. Deprez-DeCampaneere, M. Masquelier, and P. Prison, in *Targeting of Drugs* (G. Gregoriadis, J. Senior, and A. Trouet, eds.), Plenum Press, New York, 1982, p. 19.
22. K. J. Widder, A. E. Senyei, and B. Sears, *J. Pharm. Sci. 71*:379 (1982).
23. G. Gregoriadis, *Nature 265*:407 (1977).
24. G. Kohler and C. Milstein, *Nature 256*:495 (1975).
25. K. Sikora and M. Smedley, *Cancer Surv. 1*:521 (1983).
26. P. N. Kulkarni, P. H. Blair, and T. Ghose, *Fed. Proc. 40*:642 (1981).
27. G. F. Rowland, in *Targeted Drugs* (E. P. Goldberg, ed.), Wiley-Interscience, New York, 1983.
28. M. V. Pimm, J. A. Jones, M. R. Price, J. G. Middle, M. J. Embleton, and R. W. Baldwin, *Cancer Immunol. Immunother. 12*:125 (1982).
29. M. Muirhead, P. J. Martin, B. Torok-Storb, J. W. Uhr, and S. Vitelia, *Blood 62*:327 (1983).
30. P. Ehrlich, in *Chemotherapy* (F. Himmelweit, ed.), Pergamon Press, London, 1960.

31. P. E. Thorpe and W. C. J. Ross, *Immunol. Rev. 62*:119 (1982).
32. S. Olsnes and A. Pihl, *Pharmacol. Ther. 15*:355 (1981).
33. M. Yamaizumi, E. Mekada, T. Uchida, and Y. Okada, *Cell 15*:245 (1978).
34. K. Eiklid, S. Olsnes, and A. Pihl, *Exp. Cell Res. 126*:321 (1980).
35. G. F. Rowland, R. G. Simmons, J. R. F. Corvalan, R. W. Baldwin, J. P. Browns, M. J. Embelton, C. H. J. Ford, K. E. Hellstorm, I. Hellstrom, J. T. Kemshead, C. E. Newmans, and C. S. Woodhouse, in *Protides of the Biological Fluids*, Proc. Colloq. 30 (H. Peters, ed.), Pergammon Press, New York, 1983, p. 375.
36. J. M. Whiteley, *Ann. N.Y. Acad. Sci. 79*:621, (1982).
37. G. Galfre, S. C. Howe, C. Milstein, G. W. Butcher, and J. C. Howard, *Nature 266*:550 (1977).
38. H. Koprowski and Z. Steplewski, in *Monoclonal Antibodies and T-Cell Hybridomas* (G. J. Hammerling, U. Hammerling, and J. F. Kearney, eds.), Elsevier-North-Holland, 1981, p. 161.
39. K. E. Hellstrom, J. P. Brown, and I. Hellstrom, in *Contemporary Topics in Immunobiology*, Vol. II (N. L. Warner, ed.), Plenum Press, New York, 1980, p. 117.
40. S. M. Larson, J. A. Carrasquillo, K. A. Krohn, J. P. Brown, R. W. McGuffin, J. M. Ferens, M. M. Graham, L. D. Hill, P. L. Beaumier, K. E. Hellstrom, and I. Hellstrom, *J. Clin. Invest. 72*:2101 (1983).
41. R. Arnon and E. Hurwitz, in *Durable Resistance in Crops* (F. Lamberti, J. M. Waller, and N. A. Van der Graff, eds.), Plenum Press, New York, 1983, p. 23.
42. B. A. Chabner and C. E. Myers, in *Cancer, Principles and Practice of Oncology* (V. T. DeVita and R. S. Hellman, eds.), J. B. Lippincott, Philadelphia, 1982, p. 156.
43. V. Byers, M. V. Pimm, I. Z. A. Pawlucyk, H. M. Lee, P. J. Scannon, and R. W. Baldwin, *Cancer Res. 47*:5277 (1987).
44. N. R. Worrell, A. J. Cumber, G. D. Parnell, W. C. J. Ross, and J. A. Forrester, *Biochem. Pharmacol. 35*:417 (1986).
45. J. Bertram, P. S. Gill, A. M. Levine, D. Boquiren, F. M. Hoffman, P. Meyer, and M. S. Mitchell, *Blood 68*:752 (1986).
46. M. J. Embleton, V. S. Byers, H. M. Lee, P. J. Scannon, N. W. Blackhall, and R. W. Baldwin, *Cancer Res. 46*:5524 (1986).
47. P. E. Thorpe and W. C. J. Ross, *Immunol. Rev. 62*:119 (1982).
48. V. S. Byers, M. V. Pimm, P. J. Scannon, and R. W. Baldwin, *Cancer Res. 47*:5042 (1987).
49. R. W. Baldwin, and V. S. Byers, in *Gastrointestinal Cancer: Current Approaches to Diagnosis and Treatment*. (Bernard Levin, ed.), University of Texas Press, Houston, 1987.
50. P. E. Thorpe, S. I. Detre, B. M. J. Foxwell, A. N. F. Brown, D. N. Skilleter, G. Wilson, J. A. Forrester, and F. Stripe, *Eur. J. Biochem. 147*:197 (1985).
51. D. N. Skilleter and B. M. J. Foxwell, *FEBS Lett. 196*:344 (1986).
52. B. M. Simmons, P. D. Stahl, and J. H. Russell, *J. Biol. Chem. 261*:7912 (1986).

53. B. J. P. Bourrie, P. Casellas, H. E. Blythman, and F. K. Jenson, *Eur. J. Biochem. 155*:1 (1986).
54. M. E. Spearman, R. M. Goodman, L. D. Apelgren, and T. F. Bumol, *J. Pharmacol. Exp. Ther. 241*:695 (1987).
55. A. C. Perkins, M. V. Pimm, and R. W. Baldwin, *Eur. J. Cancer Clin. Oncol. 23*:1225 (1987).
56. N. L. Levin, V. S. Goldmacher, J. Ritz, J. M. Yetz, S. F. Schlossman, and J. M. Lambert, *J. Clin. Invest. 77*:977 (1986).
57. N. C. Armitage, A. C. Perkins, M. V. Pimm, M. G. Wastie, R. W. Baldwin, and J. D. Hardcastle, *Nucl. Med. Commun. 6*:625 (1985).
58. K. C. Ballantyne, A. C. Perkins, M. V. Pimm, M. C. Garnett, J. A. Clegg, N. C. Armitage, R. W. Baldwin, and J. D. Hardcastle, *Int. J. Cancer*, Suppl. 2, 1988, p. 103.
59. C. F. Scott, Jr., J. M. Lambert, A. S. Goldsmith, W. A. Blattler, R. Sobel, S. F. Schlossman, and B. Benacerraf, *Int. J. Immunopharmacol. 9*:211 (1987).
60. S. Ramakrishnan and L. L. Houston, *Cancer Res. 45*:2031 (1985).
61. J. Carlsson, H. Drevin, and R. Axen, *Biochem. J. 173*:723 (1978).
62. E. C. Petrella, S. D. Wilkie, C. A. Smith, A. C. Morgan, and C. W. Vogel, *J. Immunol. Meth. 104*:159 (1987).
63. D. M. Ziegler, *Ann. Rev. Biochem. 54*:305 (1985); F. J. Martin, W. L. Hubbell, and D. Papahadjopoulos, *Biochemistry 20*:4229 (1981).
64. G. Sivam, J. W. Pearson, W. Bohn, R. K. Oldham, J. C. Sadoff, and A. C. Morgan, *Cancer Res. 47*:3169 (1987).
65. G. Weil-Hillman, F. M. Uckan, J. M. Manske, and D. A. Vallera, *Cancer Res. 47*:579 (1987).
66. M. J. A. Daeman, J. F. M. Smits, H. H. W. Thijssen, and H. A. J. Struyker-Boudier, *Trends Pharmacol. Sci. 9*:138 (1988).
67. A. I. Freeman and E. Mayhew, *Cancer 58*:573 (1986).
68. T. Takahashi, Y. Yamaguchi, K. Kitamura, H. Suzuyama, M. Honda, T. Yokota, H. Kotanagi, M. Takahashi, and Y. Hashimoto, *Cancer 61*:881 (1988).
69. R. F. Ozols, R. C. Young, J. L. Speyer, P. H. Sugarbaker, R. Greene, J. Jenkins, and C. E. Myers, *Cancer Res. 42*:4265 (1982).
70. P. Uadia, A. H. Blair, and T. Ghose, *Cancer Res. 44*:4263 (1984).
71. H. E. Blythman, P. Casselas, O. Gros, P. Gros, F. K. Jansen, F. Paolucci, B. Pau, and H. Vidal, *Nature 290*:145 (1981).
72. F. Bergman, *Acta Obstet, Gynecol. Scand. 45*:211 (1966).
73. M. J. Mattes, *J. Natl. Cancer Inst. 79*:855 (1987).
74. R. M. Macklis, B. M. Kinsey, A. I. Kassis, J. L. M. Ferrara, R. W. Atcher, J. J. Hines, C. N. Coleman, S. J. Adelstein, and S. J. Burakoff, *Science 240*: 1024 (1988).
75. C. F. Scott, V. S. Goldmacher, J. M. Lambert, R. V. J. Chari, S. Bolender, M. N. Gauthier, and W. A. Blattler, *Cancer Immunol. Immunother. 25*: 31 (1987).

76. E. Hurwitz, R. Kashi, D. Burowsky, R. Arnon, and J. Haimovich, *Int. J. Cancer 31*:745 (1983).
77. A. Supersaxo, W. Hein, H. Gallati, and H. Steffan, *Pharm. Res. 5*:472 (1988).
78. K. M. Hwang, K. A. Foon, P. H. Cheung, J. W. Pearson, and R. K. Oldham, *Cancer Res. 44*:4578 (1984).
79. G. Gregoriadis, *Trends Pharmacol. Sci. 4*:304 (1983).
80. S. M. Sullivan, J. Connor, and L. Huang, *Medicinal Res. Rev. 6*:171 (1986).
81. A. L. Weiner, J. B. Cannon, and P. Tyle, in *Topics in Controlled Release Science* (M. Rosoff, ed.) VCH, Deerfield Beach, Fla., 1989, p. 217.
82. T. D. Heath and F. J. Martin, *Chem. Phys. Lipids 40*:347 (1986).
83. W. B. Geho and J. R. Lau, U. S. Patent No. 4,501,728 (1985); T. M. Allen and A. Chonn, *FEBS Lett. 223*:42 (1987).
84. C. A. Hunt, Y. M. Rustum, E. Mayhew, and D. Papahadjopoulos, *Drug Metab. Dispos. 7*:124 (1979).
85. M. B. Yatvin and P. I. Laelkes, *Med. Physics 9*:149 (1982).
86. K. K. Matthay, T. D. Heath, and D. Papahadjopoulos, *Cancer Res. 44*: 1880 (1984).
87. P. Machy and L. D. Leserman, *Biochim. Biophys. Acta 730*:313 (1983).
88. E. Arakawa, Y. Imai, H. Kobayashi, K. Okumura, and H. Sezaki, *Chem. Pharm. Bull. 23*:2218 (1975).
89. G. Dapergolas, E. D. Neerunjun, and G. Gregoriadis, *FEBS Lett. 63*:235 (1976); M. Arrowsmith, J. Hadgraft, and I. W. Kellaway, *Int. J. Pharm. 20*:347 (1984).
90. A. J. Jackson, *Drug Metab. Dispos. 9*:535 (1981).
91. B. E. Ryman, R. F. Jenkes, K. Jeyasingh, M. P. Osborn, H. M. Patel, V. J. Richardson, M. H. N., Tattersall, and D. A. Tyrrell, *Ann. N.Y. Acad. Sci. 308*:281 (1978); J. Khato, E. R. Priester, and S. M. Sieber, *Cancer Treat. Rep. 66*:517 (1982).
92. A. L. Weiner, S. S. Carpenter-Green, E. C. Soehngen, R. P. Lenk, and M. C. Popescu, *J. Pharm. Sci. 74*:922 (1985).
93. M. C. Popescu, A. L. Weiner, and S. S. Carpenter-Green, PCT Patent No. WO 85/03640 (1985); A. Kato, M. Arakawa, and T. Kondo, *J. Microencapsulation 1*:105 (1984).
94. R. Duncan, P. Kopeckova-Rejmanova, J. Strohalm, I. Hume, H. C. Cable, J. Pohl, J. B. Lloyd, and J. Kopecek, *Br. J. Cancer 55*:165 (1987).
95. B. Rihova, J. Kopecek, J. Kopeckova-Rejmanova, J. Strohalm, D. Plocova, and H. Semoradova, *J. Chromatogr. 376*:221 (1986).
96. B. Rihova, J. Kopeckova, J. Strohalm, P. Rossmann, V. Vetvicka, and J. Kopecek, *J. Clin. Immunol. Immunopathol. 46*:100 (1988).
97. C. Kubiak, L. Manil, and P. Couvreur, *Int. J. Pharm. 41*:181 (1988).
98. D. Carriere, P. Casellas, G. Richer, P. Gros, and F. K. Jansen, *Exp. Cell Res. 156*:327 (1985).

7

Drug–Immunoglobulin Conjugates as Targeted Therapeutic Systems

ASHIM K. MITRA and MRIDUL K. GHOSH
Purdue University, West Lafayette, Indiana

I. INTRODUCTION

A. Concept of Antibody-Mediated Drug Delivery

Cancer is still a life-threatening disease, despite the development of medical science and modern therapeutic techniques. Metastasis is the major concern in the management of cancer [1]; once cancer has metastasized, it is often incurable. At present, chemotherapy is one of the major therapeutic approaches in the prevention and therapy of metastasized cancer. Unfortunately, agents effective in killing neoplastic cells usually have detrimental effects on nonmalignant cells, particularly the rapidly proliferating ones of the gastrointestinal tract, bone marrow, lymphoid tissue, and genitourinary epithelium [2,3]. For example, adriamycin, often used alone or in combination with other drugs in the treatment of breast cancer, can cause nausea, frequent vomiting, hair loss, and a specific action on myocardium [4]. Therefore, in order to avoid severe damage to normal tissues, doses of drugs that are borderline for killing the cancer cells have to be administered, thereby diminishing the effectiveness of the treatment. A possible way to overcome this limitation is by employing affinity chemotherapy, which is based on the biological recognition between a target cell and a site-specific drug counterpart.

Paul Ehrlich [5], at the turn of the century, suggested that molecules with an affinity for certain tissues might be able to serve as carriers of cytotoxic agents, enabling them to concentrate on the appropriate target cells in vivo. The target molecule at the cell surface can be a defined antigen expressed preferentially on the tumor cell or a specific surface receptor; it could constitute an integral membrane component or a compound produced by the cells and only transiently expressed on its membrane. For targeted delivery the most important aspect of the study is to select a water-soluble macromolecule that can be recognized for pinocytic ingestion by one cell type only—the cell type in which the drug's action is desired. The drug molecule is then attached to this macromolecule in such a way that the linkage is susceptible to hydrolysis by one of the lysosomal enzymes. Administration of such conjugates should lead to drug release and action in the target cells only. During the past decade, a wide variety of macromolecules have been designed and evaluated, including lectins, hormones, erythrocytes, glycoproteins, liposomes, polynucleotides, and polymers [6,7]. Unfortunately, none of them seems to have a high degree of specificity necessary for selective delivery of the cytotoxic agents to target cells. The factors affecting the selection of a macromolecule optimum for drug targeting are likely to be (1) recognition of the conjugate by only one type of cell, (2) stability in the bloodstream, (3) possibilities for linking drugs and targeting moieties without significant loss of drug and antibody biochemical properties, (4) less immunogenicity, and (5) amenability to a desired specification.

The idea of addressing a cytotoxic drug to a particular tissue (tumor) destination and posting it on a tissue-specific antibody has existed for many years. Ehrlich [8], even in 1900, pointed out the possible use of diphtheria toxin bound to antitumor antibodies as a "magic bullet" against malignancies. With the development of tumor immunology, several investigators have sought to use the antibodies to antigenic determinants expressed preferentially on tumor cells as carriers of cytotoxic agents. The first report of drug-antibody conjugate was reported in the 1950s, when Mathe et al. [9] linked the drug methotrexate (MTX) by diazotization to antibodies raised against L1210 leukemia cells and used the conjugate in the successful treatment of L1210-bearing DBA/2 mice. A similar approach was adopted by Calendi et al. [10], who linked MTX by diazotization to the globulin fraction of a rabbit antiserum raised against human sarcoma tissue. The conjugate was found to retain specificity when assessed by indirect immunofluorescence, but no attempt was made to assess drug activity in the complex. Although the initial results with drug-antibody conjugates appeared promising, no serious efforts toward therapeutic modalities were made until the 1970s, when Ghose and colleagues developed chlorambucil coupled to goat antitumor antibodies [11]. These early experiments led to the clinical use of chlorambucil-antibody preparations in humans [12], although the exact nature

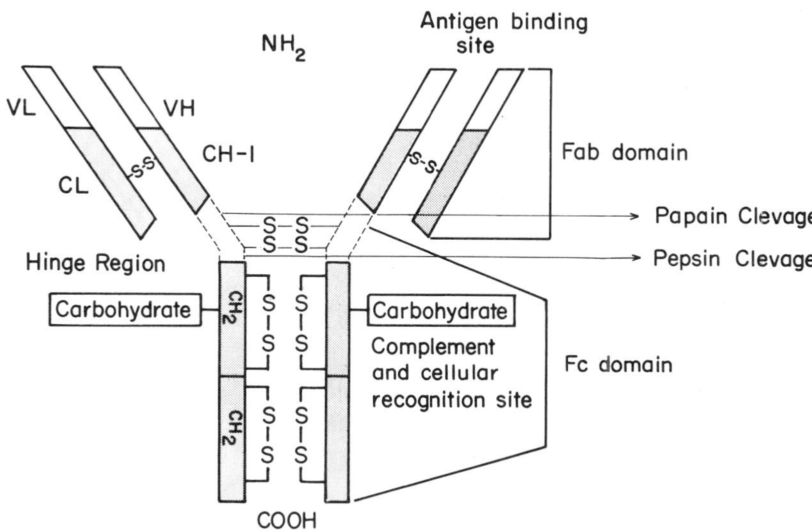

Figure 1 Schematic diagram of an antibody molecule. Two heavy (H) and two light chains (L) are joined together by disulfide bridges, giving the molecule a Y-like configuration. The Fab domain, which consists of the variable (V) and the constant (C) regions of both chains, confers antigen-binding specificity on the molecule. The Fc domain, on the other hand, is responsible for the overall structure of the molecule and for its recognition by other components of the immune system.

of the complex (covalent or noncovalent coupling of drug to antibody) was not clear, nor were the results. In recent years several laboratories have been engaged in the utilization of antibodies specific to tumor cells as carriers of various drug complexes [13-18]. Drugs have also been covalently linked to antibodies [9, 19-22] and studied in a variety of experimental tumor systems.

The best-characterized and best-understood immunoglobulin is immunoglobulin G (IgG); other classes of immunoglobulin are IgA, IgM, IgD, and IgE. Structurally, an IgG molecule has a tetrameric arrangement of pairs of identical light and heavy polypeptide chains held together by noncovalent forces and disulfide bridges. This arrangement renders the molecule a characteristic Y configuration (Figure 1). The arms, called the Fab fragments, contain the variable regions that create the unique binding sites for the antigen. The base of the molecule, the Fc fragment, retains the antigenicity and is responsible for recognition of the molecule by other components of the immune system.

Several advantages may be seen when drugs are delivered as antibody conjugates. The conjugate can specifically reach the target cells without significant distribution to normal tissues, thereby reducing the chances of detrimental side effects associated with free drugs. The binding of drugs with antibody might impart enhanced stability of drug molecules against enzymatic degradation and prevent rapid excretion, thereby increasing the drug half-life. In addition, the antibody molecules are fairly rigid and can withstand conditions employed in labeling. The molecules are little affected by the chemical manipulations required for conjugate synthesis [23]. The antibodies can also be broken down to given Fab fragments [24]. The consequent decrease in molecular weight might aid diffusion, and the removal of Fc binding may improve specificity of action. Selective delivery to the target site may reduce the doses of drugs necessary for treatment. Thus, the advantages are clear—less side effects, smaller doses, selective targeting, and so on. Therefore, drug-antibody conjugate could be expected to be the ideal agents for drug targeting in chemotherapy.

B. Strategies for Linking Drugs to Antibodies

Most of the antineoplastic agents have complex chemical structures, and their cell-killing properties are sensitive to any major chemical modification. Nevertheless, these compounds need to be finely tuned to act on the target cells so that they can interfere with a particular biological function in the processes of cell binding to cell death [25]. These low-molecular-weight compounds also provide very few (if any) reactive groups, which often are not amenable to any chemical modification owing to their involvement in binding to the target sites. For example, the intact pteridine moiety (with its free amino group in position 4) of methotrexate (MTX) must be preserved in MTX-antibody conjugates in order for it to retain antifolate and subsequent cytotoxic activity [26]. Neverthless, γ-carboxyl groups of the glutamate moiety at the other end of the molecule are available for conjugation to immunoglobulin (IgG) and other carriers [27]. Similarly with chlorambucil, any modification to the chloroethyl group will yield an inactive drug; however, the carboxyl group can be used for coupling, as it is remote from the chloroethyl group [28].

In designing drug-antibody conjugates, consideration must be given to the studies on the structure-activity relationship of the drug as well as to the methods of conjugation. An ideal method of conjugation should provide maximum number of incorporation without altering much of the antibody activity and acceptable protein recovery. In general, substitution ratios of greater than 10:1 with respect to IgG produce a marked loss of antibody reactivity; and in many cases, substitution of as few as four drug residues per IgG molecule renders unacceptable antibody damage [29]. The method should not also cause any aggregation and/or precipitation during the reaction. Most of the anticancer agents have relatively large hydrophobic structures, which upon conjugation with the antibodies through carboxylic or amino groups of the antibody molecule, may give rise to a conjugate with reduced aqueous solubility leading to aggregation and precipitation. However, with chlorambucil or N-acetyl meephalan, even a substitution ratio of 30:1 resulted in substantial protein recovery and good antibody activity [28,30]. Care should be taken during conjugation so that the process does not yield a homopolymer. Since proteins (or peptides) share common functional groups with antibodies, chances of forming homopolymer is more frequent. Finally, the conjugation method should be simple and straightforward, and must produce covalent linkage between drug and antibody. Although the established chemical reactions are supposed to produce covalent linkage between drug and antibody, sometimes noncovalent interactions such as ionic, hydrogen, or van der Waals bonds also occur [31,32]. If action of the drugs needs dissociation from the carrier antibody, as it often does, binding by such noncovalent forces would be useful if the binding is stable enough to prevent rapid dissociation of the drug in blood and/or other body fluids. Some effort, therefore, should be directed toward finding the proportion of drug bound covalently versus that bound noncovalently.

Selection of a method for linkage of a drug to immunoglobulins should be based on the following critiera.

1. Appropriate consideration must be given to the nature of reactive groups in the drug molecules that can be used for coupling reaction. However, the functional group selected for such conjugation reactions should yield immunoconjugates that upon hydrolysis are capable of regenerating the original group or a derivative of it that is capable of cytotoxic action.
2. Consideration of the type of linkage that is needed for maximum biological activity is also necessary—i.e., one that will show cytotoxic action in the intracellular milieu either by maintaining a favorable steric orientation or by allowing the release of active agent.
3. The stability of the immunoconjugate during transit from the site of application to the target organ should be considered. Instability in the plasma would lead to modification of the carrier and/or premature release of the drug.
4. The possibility of forming homopolymers of either agent during conjugation should be stressed.
5. Assessment should be made regarding alteration in the drug or immunoglobulin molecule for preparing effective biological conjugates, such as the removal of the "Fc" fragment of IgG or the cell surface binding subunit of Ricin (B chain).

6. Introduction of a spacer molecule between the drug and antibody, to increase the drug load and also to provide a favorable steric orientation, may be prudent under certain circumstances.
7. The possibility of biodegradation of the spacer molecule by lysosomal enzymes should be considered. The molecule should not remain as an indigestible residue in the lysosomes for the entire life span of that cell.
8. Finally, possible adverse effects of conjugating chemical reactions on the activities of the drug and antibody and/or on the biological half-life of the immunoconjugate must be carefully considered.

C. Choice of Antibody

Irrespective of whether the antibodies are used clinically as therapeutic or diagnostic entities, they must possess the ability to bind with the specific antigen. Before the discovery of monoclonal antibody technology, the only option was to isolate antibodies from the serum of an immunized animal. In most cases, if not all, the use of antibodies as carriers for therapeutic agents was limited by the availability of specific antibodies. In an experiment with affinity-purified polyclonal antibodies raised against membrane antigen of the "yac lymphoma" tumor cells, Ruth et al. [33] demonstrated that these antibodies were not just specific toward Yac cells but bound to normal splenocytes as well. In a similar type of experiment, they observed that daunomycin conjugated to polyclonal antibodies (against lung-associated antigen) showed prominent efficacy in vitro, but failed to show a specific distribution factor (less than twofold) in the lung as compared to other organs [33]. Therefore, to become a suitable drug carrier the antibody molecule must have high specificity and should possess high binding affinity and avidity toward the target cells. Köhler and Milstein [34] introduced novel methods for production of hybrid myelomas (hybridomas) that can synthesize monoclonal antibodies against single antigenic determinants. This method has allowed the production of apparently inexhaustible supplies of pure, specific, standardized antibody. Using this technology, the monoclonal antibody (791T/36) was initially prepared against a human osteogenic sarcoma cell line [35]. Subsequently it has been used for successful clinical radioimaging of a variety of both primary tumors and their metastases, such as the osteogenic sarcomas [36], colorectal carcinomas [37,38], ovarian carcinomas, and primary and metastatic breast carcinoma [39]. The preparation of tumor-reactive monoclonal antibodies has stimulated much interest in their use for the targeting of cytotoxic agents to tumor cells or tissues. In fact, several groups have prepared monoclonal antibodies against tumor-associated antigens and have used them as carriers of plant or bacterial toxins [40–44] and, recently, of adriamycin [45].

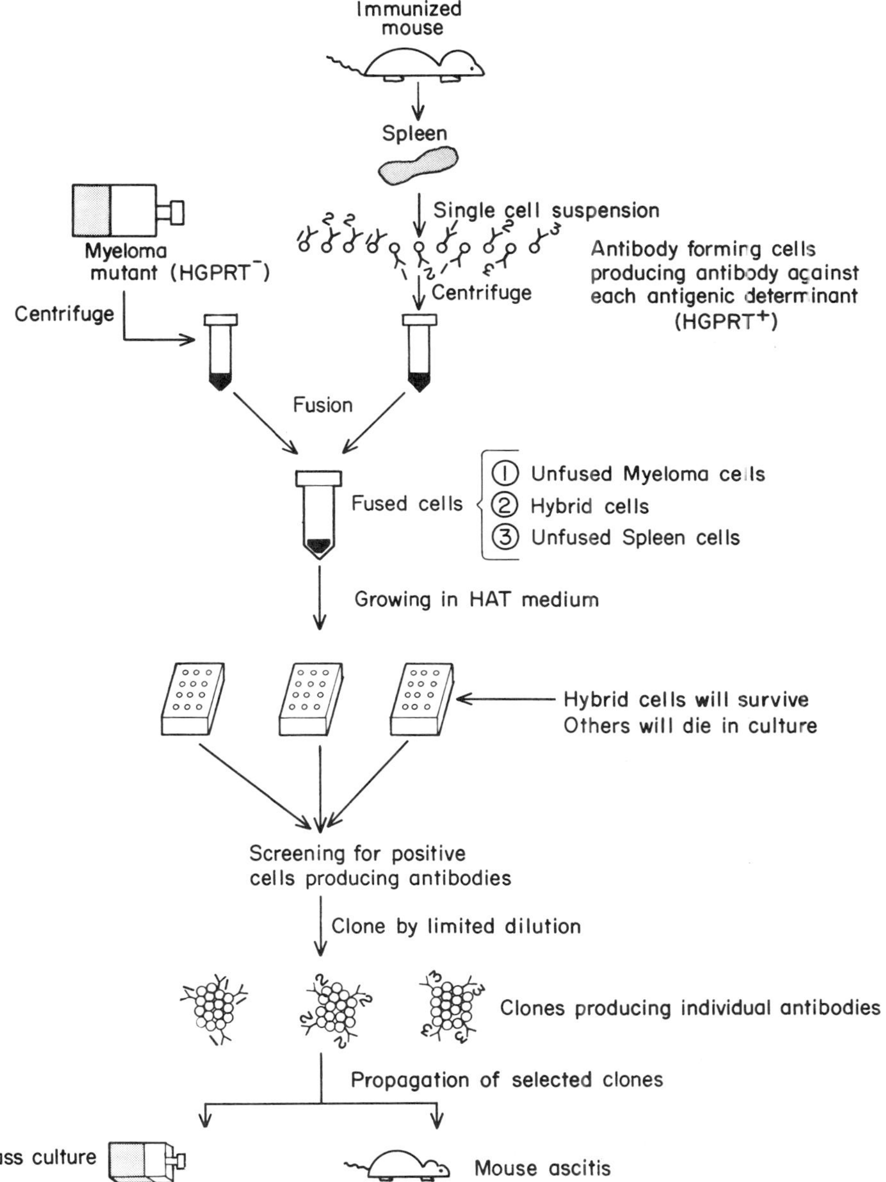

Figure 2 Basic protocol for the development of monoclonal antibodies by in vivo immunization. Antibody-forming cells, eg., spleen cells from a mouse immunized with desired antigen, are fused with cultured myeloma cells lacking HGPRT (hypoxanthine guanine phosphoribosyl transferase). The hybrid cells growing in HAT (hypoxanthine-aminopterine-thymidine) medium are then screened for their ability to produce and secrete the desired antibody. Hybridoma cells, producing antibodies, are then subcultured to produce population derived from a single hybridoma.

The entire scheme of monoclonal antibody (MoAb) production is illustrated in Figure 2. The spleen cells from a BALB/c mice (immunized with desired antigen) are fused with mouse myeloma cells. Each sensitized spleen B cell carries the genetic information and has the capacity to produce antibody against a single antigenic determinant, but is not able to survive in culture for long periods of time. The other cell partner is a myeloma cell, which retains the capacity to produce immunoglobulins and is also capable of surviving indefinitely in tissue culture. Thus, hybridomas resulting from such fusions conserve the properties of continuous growth and antibody secretion. Following an initial screening, the hybrid cells secreting the desired antibody are cloned by limited dilution to one cell per well. The positive cell obtained after cloning are the source for monoclonal antibodies. Such cells can be cryopreserved and, for the production of large amounts of MoAb, can be grown in mass cultures, where they secrete 10-50 μg antibody/ml, or they can be propagated as ascites tumor, where they usually produce 1-10 mg/ml of MoAb into the ascitic fluid. Thus, by immunizing a mouse with a chosen antigen, it is possible to generate a monospecific antibody in amounts sufficient for therapeutic purposes. The principal advantages of monoclonal antibodies produced by hybridoma technology over the polyclonal antibodies obtained by "conventional" immunization are reflected in the defined specificity, homogeneity, availability of practically unlimited quantities in a reproducible manner, and the ability to immortalize the production of such monospecific reagents by freezing the hybridomas for unlimited periods. Many improvements have been made since hybridoma technology was first introduced, such as the fusion of cells in the presence of electric current, hybridoma selection by flow cytometry [46], in vitro immunization [47,48], and the use of genetically altered mice for immunization [49].

An elegant example of specific cytotoxicity is provided by the monoclonal antibodies coupled to Ricinus toxin A chain, with specificity for one of the allotypes of the delta (δ) heavy chain of murine immunoglobulin D. In toxicity experiments employing these immunotoxins it was observed that conjugates specific for δa allotypes were cytotoxic to cells expressing the δa marker but not those expressing the δb allelic form; the reciprocal experiment using anti-δb antibodies demonstrated the same differential specificity [50]. Similar data have been obtained from the two alleles of the Thy1 marker [51].

While the specificity of the antibodies is of paramount importance for successful antibody-mediated drug delivery, the biochemical properties of a particular monoclonal antibody also need to be considered. Polyclonal antibodies show high affinity and avidity, and the molecules are capable of withstanding fairly rigorous chemical procedures. Monoclonal antibodies, on the other hand, tend to display weaker affinities and avidities, and their immunological specificity may be profoundly influenced by the local chemical environment [52].

Early work on drug targeting with polyclonal antibodies suggested very little effect of conjugation methodology on the binding affinity, which may be due partly to the heterogeneity of such antibodies. It has been noted by Embleton and Garnett [53] that individual monoclonal antibodies vary in their ruggedness toward substitution. Some antibodies completely lose their antibody binding properties even after mild procedures such as iodination or labeling with fluorescein, while others withstand a relatively high degree of substitution with no or minimal loss activity. Rowland et al. [54] conjugated vindesine to three different monoclonal antibodies by the same procedure. The conjugates produced the following antibody activities relative to the unconjugated antibody: 3:1, 84%; 7:1, 76%; and 9:1, 70%. Embleton and Garnett [53] have also reported that a particular antibody is inactivated to a different extent when different drug molecules are coupled by an equally mild reaction, e.g., 6 moles of vindesine or 8 moles of 14-bromodaunomycin can be coupled to 1 mole of 791T/36 monoclonal antibody with hardly any loss of activity. But substitution with 3 moles of MTX produced 70% loss of activity. The degree of retention of binding activity decreases with increasing substitution for any one drug molecule. Therefore, it is necessary to choose an antibody that withstands chemical manipulations and that allows adequate substitution without appreciable damage to antibody binding property.

Although Fab dimer has several advantages (listed below), its use has been mostly restricted to preparing immunoliposomes.

1. Antitumor antibodies may still show toxicity to normal cells even after exhaustive adsorption of the antiserum on normal cells from several tissues. This toxicity is either residual or due to a shift of specificity [55] and is most probably complement-mediated. Complement system binds with the antibodies via the "Fc" fragment of the IgG, and its removal will overcome this undesired toxic effect induced by complement.

2. The immunoconjugate needs to react with the cellular antigen to deliver the drug to the target site. It is highly desirable that the fragment attached to the cellular antigen be cleared from the circulation as quickly as possible. Since Fab dimer is known to be cleared from the circulation two to three times faster than the intact IgG [56-58], the use of Fab dimer derived from an antibody is certainly superior to an intact IgG molecule in this regard.

3. The use of intact IgG for local delivery of drug might elicit unfavorable reaction because of its immunogenicity. The "Fc" fragment of IgG is the most immunopotent region, and its removal will render the molecule less immunogenic. However, the remaining fragment is still likely to produce antibodies capable of reacting with determinants on the Fab′ fragment.

4. The Fab dimer may help the conjugate in transcapillary passage and confer enhanced diffusive properties compared to intact IgG because of the smaller

size of the immunoconjugate. Moreover, the Fab dimer-drug conjugate may also achieve superior tumor:blood ratio [59,60].

5. The removal of the "Fc" fragment from the IgG molecule may also reduce the nonspecific interactions with the cells having "Fc" receptors (e.g., phagocytes). Such interactions are common problems exhibited by both monoclonal and polyclonal antibodies.

Hurwitz and colleagues [61] were the first group to explore the possibility of Fab dimer utilization for drug delivery. In that study, daunomycin was coupled to Fab dimers obtained by pepsin digestion of an IgG fraction of a rabbit antiserum against a moloney virus-induced lymphoma (YAC). The conjugate was found to retain 70% of the binding activity of the uncomplexed dimer and had twice the toxicity to YAC tumor cells as that of the conjugate prepared with unrelated $(Fab)_2$ of anti-BSA. The most interesting observation was the ability of the conjugate to withstand a higher degree of drug substitution without significant loss of antibody activity than with intact IgG. Similar results were obtained in the study reported by Arnon and Sela [62] in which specific cytotoxicity of daunomycin linked to anti-YAC $(Fab')_2$ was assessed by the inhibition of $[^3H]$-uridine incorporation into the YAC cells following exposure to the conjugate for 5 min. Sela's group also performed another experiment to show the superiority of Fab dimer over intact IgG. Mice in the control groups injected with a 5-mg dose of the antitumor IgG showed high mortality within 24 h and the surviving animals appeared rather sick, whereas mice injected with a similar dose of the Fab′ dimer preparation appeared normal and healthy.

Well-established procedures are currently available for obtaining the immunologically active fragments from intact immunoglobulins. The methods entail controlled pepsin digestion followed by gel filtration and affinity chromatography to remove the "Fc" fragments [63–65].

II. DRUG CONJUGATION TO IMMUNOGLOBULIN

A. Molecular Modification of Drugs to Generate Reactive Groups

Conjugation of a drug molecule with an antibody requires the presence of a free, chemically reactive functionality such as carboxyl, amino (either aromatic or aliphatic), hydroxyl, phenol, sugar moieties, ketones, aldehydes, or sulphydryl groups in the drug molecule. The most useful among them are carboxyl and aliphatic amino groups, since information regarding the chemistry of peptide bonds and their susceptibility to lysosomal enzymes are readily available [66–71]. The lack of necessary functional groups in a large number of potent antineoplastic agents may restrict their direct covalent attachment with antibody. However, these compounds can be modified, as shown in Figure 3, to form appropriate derivatives for direct conjugation with antibody. Different pathways of drug modification rendering them suitable for conjugation to immunoglobulins are described in the following paragraphs.

1. The esterification of a primary alcohol group with succinic anhydride to form hemisuccinates [72] is a simple mechanism to generate a carboxyl group that can be used for further reaction with immunoglobulin/protein by any of the procedures described later. Reaction with succinic anhydride in dry pyridine yields hemisuccinate, wich can be isolated by silica gel chromatography. A representative procedure can be found in the studies on steroid-protein conjugates [73].

2. Formation of O-(carboxymethyl) oximes from the reaction of ketone or aldehyde groups with carboxymethyl hydroxylamine [74] is another way of introducing a carboxyl group into the drug molecule, which can be further exploited for coupling to protein/immunoglobulin by any of the procedures described later. Such reactions are commonly applied for conjugating steroid hormones to bovine serum albumin (BSA) for raising antibodies against them.

3. The ketone groups of aldosterone, corticosterons, and cortisol can be derivatized with p-hydrazinobenzoic acid [75] and the resulting carboxyl acid derivatives can be linked to proteins/immunoglobulins.

4. A carboxyl group can be introduced into a phenolic residue by reaction with chloroacetate, and the resulting carboxyl acid derivative can be used for coupling to protein/immunoglobulin as described later. An example of this methodology can be found in the coupling of morphine to BSA [76].

5. Introduction of a carboxyl group into a drug molecule containing a phenolic group by reaction with diazotized p-aminobenzoic acid is another type of reaction that has been carried out successfully with 17β-estradiol]77]. Recently, similar procedures have also been applied to the preparation of A_7-tetrahydrocannabinol [78] and reserpine conjugates [79].

6. The classical procedure for the coupling of sugars to proteins involves the formation of p-nitrophenylglycosides, subsequent reduction of the latter to p-aminophenylglycosides, followed by attachment to the protein by diazotization. Landsteiner [80] used this methodology for a number of cases, but later on a variation of this method was developed by Goebel [81,82] and Goebel and Hotchkiss [83] in which p-aminophenylglycosides were converted to aminobenzyl ether prior to diazotization.

7. The oxidation of a primary alcohol group into a carboxyl group prior to conjugation is another approach. The 5′-hydroxyl groups of uridine [84,85], pseudouridine [84], and other nucleosides [86] can be oxidized to corresponding carboxylic acid derivatives, which can subsequently be linked to proteins/immunoglobulins by already-available methods.

While it is fairly simple to introduce a reactive functional group to any drug molecule for conjugation to an antibody, susceptibility of the generated drug-

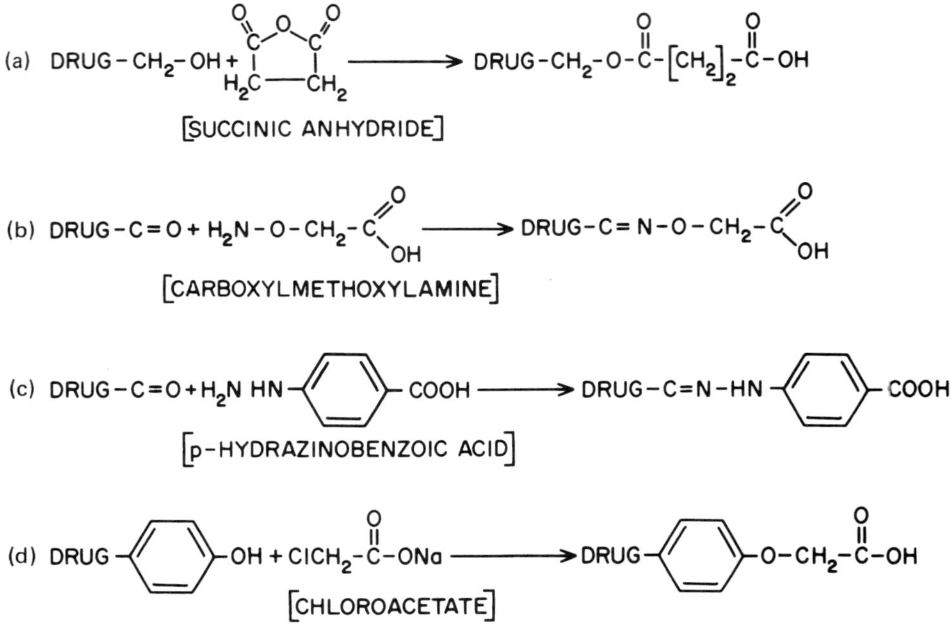

(a) $\text{DRUG-CH}_2\text{-OH} + \text{(succinic anhydride)} \longrightarrow \text{DRUG-CH}_2\text{-O-}\overset{O}{\overset{\|}{C}}\text{-[CH}_2]_2\text{-}\overset{O}{\overset{\|}{C}}\text{-OH}$

[SUCCINIC ANHYDRIDE]

(b) $\text{DRUG-C=O} + \text{H}_2\text{N-O-CH}_2\text{-C(=O)OH} \longrightarrow \text{DRUG-C=N-O-CH}_2\text{-C(=O)OH}$

[CARBOXYLMETHOXYLAMINE]

(c) $\text{DRUG-C=O} + \text{H}_2\text{N HN-C}_6\text{H}_4\text{-COOH} \longrightarrow \text{DRUG-C=N-HN-C}_6\text{H}_4\text{-COOH}$

[p-HYDRAZINOBENZOIC ACID]

(d) $\text{DRUG-C}_6\text{H}_4\text{-OH} + \text{ClCH}_2\text{-}\overset{O}{\overset{\|}{C}}\text{-ONa} \longrightarrow \text{DRUG-C}_6\text{H}_4\text{-O-CH}_2\text{-}\overset{O}{\overset{\|}{C}}\text{-OH}$

[CHLOROACETATE]

Figure 3 Formation of a drug derivative suitable for conjugation to immunoglobulin. (a) Esterification of a primary alcohol with succinic anhydride to form hemisuccinates. (b) Formation of an oxime from the reaction of a ketone group with carboxymethyl hydroxylamine. (c) Introduction of a carboxyl group by derivatization of ketone group with p-hydrazinobenzoic acid. (d) Introduction of a carboxyl group into a phenolic residue by using chloroacetate. (e) Diazotization with p-aminobenzoic acid. (f) Reduction of p-nitrophenyl group to p-aminophenyl group followed by attachment with immunoglobulin by diazotization. (g) Oxidation of 5′-hydroxyl group of the nucleosides to corresponding carboxylic acid derivative.

antibody linkage to lysosomal enzymes may be variable and may require more basic chemical research and biological testing before a useful conjugate can be developed. Conjugation of adriamycin highlights the difficulties in this area, in particular, the problems associated with the derivatization of the drug and the difficulty of obtaining an effective drug-antibody conjugate. A number of methods are available for coupling daunomycin, an analog of adriamycin, to monoclonal antibodies [87], but work with adriamycin has had little success due to its lower solubility and stability. Several attempts have been made to couple adriamycin to antibody, including oxidation of the sugar moiety coupled

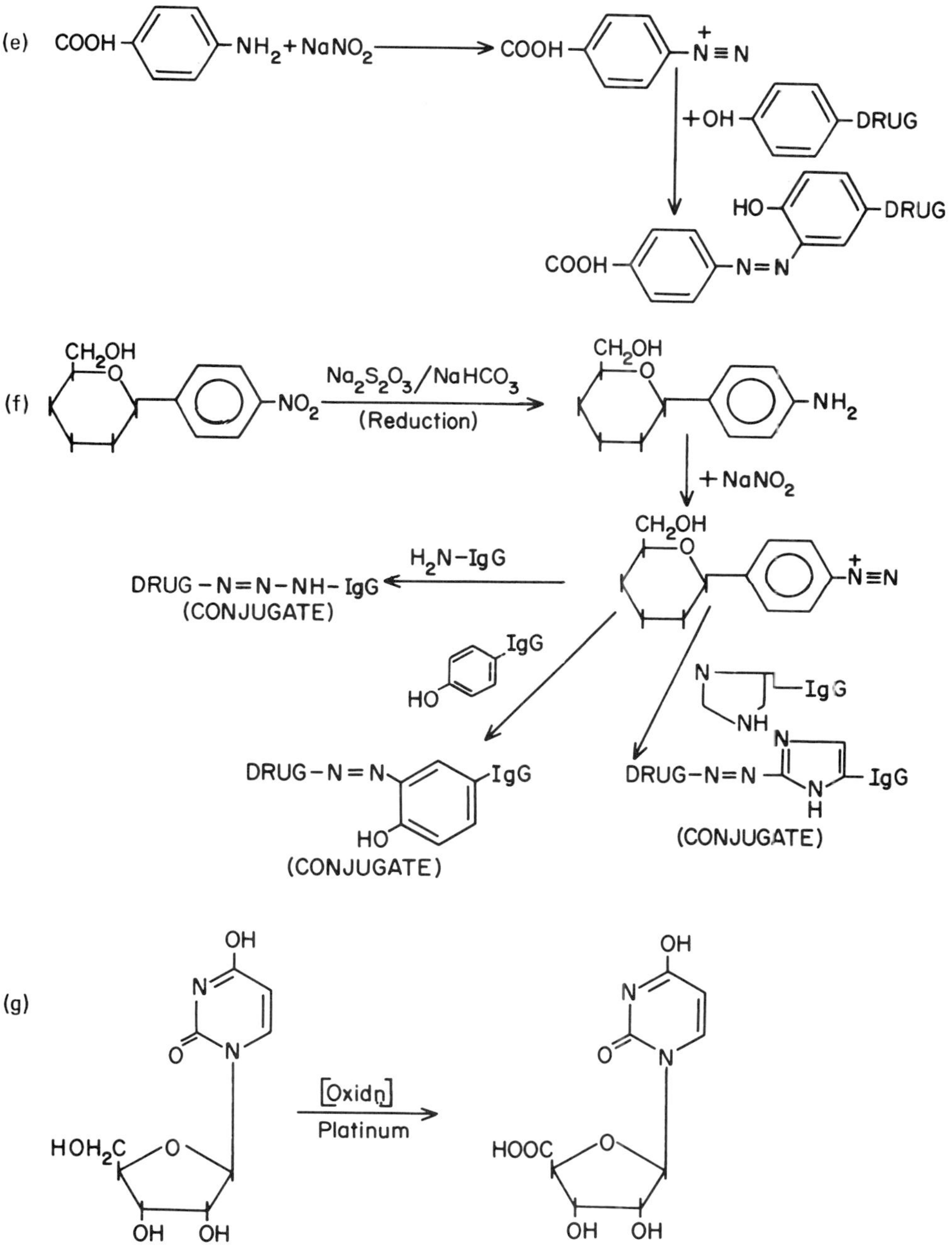

(e) COOH–C$_6$H$_4$–NH$_2$ + NaNO$_2$ ⟶ COOH–C$_6$H$_4$–$\overset{+}{N}\equiv N$

+OH–C$_6$H$_4$–DRUG

HO– –DRUG

COOH–C$_6$H$_4$–N=N–

(f) CH$_2$OH –NO$_2$ $Na_2S_2O_3/NaHCO_3$ (Reduction) ⟶ CH$_2$OH –NH$_2$

+ NaNO$_2$

CH$_2$OH –$\overset{+}{N}\equiv N$

H$_2$N–IgG

DRUG – N=N–NH– IgG
(CONJUGATE)

IgG

HO

N –IgG
NH

DRUG–N=N– –IgG
HO
(CONJUGATE)

DRUG–N=N– N –IgG
N
H
(CONJUGATE)

(g) OH N O N HOH$_2$C O OH OH

[Oxidn]
Platinum

OH N O N HOOC O OH OH

Figure 3 (continued)

(a)

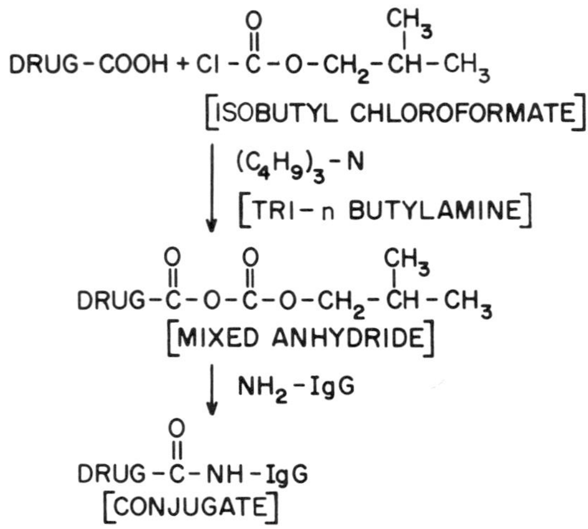

$$\text{DRUG-COOH} + \text{Cl}-\overset{\overset{\text{O}}{\|}}{\text{C}}-\text{O}-\text{CH}_2-\overset{\overset{\text{CH}_3}{|}}{\text{CH}}-\text{CH}_3$$

[ISOBUTYL CHLOROFORMATE]

$$\downarrow (C_4H_9)_3-N$$

[TRI-n BUTYLAMINE]

$$\text{DRUG}-\overset{\overset{\text{O}}{\|}}{\text{C}}-\text{O}-\overset{\overset{\text{O}}{\|}}{\text{C}}-\text{O}-\text{CH}_2-\overset{\overset{\text{CH}_3}{|}}{\text{CH}}-\text{CH}_3$$

[MIXED ANHYDRIDE]

$$\downarrow NH_2-IgG$$

$$\text{DRUG}-\overset{\overset{\text{O}}{\|}}{\text{C}}-\text{NH}-\text{IgG}$$

[CONJUGATE]

Figure 4 (a) Formation of a mixed anhydride by reaction with isobutyl chloroformate and tri-n-butylamine followed by reaction with amino groups in immunoglobulin molecule. (b) Formation of an active ester by reaction with N-hydroxy succinimide and dicyclohexyl carbodiimide followed by reaction with amino groups in immunoglobulin molecule. (c) Carbodiimide activation of carboxyl group followed by reaction with amino groups in immunoglobulin molecule.

through carbodiimide and glutaraldehyde as well as through a dextran intermediary, none of which produced potent complexes. Recently, Pietersz et al. [88] prepared three different derivatives by reacting adriamycin with succinic anhydride, succinimidyl maleimidophenylbutyric acid, and the active ester of iodoacetic acid. Of these derivatives, iodoacetyl adriamycin was found to possess the optimum effectiveness in terms of cytotoxicity and retention of activity.

B. Coupling Reactions

Once the necessary functional group has been introduced into the drug molecule, the following reactions can be performed to attach them to antibodies.

Drug Molecules with Carboxyl Groups

Mixed-anhydride Method. The method is simple and straightforward and does not require isolation of an active intermediate. The mixed anhydride procedure was originally developed for peptide synthesis [89,90], but it has since

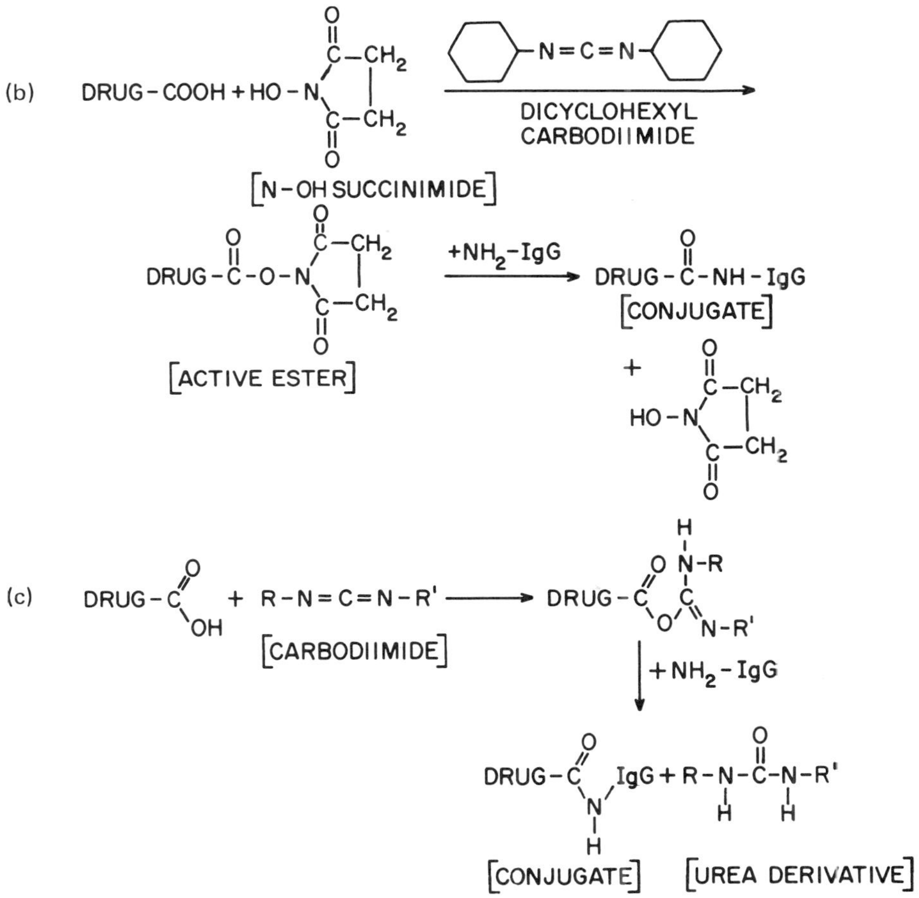

(b) DRUG − COOH + HO − N(succinimide) → [DICYCLOHEXYL CARBODIIMIDE: $C_6H_{11}-N=C=N-C_6H_{11}$]

[N-OH SUCCINIMIDE]

DRUG − C(=O) − O − N(succinimide) [ACTIVE ESTER] $\xrightarrow{+NH_2-IgG}$ DRUG − C(=O) − NH − IgG [CONJUGATE] + HO − N(succinimide)

(c) DRUG − C(=O)OH + R − N = C = N − R' [CARBODIIMIDE] → DRUG − C(=O) − O − C(=N − R') − NH − R

$\downarrow$ + NH_2 − IgG

DRUG − C(=O) − NH − IgG [CONJUGATE] + R − NH − C(=O) − NH − R' [UREA DERIVATIVE]

Figure 4 (continued)

been used extensively in immunochemistry for the conjugation of a variety of carboxyl-containing haptens to proteins, either for antigen preparation [91,92] or for labeling enzymes in enzyme immunoassays [93,94]. The procedure has also proven useful for preparing conjugates with a high ratio of haptens to protein in which the hapten is immunoreactive [95]. The reaction proceeds in two steps (Figure 4a). First, the hapten containing a carboxyl group is conjugated to an alkylchlorocarbonate, usually isobutylchloroformate in the presence of a tri-N-alkylamine, such as triethylamine or tri-butylamine, at low temperature and under anhydrous conditions. The reaction results in the formation of a mixed anhydride. The mixed anhydride is then added to a protein solution where the hapten reacts with free amino groups, usually ϵ amino groups of lysine side chains, linking hapten to protein via an amide bond [96]. Drug molecules having carboxyl groups can be coupled to amino groups of the antibodies by this method as long as amino or other sensitive groups are not present in the drug molecule itself or those groups have been suitably protected—otherwise, extensive autocoupling may occur. Burnstein and Knapp [97] synthesized active conjugates of MTX and an anti-mouse ovarian carcinoma antibody by reacting immunoglobulin with a product formed by heating MTX with acetic anhydride. Several groups, however, have not found any antitumor activity in such conjugates and have failed to recover active MTX under conditions appropriate for hydrolysis of a mixed anhydride [98,99].

Active Ester Method. This is another simple and direct method of coupling drug molecules with proteins [100]. The procedure, like the mixed anhydride method, does not require isolation of the active intermediate. Moreover, the intermediate can be stored under nitrogen in a dry condition for a longer period of time. The carboxyl group of the drug molecule is converted to an ester by reacting with equimolar amounts of N-hydroxy succinimide and dicyclohexylcarbodiimide in a small volume of organic solvent for several hours. The precipitate formed during the reaction is removed by centrifugation, and the clear supernatant is added dropwise directly to a stirred aqueous-acetone solution of proteins, thus forming an amide (or peptide) bond between the drug and immunoglobulin (Figure 4b). Methotrexate has been coupled to anti-EL4 IgG by this method. Seventy percent of antibody activity was retained at a drug incorporation level of 12 moles of drug per mole of IgG, but there was dramatic loss of antibody activity beyond 15 moles of drug per mole of antibody. Antitumor activity of the conjugates were also retained under both in vitro and in vivo conditions [98].

Carbodiimide Method. This direct method of drug coupling has been used extensively in preparing drug-protein conjugates [101,102] or protein-protein conjugates [103,104]. The procedure can be carried out in aqueous media and results in the formation of amide bonds between carboxyl and amino groups present on either of the molecules (Figure 4c). Although dicyclohexyl carbodiimide was first utilized for such reactions, a number of water-soluble carbodiimides—1-ethyl-3-(3′-dimethylamino propyl) carbodiimide, 1-cyclohexyl-3-(2-morpholine ethyl) carbodiimide, metho-p-toluene sulfonate, and others [105]—have also been tried in linking a variety of drugs including chlorambucil, N,N-bis(2-chloroethyl)-p-phenylenediamine (PDM), adriamycin/daunomycin, and MTX directly to immunoglobulins or via appropriate spacers. If the drug molecules do not have appreciable aqueous solubility, the compounds can be dissolved in water-miscible solvents such as dimethyl formamide and added to the aqueous protein solution [73]. The degree of coupling is dependent on the density of the reactive groups. It is important that the pH of the reaction medium be kept at or near the neutral region. At pH lower than the pK_a, the amine is protonated and does not react readily, whereas at a higher pH the carbodiimide may decompose. Since proteins contain both amino and carboxyl groups, it is possible that unwanted cross-linkage and polymerization of the antibody molecule can occur [106].

Drugs with Amino Groups

Drug molecules with available amino groups can be characterized as aromatic and aliphatic amines.

Aromatic Amines. Molecules having aromatic amino groups can be converted to diazonium salts by reaction with nitrous acids. The dizonium salts are subsequently reacted with proteins at alkaline pH (approximately 9). Histidine, tyrosine, and tryptophan residues of the protein molecule are the primary sites of conjugation. The method was applied for the preparation of a chloramphenicol-protein conjugate to raise antibodies specific for chloramphenicol [107]. As early as 1937, carcinogenic compounds were conjugated to protein carriers by means of their isocyanate derivatives derived from amines [108].

Haptens containing azophenyl groups can also be coupled to antibodies by means of amidination reactions [109]. The method permits attachment of a large number of drug residues with IgG without much altering antibody activity. As shown in Figure 5a, the reaction proceeds in two steps and the procedure does not require the isolation of active intermediate. In the first step, for azo formation, drug molecules are reacted with an amidinating reagent such as p-hydroxy benzimidate or 3,5-dihydroxy benzimidate in an alkaline solution at about pH 9.2. When the diazonium reagent is exhausted, the reaction solution is adjusted to pH 8.5 and is later added to the antibody solution, which converts protein amino groups to drug-linked amidine groups.

Aliphatic Amines. Drug molecules having aliphatic amines can be coupled to antibodies following reactions with water-soluble carbodiimides. Several drugs, including bradykinin [104], tobramycin [110], gentamicin [111], adriamycin [112], 5-OH tryptamine [113], and spermidine [114], have been coupled to proteins using this method.

Aliphatic amines can be converted to p-nitrobenzoyl amide by reaction with p-nitrobenzoyl chloride, which upon reduction yields the corresponding p-aminobenzoyl derivative. Such amino derivatives can be coupled to proteins by diazotization as described earlier.

Compounds having aliphatic secondary amine groups can also be coupled to proteins by the method described for coupling nortryptyline to BSA [115]. In

(a)

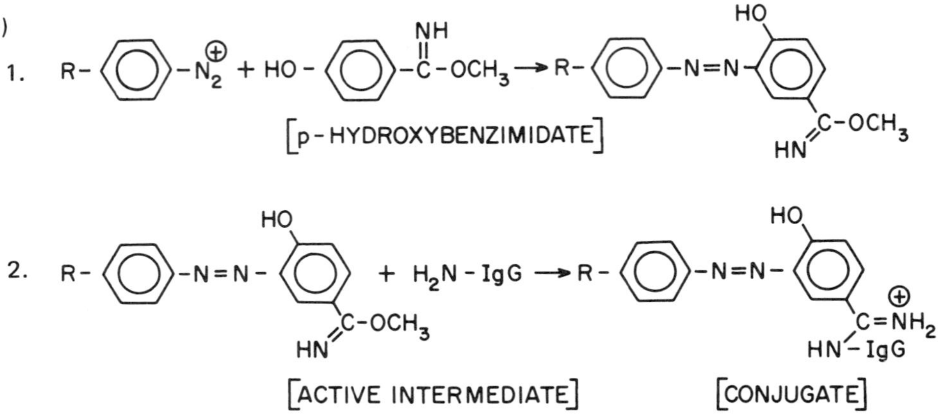

[p-HYDROXYBENZIMIDATE]

[ACTIVE INTERMEDIATE]

[CONJUGATE]

Figure 5 (a) Diazotized phenyl hapten is converted to an active intermediate by reaction with p-hydroxybenzimidate at pH 9.2 followed by reaction with immunoglobulin preparation at pH 8.5. (b) Drug molecules containing amino groups conjugated to immunoglobulin molecule using glutaraldehyde. (c) Coupling of drug molecules containing amino groups to immunoglobulin using SPDP. The reagent reacts with the amino group via its N-hydroxysuccinimide moiety and introduces a 2-pyridyldisulfide group, which is subsequently reacted with the thiol group of the Fab fragment.

this method, a succinamic acid derivative of the compound is made, which is then coupled to the protein using water-soluble carbodiimide.

Bifunctional Reagent. Compounds containing amino groups can be coupled to proteins using bifunctional reagents; the drugs conjugated in this manner are angiotensin [116], glucagon [117], normetanephrine [118], and adrenocorticotropic hormone [119].

In recent years, the use of bifunctional reagents has gained much popularity for making protein-protein and protein-drug conjugates. Both homobifunctional and heterobifunctional reagents have been actively used for this purpose. Selection of the reagent depends on the reactive groups so that a controlled sequential activation of each group is possible. For a reagent to be considered effective, it must possess a relatively unreactive group that can be photoactivated to a highly reactive group, e.g., nitrenes and carbenes. The coupling reaction with such reagents involves a two step process: an initial step in the dark followed by one in the presence of activating light. However, the extreme indiscrimination of such reagents makes their use unsuitable for controlled linkage of drug molecules to immunoglobulins. Table 1 lists the commonly used bifunctional reagents and the

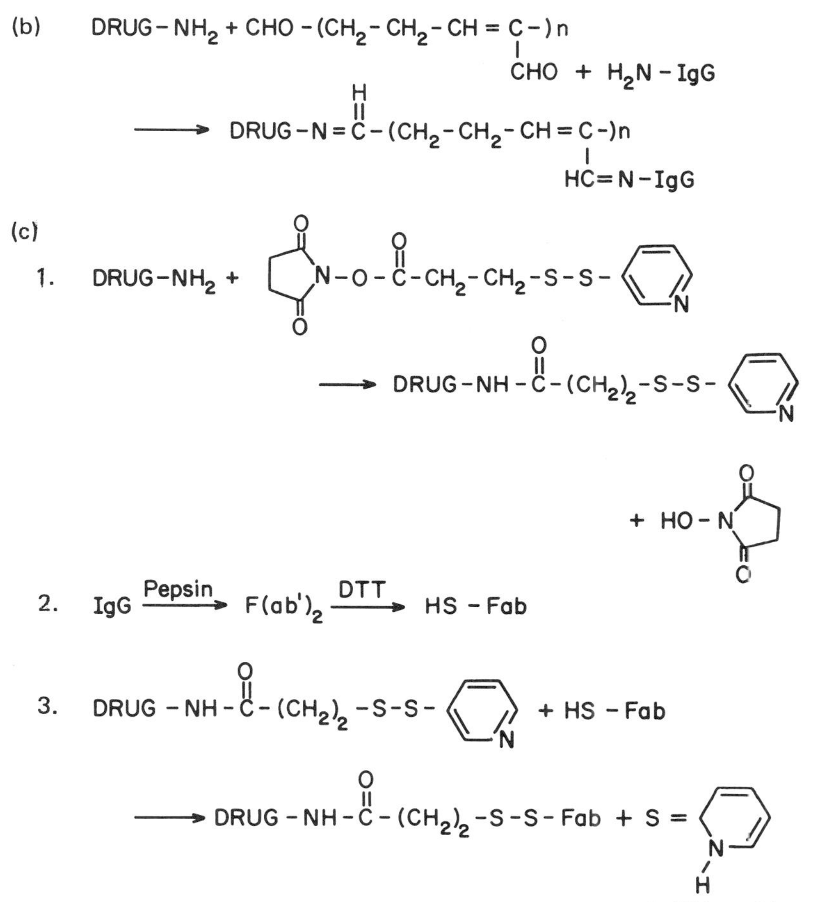

(b) $DRUG-NH_2 + CHO-(CH_2-CH_2-CH=\underset{\displaystyle CHO}{C}-)_n + H_2N-IgG$

$\longrightarrow DRUG-N=\overset{H}{C}-(CH_2-CH_2-CH=\underset{\displaystyle HC=N-IgG}{C}-)_n$

(c)

1. $DRUG-NH_2$ + [N-succinimidyl]$-O-\overset{O}{\overset{\|}{C}}-CH_2-CH_2-S-S-$[2-pyridyl]

$\longrightarrow DRUG-NH-\overset{O}{\overset{\|}{C}}-(CH_2)_2-S-S-$[2-pyridyl]

+ HO-N[succinimide]

2. $IgG \xrightarrow{Pepsin} F(ab')_2 \xrightarrow{DTT} HS-Fab$

3. $DRUG-NH-\overset{O}{\overset{\|}{C}}-(CH_2)_2-S-S-$[2-pyridyl] + HS-Fab

$\longrightarrow DRUG-NH-\overset{O}{\overset{\|}{C}}-(CH_2)_2-S-S-Fab + S=$[pyridine, N-H]

2 Thiopyridone

Figure 5 (continued)

functional groups involved. Moreover, examples of one heterobifunctional and one homobifunctional reagent are addressed in the following paragraphs.

1. *Heterobifunctional reagent.* Heterobifunctional reagents [120] are becoming extremely common reagents for the preparation of protein-protein conjugates [41,121], protein-hapten conjugates [122], and immunoliposomes [123]. The most versatile heterobifunctional reagent is N-succinimidyl-3(2-pyridyldithio) propionate (SPDP), which reacts with amino groups via its N-hydroxy succinimide moiety and with thiol groups via its 2-pyridyldisulfide

Table 1 Bifunctional Reagents Used in Covalent Linkage

	Reactive groups	Comments
Homobifunctional reagents		
Dimaleimide	Sulfhydryl	Relatively insoluble in water with the exception of N,N′-(oxydimethylene) dimaleimide; mild reagent
Alkyl/aryl halides	Sulfhydryl, amino, imidazole, phenol	Highly reactive; and reaction with different groups is pH-dependent
Diisocyanates	Amino	Mostly insoluble in water; reaction may lead to aggregation and also inactivation of IgG
Acylating agents	Amino	Highly reactive
Diimido esters	Amino	Soluble in water; reaction proceeds under mild conditions
Dialdehydes	Amino	Highly reactive but produces intra- or intermolecular cross-linking
p-Benzoquinones	Amino	Soluble in water; reaction does not interfere with antigen-binding specificity
Heterobifunctional reagents		
N-Succinimidyl-3-(2-pyridyldithio) propionate (SPDP) or Succinimidyl-4-(p-maleimidophenyl) butyrate (SMPB)	Amino, sulfhydryl	Soluble in water; reaction is controllable and less chance of forming homopolymers; reagents are also mild
m-Maleimido benzoyl-N-hydroxysuccinimide	Amino, sulfhydryl	Reaction proceeds under mild conditions

group. Reaction conditions for SPDP coupling are mild, side reactions are limited, and chances of forming homopolymers are very low. The scheme illustrating SPDP reaction with an antibody has been depicted in Figure 5c.

2. *Homobifunctional reagent.* The most popular homobifunctional reagent among aliphatic dialdehydes is glutaraldehyde, which has been used extensively in the preparation of protein-protein conjugates. However, drug molecules containing amino groups can also be conjugated by glutaraldehyde, as illustrated by Hurwitz et al. [87] in the coupling of daunomycin to IgG. Although the dialdehyde-mediated linkage is believed to occur via Schiff base formation between an aldehyde and an amino group, the reaction mechanism is as yet not fully understood. Richards and Knowles [24] proposed a mechanism in which the reaction with amino groups is believed to occur through the α,β-unsaturated aldehyde forming Michael-type adducts (Figure 5b).

Drugs with Hydroxyl Groups

The hydroxyl group includes alcohols, phenols, sugars, polysaccharides, and nucleosides. In most cases, a carboxyl group can be introduced into these compounds by derivatization with an appropriate reagent (described earlier), and then the newly introduced carboxyl groups can be derivatized for preparing conjugates with protein as described in the section on drugs with carboxyl groups. In the case of periodate oxidation, the compounds containing two vicinal hydroxyl groups are oxidized to dialdehydes which, without further isolation, capable of forming a Schiff base with the amino groups of the proteins.

Periodate Oxidation. The periodate oxidation method involves a simple reaction that can be applied to link (1) a carbohydrate moiety in the drug molecule with an amino group in the immunoglobulin; (2) an amino group in the drug molecule with carbohydrate moieties of the immunoglobulin; and (3) via carbohydrate-containing spacers. The dialdehydes, formed in situ during the reaction between excess periodate and vicinal hydroxyl groups, can further react with the amino groups of the protein molecules in the absence of light and at pH 9.5. The resulting aldimines, following reduction with sodium borohydride, give rise to a desired stable product (Figure 6). The reaction is applicable to any compounds having vicinal hydroxyl groups, such as glycols, glycerol derivatives, and glycosides. Although the method is capable of incorporating a large number of drug molecules [125], particularly when dextran intermediaries are used [126,127], it is difficult to control the reaction to obtain the desired extent of drug incorporation, and there is usually a substantial loss of antibody activity in the conjugate. Secondary amines, generated from the reduction of Schiff bases with sodium borohydride, are not easily cleaved in vivo and results in the loss of activity of the conjugate. The main disadvantage of periodate as the coupling re-

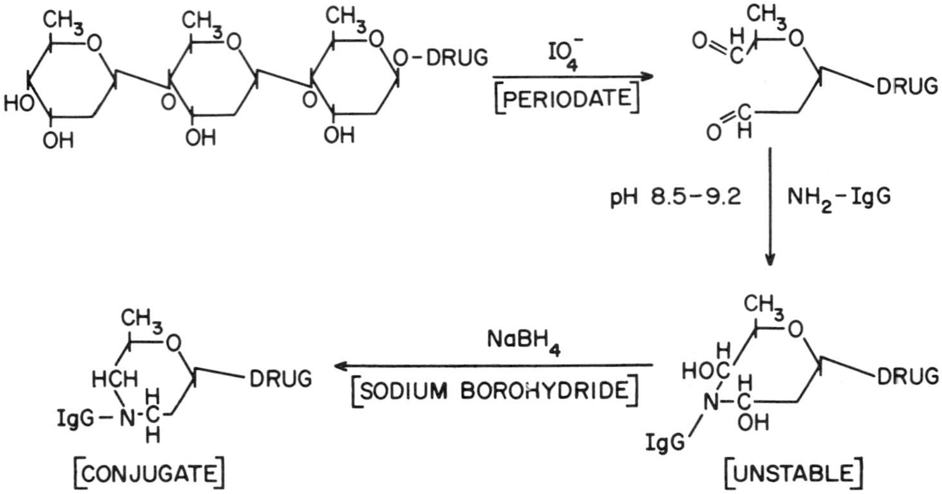

Figure 6 The dialdehydes, formed in situ between periodate and vicinal hydroxyl groups, can react with the amino groups in the immunoglobulin molecule to form aldimines. Subsequent reduction of the aldimines with sodium borohydride yields a stabe immunoconjugate.

agent is the formation of intra- and intermolecular cross-linked immunoglobulinks between carbohydrate moieties and amino groups [125].

C. Noncovalent Linkages

For antibody-mediated drug delivery, most of the conjugates are prepared by methods aimed at establishing a covalent linkage between the two components. An alternative approach, however, based on noncovalent association, has been explored primarily by Ghose and his colleagues with the bifunctional alkylating agent chlorambucil [126,127]. Several reports have been published of research in which noncovalent complexes of alkylating drugs with immunoglobulin [13,16,126–130] and other macromolecules [130] have been studied. For example, chlorambucil linked noncovalently to antitumor antibodies was found to be more effective in killing the target tumor cells than either the free drug or the antibody alone [13,16,126,129]. It has been established from these studies that an immunochemotherapeutic approach can be applied successfully with conjugates having drugs physically adsorbed onto the globulin fraction from tumor-specific antiserum. The mechanism of physical adsorption is not known, but temperature, time of reaction, and concentration of the reactants have definite roles in the formation of such complexes. The effects of pH and ionic strength are not highly critical, but low ionic strength and alkaline pH favor the complexation between the two components. In the course of many experiments with chlorambucil-immunoglobulin system, reproducibility of binding under a given set of conditions has been generally good but has never been exact; the number always varies fron 60 to 70 molecules per molecule of IgG. The physically bound chlorambucil is generally measured by ethanol extraction of the complex followed by spectrophotometric determination at 258 nm [131].

A problem in noncovalent linking of chlorambucil to antibody is the possible formation of aggregates of chlorambucil [132]. The presence of these aggregates might lead to overestimation of the amount of active chlorambucil physically linked to the antibody. Some valid questions regarding the use of noncovalent bonding in conjugate construction have been raised by Ross et al. [23], particularly with regard to the stability of the conjugate in body fluids. In at least two different studies [13,129] it has been shown that similar effects could be obtained by administering the free drug and the antibody separately. The possibility exists that the drug administered as a noncovalent complex with antibodies dissociates in vivo and acts separately and synergistically with the antibody, as suggested by Segerling et al. [133].

D. Use of Carriers and Spacers in the Design of Immunoconjugates

Use of Carriers

To deliver a substantial amount of drug to a target organ having a limited number of available antigenic sites, the antibody molecule should have a large number of drug residues attached to it. The extent of direct covalent binding of drug molecules onto immunoglobulin is likely to be limited by progressive loss of antibody activity and/or solubility, as already discussed [87,134]. Such limitations could constitute major barriers to the emergence of effective drug-antibody complexes. The best approach would be either to use a robust antibody or to use a drug carrier that is then attached to the antibody. The problems in obtaining monoclonal antibodies with desired specificity and consistent biochemical stability may encourage researchers to use a drug carrier.

The introduction of a carrier has several advantages over that of direct linking. First, conjugates constructed in this manner can be produced with a "drug-loading" capacity at least 10 times greater than the produced by direct conjugation without any appreciable loss of antibody activity. Another advantage of drug carrier introduction is the enhanced stability of the protein to any relatively harsh coupling reactions. The major drawbacks of carrier use are likely to be seen in vivo relative to the stability and altered biodistribution of the conjugate due to the considerable increase in molecular weights. Although several articles reported the use of carriers in targeted and nontargeted delivery, the selection of an appropriate carrier still needs thorough study. Poly-L-lysine has been used as a carrier for several drugs such that free drugs can be released in the lysosomal milieu [70,135]. The advantage of polylysine as a drug carrier stems from the fact that its reaction with antibody needs a relatively small linkage site, and a large drug load can be incorporated as well. The disadvantages of such carriers are the extreme toxicity and the stickiness of the compounds, rendering antibody assay difficult.

Other carriers that have been reported so far include albumin [136], fibrinogen and α-globulins [130,137], melanotropin [138], wheat germ agglutinin [69], chymotrypsinogen [139], concanavalin A [140], dextrans [141,142], polyglutamate [22], polyaspartate [143], carboxymethylcellulose [144], and N-(2-hydroxypropyl)methylacrylamide copolymers [67,68]. The criteria for selecting a drug carrier include molecular size, numbers of functional groups, homogeneity, ease of handling, ease of coupling stability, toxicity, and possibly molecular shape (whether globular or random or straight chain).

Two types of drug carrier-antibody conjugate have been discussed in the literature. In one type of conjugation, the drug is coupled to the carrier by one method followed by coupling to the antibody by a second method [22,142, 145]. In the other type, the carrier has been used as a multifunctional agent linking the drug to the antibody [146]. The first type is usually preferred as it confers more control to the ratio of drug to carrier, and carrier to antibody, and degree of polymerization of the final product. It also permits the selection of different types of linkages between the drug and the carrier and between the carrier and the antibody.

Use of Spacers

It has become apparent that drug molecules linked directly to antibodies are often poorly released from their conjugates by the action of lysosomal enzymes. Such slow release rates might be due to a simple steric effect, since success has been achieved by introducing a spacer molecule between the two components. Thus, the advantage of a spacer is twofold: It reduces the steric effect and it also helps to increase the drug load per immunoglobulin molecule.

The successful application of spacers in the preparation of drug-antibody conjugate has been illustrated mainly with adriamycin and daunomycin. The cytotoxic agent daunorubicin, when conjugated to serum albumin by a tetrapeptide (Ala-Leu-Ala-Leu), exhibited release by lysosomal enzymes [147]. Similarly, the linkage of daunomycin to wheat germ agglutinin via a Glc-S-Et-Arg-Leu arm [69] or to serum albumin via N-L-leucyl-L-alanyl-L-leucyl or N-

L-leucyl-L-alanyl-L-leucyl-L-alanyl arms [71] provided active cytocidal conjugates. The spacers are designed in such a way that the conjugates remain stable during their transit through blood, but release free drug in the lysosomal milieu. Most often, amide linkages between the components in the ternary conjugates are chosen because of the presence of higher amidase activity in lysosomes and the absence of similar activity in the serum.

An alternative approach has been to achieve drug release within the lysosomes based not on the enzymatic attack but on the acid pH within the lysosomes. An acid-labile spacer is used in such cases to link the drug with the antibody molecules. Such linkage undergoes cleavage within the lysosome and consequently releases the drug. For example, daunorubicin was conjugated to poly-D-lysine by a pH-sensitive cis-aconityl spacer, and the cytotoxicity of the conjugate was attributed to the release of the drug under the mildly acidic conditions of lysosomes [70]. The major limitations with this type of drug delivery system is the decrease in water solubility of the conjugate resulting from the attachment of a high payload of certain hydrophobic drugs.

E. Comparison of Different Coupling Methods

Comparison of different methods of coupling in the same ligand-and-immunoglobulin system has not been carried out in great detail. The first attempt in this area was made in 1975 by Sela's group, who coupled individually two potent anticancer drugs, viz., adriamycin and daunomycin, with anti-BSA antibodies by three different coupling methods; using glutaraldehyde (Michael-type adduct), periodate borohydride (secondary amine), and a water-soluble carbodiimide (amide) as coupling reagents. The periodate-borohydride method (which cleaves the bond between C-3 and C-4 of the amino sugar groups) was reported to be the best coupling method in terms of retention of drug and antibody activities in the conjugate [87,148]. Hurwitz and colleagues also suggested the effectiveness of this method in their preparation of daunomycin conjugate with anti-mouse lymphoma IgG via a dextran bridge, utilizing periodate-oxidized dextran. The ternary conjugate exhibited more antitumor activity than the free drug [146]. However, in light of more recent studies, this conclusion appears to be questionable. In 1980, Hurwitz et al. failed to produce a soluble complex while conjugating daunorubicin with goat anti-mouse lymphoma IgG by periodate oxidation [144]. In another experiment, Ghose et al. [149] prepared a ternary conjugate (adriamycin-dextran T40-anti BSA IgG) by the periodate oxidation method, which caused a significant amount of antitumor activity retention both in vivo and in vitro. Upon further reduction with borohydride, however, the ternary conjugate completely lost the antitumor action of adriamycin. Moreover, direct conjugation of adriamycin with anti-BSA IgG by the periodate borohydride method did not produce an effective conjugate.

Burnstein and Knapp [97] compared the cell-killing properties of methotrexate in ovarian carcinoma when it was conjugated by two different coupling methods. Reaction with water-soluble carbodiimide, in one case, resulted in a complex with greater antitumor activity than the free drug or antiserum alone, but left a large fraction of insoluble material—a problem encountered by other investigators [13]. An alternative coupling reaction, based on a mixed-anhydride reaction, produced a more soluble complex that retained most of the antitumor activity. Shigeru et al. [150] coupled adriamycin to anti-α-fetoprotein IgG using a water-soluble carbodiimide as well as a bifunctional reagent such as glutaraldehyde. Each of the conjugates, with drug-antibody molar ratios of 4–5 to 1, retained 65–75% of the original antibody activity and a substantial fraction of antitumor action. Pietersz et al. [88] failed to produce a potent immunoconjugate when adriamycin was linked to monoclonal antibodies by various methods including oxidation of the sugar moiety followed by direct coupling reaction with carbodiimide and glutaraldehyde, as well as through a dextran intermediary.

In a recent publication [151] we have compared three different coupling reactions (all predicted to produce amide bonds) for conjugating MTX to IgG. The first reaction involved coupling via water-soluble 1-ethyl-3(3′-dimethyl aminopropyl) carbodiimide (EDCI), whereas two other methods required prior activation of MTX carboxyl groups through active ester formation by reaction with N-hydroxysuccinimide or through mixed anhydride formation by reaction with isobutyl chloroformate. When these methods were compared with respect to protein recovery, degree of drug substitution, and drug and antibody activities in the conjugate, none of them showed absolute superiority except the simplicity of the reaction offered by the active ester method and the relative stability of the ester under nitrogen for a long time.

In a study comparing covalent and noncovalent attachment, Tai et al. [152] observed that superior tumor inhibition was caused by chlorambucil linked to an antitumor IgG by noncovalent bonds. Warzynski et al. [153] reported that coupling of triaziquinone (Trenimon) to IgG by a thiolation procedure gave much more reliable and reproducible results than that involving dithiothreitol-induced reduction of IgG.

F. Purification and Characterization of Immunoconjugate

Following conjugation, the purification and characterization of the immunoconjugate are the two other important steps in preparing drug-antibody conjugate. The high-molecular-weight immunoconjugate can be purified from low-molecular-weight unbound drugs, reactants, or side products by standard techniques such as dialysis, gel filtration, affinity chromatography, or ammonium sulfate fractionation. Although these methods will remove most of the low-molecular-weight compounds, the preparation will always contain some unreacted immunoglobulin. In the case of low-molecular-weight drugs, the resolution of unbound immunoglobulin from the immunoconjugate is difficult, and further purifications are not usually attempted. However, affinity chromatography using immobilized ligands with affinity for drug molecules may be applied to remove unreacted immunoglobulin from the drug-immunoglobulin conjugate. For example, antibody for toxins or low-molecular-weight drugs, dihydrofolate reductase for MTX [154], and Poropak Q for adriamycin and daunomycin have been employed. When protein, peptide, or enzyme is coupled to immunoglobulin, the choice of an appropriate gel-filtration media can allow resolution of unreacted protein, unreacted immunoglobulin, and the conjugate.

The number of drug residues in the immunoconjugate can be determined by either of the standard methods, i.e., isotope incorporation and absorption spectrophotometry, provided that the drug molecule has a suitable chromophore [63,155]. Even if the spectrum of the drug overlaps that of immunoglobulin, the difference in absorbance values between the conjugated antibody and the same amount of free antibody will constitute a reasonably accurate determination of the number of drug molecules per immunoglobulin molecule [152,156]. Certain compounds (azo derivatives) have an absorbance spectrum in the visible region that may enable one to differentiate the drug molecules from the protein carrier. The determination of drug incorporation by the isotope technique was first developed by Abraham et al. [157], who calculated a direct ratio of substitution by counting undialyzable radioactive material.

In indirect techniques, the molar ratio can be roughly estimated by measuring the residual functional group in the immunoglobulin molecule following the coupling reaction (e.g., free amino groups for trinitrobenzene sulfonic acid, or carboxyl groups for a carbodiimide-based method). This indirect method was found to be convenient and entirely satisfactory in the case of insect juvenile hormone-protein conjugates [158-160]. Another indirect method introduced by Erlanger et al. [70] requires the estimation of the residual free amino groups with the dinitrophenylation technique of Sanger [161]. Dinitrophenyllysine is not isolated, but is estimated directly by spectrophotometry after ether extraction of the acid hydrolysate. A control experiment with unsubstituted carrier is always run simultaneously, and the difference between the two readings is taken to be the extent of substitution by drug.

III. BIOLOGICAL PROPERTIES OF DRUG-IMMUNOGLOBULIN CONJUGATES

A. Retention of Antibody Activity

The immunoconjugate resulting from the conjugation of drugs with antibodies should have high residual antibody activity. Conjugation reactions, however, may bring about some conformational changes in the antibody molecule, which may sometimes lead to the loss of specificity and antigen-binding abilities. Monoclonal antibodies raised against carcino embryonic antigen lost some of their antigen-binding property following conjugation with vindessine [54]. However, this is not always true, since many antibodies were found to resist conformational changes following conjugation [54]. In some cases, the loss of antibody activity may arise from the extent of drug incorporation into antibody molecule. In our early experiments with polyclonal antibodies and methotrexate, we found that the ratio between MTX and anti-BSA antibodies also plays an important role in the retention of antibody activity [151]. In general, increased incorporation of drugs to antibody is associated with a decrease in antibody activity. Thus, measurement of antigen-binding properties of the conjugate becomes an important parameter for developing drug-antibody conjugate.

Currently, a number of assays are used to quantitate antibody activity, such as rosetting assays [162], radioimmunoassay [163], flowcytometry [164], radial immunodiffusion [98], and other serological assays. When the conjugates are examined for their antigen-binding properties, a comparison is always made with the same amount of unlabeled antibody as well as with the antibodies exposed to the same environment. Radial immunodiffusion is the simplest and easiest among all the methods used for quantitation of antibody activity. Under conditions where antigen-antibody reactions are less likely to develop visible precipitates, including the assay of monovalent fragments (Fab), other suitable procedures should be adopted. In radioimmunoassay, the immunoconjugates are allowed to react with the labeled antigen and the immunocomplex is precipitated with a second antibody directed against the first antibody. The radioactivity in the pellet measures the antigen-binding property of the conjugate. In flow cytometry, the immunoconjugate is added to an excess of fluorescein isothiocyanate (FITC)-labeled antibody, and the mixture is reacted with tumor cells. At equilibrium, the relative amounts of FITC-antibody bound to tumor cells are estimated by quantitative fluorescein measurements to give a measure of residual activity. The assay is simple, quick to perform, and produces very accurate and reproducible results.

To examine the specificity of the antibody molecule in the immunoconjugate, two different cell lines (one of them antibody-reactive) are separately incubated with the conjugate for 30 min. Following a thorough wash to remove the unbound conjugate, the cell lines are grown for 24 h under specified conditions. If the specificity is retained, only the antigen-bearing cells will be killed.

B. Retention of Drug Activity

Like antibody, the drug molecule should also have high residual activity following conjugation. A number of assays are used to assess the activity of the drug, the most common being the inhibition of incorporation of radiolabeled precursor into either DNA or RNA [165]. Other methods, including dye exclusion [166] and sensitive clonogenic assays [145], are also available to measure drug activity in the conjugate. Some of the assays are also designed on the basis of the drug's biochemical reaction. For example, conjugates containing methotrexate are assayed by its ability to inhibit dihydrofolate reductase [151]. Other conjugates containing alkylating agents (e.g., chlorambucil and Trenimon) are assayed by a colorimetric procedure based on the alkylation of 4-nitrobenzyl pyridine [167]. Whatever the method of choice for assaying drug activity in the conjugates, a comparison is always made with the same amount of free drug.

Although the success of drug targeting with antibodies is based entirely on the specificity of the directing agent, the role of drug molecule in the conjugate should not be overlooked. Some drugs, e.g., MTX [168] and some alkylating agents [169], are taken up selectively by cells, and other drugs, e.g., adriamycin/daunomycin, would be expected to have interactions with cell membranes and thus have nonspecific interaction with the antigen-negative cells. Similar results have also been observed with some carriers, where nonspecific interactions are facilitated by the presence of strong positive charge in the carrier molecule, i.e., poly-L-lysine [135,170]. For these reasons, it is difficult to obtain absolute specificity of action for targeted drug delivery, but an improvement in the discrimination property of the conjugate between tumor and nontumor cells would be of considerable benefit.

In order to examine the specificity of the drug molecule in the immunoconjugate, a single cell line (antibody-reactive) is incubated similarly with two different drug-antibody conjugates (one from specific antibody and the other from nonspecific antibody) as described earlier. If there is no nonspecific interaction between the drug molecule and the cell line, the conjugate containing only the specific antibody will cause cell death. These in vitro assays are similar to the in vivo methods except that a brief contact takes place between the conjugates and the cell lines in which binding must occur.

C. Mechanism of Action

Since considerable evidence (in experimental systems) of the therapeutic effectiveness of drug-antibody conjugates has been accumulated, their mode of action is currently a topic of considerable interest. The synergistic action between free drugs and tumor-specific antibodies may be responsible for such effectiveness. Davies and O'Neil [13], using the EL4 and SB1 mouse lymphomas, showed that antibody-drug complexes were better able to protect mice against death in these model systems than either the drug or the antibody alone. However, the conjugates were not more effective than the same amounts of antibody or drug injected 1 h apart. This phenomenon, in fact, has been described by several other studies [14,129,171,172]. It is possible that the weakly bound drugs might dissociate from the immunoconjugate and act in this manner. Even if such separation does not occur, the presence of unreacted antibody (as described earlier) may act synergistically with drug-conjugated antibodies [142]. According to Ghose et al. [11], even though the synergistic action may be responsible for the improvement seen with the noncovalent conjugates, it does not appear to explain all cases. Hirschberg et al. [173] also ruled out the possibility of synergism as the only explanation for their findings.

Binding of the drug to antibody gives rise to a macromolecular complex that can act either on the cell surface or intracellularly. If the mechanism of drug action is intracellular, and if it can act only in its free form, then the conjugate must first be endocytosed and then digested by lysosomal enzymes before it can exert pharmacological action. Recently, Smith et al. [174] and several other groups [175,176] have coupled methotrexate (MTX) with various monoclonal antibodies to investigate their mode of entry into cells and also their mechanism of action. It is clear from these studies that the antibody-drug conjugates are not degraded at the cell surface but bind to their receptors and then enter the cell by endocytosis as one entity. The drug-antibody conjugates are then degraded within the lysosome by one of the lysosomal enzymes, resulting in the release of free drug into the cytoplasm. The above inferences have been drawn based on the results from the following studies.

1. Chloroquine and ammonium chloride inhibit the action of MTX-antibody but not MTX (both are known to inhibit lysosomal function).
2. p-Chloromercuribenzene sulfonate and folinic acid (transport inhibitor of MTX) affect the toxicity of free MTX but not the conjugated form.
3. Alterations in temperature had a greater effect on the toxicity of MTX-antibody conjugate than on free MTX.
4. Concentration of various divalent cations (Ca^{2+}, Mg^{2+}, Mr^{2+}) effected the entry of MTX-antibody conjugates but had no effect on free MTX.

In summary, once the drug-antibody conjugate comes in contact with the specific antigen/receptors, it follows either of two entry pathways. One is the classical receptor-mediated pathway as described by Willingham et al. [177]. The other is direct fusion of the vesicle containing the drug-antibody conjugate with lysosome. In the second case, the entire complex is degraded and the receptor/antigen would not be able to transport back to the cell surface. Whatever the mechanism might be, once the conjugate reaches the acid environment of the lysosome, the bonds between the drugs and antibodies are broken, thereby releasing the drug which is then free to interact in the cytoplasm.

IV. CONJUGATION OF IMMUNOGLOBULINS TO DRUG-LOADED CARRIERS

The major problem associated with antibody-mediated drug delivery is the limited number of functional groups in the antibody molecule that can be successfully modified without significant loss of antibody activity [54]. Such lack of functional groups limits the molar drug-to-antibody ratio to about 10:1. The introduction of a drug carrier such as dextran [141,142] or HSA [145,178] might allow better drug incorporation, but the larger size of the conjugate will alter its entry into tumor cells and also may affect elimination via the reticuloendothelial system (RES).

The utilization of insoluble drug carriers for prolonged and controlled delivery of therapeutic agents has generated growing interest [179-181]. Various carrier systems have been studied, including synthetic liposomes, erythrocyte ghosts, permeable polymeric microcapsules, and solid microspheres [182-184]. Both microsphere particles and liposomes have created a great deal of interest because they are nontoxic, biodegradable, and well tolerated.

The biggest problem in using these carriers is the lack of target specificity. Following parenteral administration, these carriers are promptly and almost quantitatively cleared from the general circulation by the phagocytic cells of the RES, which are usually localized in the liver, spleen, bone marrow, and lung. In fact, rapid clearance by the RES has proved to be a major obstacle in targeting these carriers to other non-RES cells such as solid tumors [185]. It has been demonstrated that specificity can be achieved by modifying the surface of these carriers with highly specific ligands such as antibodies [123,186,187] and carbohydrates [188]. The ligands can be linked either covalently or noncovalently to the microspheres/liposomes in such a way that the carriers are able to bind to accessible cells expressing a surface molecule with an affinity for the ligand. Monoclonal antibodies, because of their unique specificity, will be ideal molecule to confer target specificity on these carriers. The goal of antibody-mediated targeting of these carriers is to add the microspheres (or liposomes) to the antibody with minimal loss of antibody specificity and bioavailability. Total saturation of the particle surface, however, seems to be important in preventing denaturation of the antibody, as suggested by two different groups of workers [189,190].

A. Use of Microspheres

Polymeric particles with bound antibody have been widely used as immunochemical reagents. Microspheres that have been investigated so far can be classified into two groups.

Nonactivated Hydrophobic Microspheres

Antibodies immobilized by physical adsorption on polystyrene or acrylic particles belong to the first category. However, the polystyrene particles have seen limited use because their hydrophobic surfaces make them adhere nonspecifically to cell surfaces and molecules. Furthermore, reliance on weak adsorption forces to hold the antibody on the polymer particles is not always satisfactory [191], and chemical bonding of the antibody to polystyrene particles under mild conditions is not a well-established procedure.

Hydrophilic Microspheres with an Additional Activation Step

Agarose and poly (hydroxyl ethyl methacrylate) spheres with new functional groups generated by cyanogen bromide or toluenesulfonyl chloride and carboxylated polystyrene activated with carbodiimide are examples of the second group of microspheres [104,192,193]. Hydrophilic microspheres are considered to be more suitable for cell targeting. However, these hydrophilic particles can be easily coated with immunoglobulins through physical adsorption, and as such are likely to be rendered even more hydrophilic.

The adsorption of an antibody molecule onto a microsphere particle may occur via either its antigenic binding sites (Fab) or the "Fc" portion of the molecule. If the surface of the microsphere is hydrophilic, the antibodies tend to bind to the surface through the (Fab) groups with the "Fc" portion protruding outward, thus forming a more hydrophobic surface. On the contrary, if the surface is hydrophobic, the "Fc" portions bind to the surface, leaving the Fab binding sites free to interact with antigenic cells [194]. Thus, the adsorption of the antibody molecule depends on the nature of the microspheres surface.

Illum et al. [195] coated polycyanoacrylate microspheres with a monoclonal antibody and studied the immunospecific targeting of these particles to tumor cells in vitro. The microspheres ($\sim 10^{13}$ particles in phosphate buffer saline with 0.02% sodium azide), incubated overnight at 4°C in the presence of an excess of antiosteogenic sarcoma monoclonal antibody (1 mg/ml), were able to adsorb an average of 2000 monoclonal antibodies on their surface via the "Fc" portion of the immunoglobulin molecules. The reactivity of these antibody-coated microspheres with antigenic cells lasted for at least 4 days at 4°C in vitro. Failure to remove any bound antibodies following physical washings indicate that the antibody molecules are strongly bound on the surface of the microsphere. However, competitive displacement of adsorbed proteins has been shown to occur in another situation [196], which may limit the usefulness of the technique for in vivo targeting. Illum et al. [197] have also observed that in the presence of serum, the antibody-coated microspheres fail to bind to their designated target cells in vitro or to localize in vivo.

Covalent attachment of monoclonal antibodies to the surface of microspheres may be an alternative approach to improve the efficacy of antibody-mediated delivery of microspheres. For that attachment to occur, the microspheres need to possess functional groups on their surfaces that are capable of reacting with proteins, e.g., aldehyde groups. The linking of microspheres with carboxyl, hydroxyl, amide, and/or pyridine groups on the particle surface can be achieved by methods already described. These groups can be directly linked to antibody molecules or can be modified to yield reactive aldehyde groups. Microspheres with native aldehyde functional groups are particularly desirable, since the chemical derivatization procedure is simplified to a one-step reaction between the aldehyde group and a primary amino group of the protein molecule. When the microspheres of choice lack the necessary groups for direct coupling to monoclonal antibodies, it is possible to obtain antibody attachment under a variety of mild conditions such as those of the carbodiimide method, the cyanogen bromide method, and the glutaraldehyde method. Table 2 lists various methods of linkage of monoclonal antibodies to various microspheres and the corresponding reactive groups involved.

B. Use of Liposomes

Liposomes are an assembly of phospholipids held together by noncovalent forces. Use of liposomes as drug carrier has attracted considerable interests in recent years because of their ability to buffer the toxicity of encapsulated material while maintaining efficacy [198]. Also, intravenously administered liposome-entrapped material does not come into direct contact with blood, and their tissue distribution and plasma clearance are controlled by their carriers [199,200]. In treating experimental visceral leishmaniasis, a parasitic disease involving hepatic and splenic macrophages, liposome-entrapped drug has been found to be 700 times more effective than the free drug [201-205].

Early studies by Gregoriadis et al. [7,206] have shown that the liposome-associated antibody is able to mediate a 3- to 25-fold enhanced binding to the relevant target tissue. Antibody molecules, because of their hydrophilic nature, do not strongly associate with liposomal surface. Consequently, an early method enhanced association by modifying the antibody molecule chemically or thermally. The enhanced interaction of heat-treated antibody with liposomes, however, is associated with poor stability of the antibody-liposome conjugate, a destabilization of the liposomal membrane, and an increase in liposome permeability [207,208]. Heat-treated antibody, when attached to liposome, might orient itself with some of its "Fc" regions protruding outside, thus facilitating the nonspecific interaction with "Fc" receptor-bearing cells, including macrophages. Sonication of the phospholipid and antibody mixture for 20-30 min may sometimes help the antibody molecules to remain associated with liposomes

Table 2 Methods of Linkage of Monoclonal Antibodies to Microspheres

Method of attachment	Reactive group required on microspheres	Reagents used	Antibody-microsphere system	Comments
Adsorption				
Direct adsorption	None	None	Poly(alkyl cyanoacrylate) nano particles-antibody	Simple procedure; hydrophobic surface facilitates adsorption; competitive displacement possible
Indirect adsorption via protein A	None	None	Poly(alkyl cyanoacrylate) nano particles-antibody	Simple and efficient procedure; spacer effect
Indirect adsorption via avidin-Biotin	None	None	Methacrylate microspheres-antibody	Avidin needs to bind covalently on microspheres and biotin to antibody
Covalent coupling	Aldehyde	None	Poly-acrolein microspheres-antibody Poly-aldehyde microspheres-antibody	Simple procedure but chances of forming aggregates
	Carboxylic	Carbodiimide	Methacrylate microspheres-antibody	Simple procedure but chances of forming aggregate
	Hydroxyl	Cyanogen bromide	Methacrylate microspheres	Very toxic material, reaction is dependent on pH
	Amino	Glutaraldehyde	Methacrylate microspheres antibody	Efficient but requires derivatization of microspheres
	Dextran	Periodate reaction	Poly(alkyl cyanoacrylate)-dextran-nanoparticles-antibody	Efficient but lengthy procedure

for a longer time [209,210]. The formation of large aggregates and denaturation of the bound antibody, however, limit the application of this methodology to the preparation of antibody-bound liposomes.

Various covalent coupling reactions have been described whereby antibody molecules are chemically coupled with activated liposomes [211,212] or are incorporated inside the liposome lamellae following preliminary chemical modification with hydrophobic residues [180,213,214]. A priori, the method of conjugation selected must be efficient in aqueous solution under mild conditions, and should not allow any degradation of the lipids or the proteins. The formation of homopolymers of either the liposome or the protein is also undesirable, since it will reduce the yield of the desired product and may also produce a mixture of products that may prove to be difficult to separate. It must be possible to establish conditions for covalent attachment in which liposome-protein complexes are not formed by any noncovalent interactions; otherwise it may be too difficult to assess the extent of the covalent bonding between the liposomes and the proteins.

Sinha and Karush [213] used a phosphatidylethanolamine (PEA) derivative, [N-Nα-iodoacetyl, N_E dimethylaminonaphthalene sulfonyl)-PEA], to modify the reduced light chains of human myeloma antibody and to attach it to preformed liposomes. The association of the immunoglobulin with the liposome was verified by gel-exclusion chromatography and by the agglutination of the liposomes by the antibody to human myeloma light chain. In other studies, Huang et al. [215-217] and Harsch et al. [218] modified the antibody molecules by attaching to palmitic acid using the N-hydroxy succinimide method and incorporated such modified antibody into liposomes by a detergent dialysis method. Some techniques may require the association of protein ligands with the hydrophobic anchor prior to liposome formation. However, organic solvents cannot be used to maintain the hydrophobic moiety in solution due to the denaturation of the protein in those solvents. Although detergents help to maintain the hydrophobic moiety in solution and also can be used for the incorporation of the ligand-lipid complex into the liposomes, removal of detergent is difficult when liposome is formed by the detergent dialysis method. The use of detergent in liposome formation is thus restricted to the methods where antibodies are attached to the lipid prior to liposome formation.

Several investigators have attempted to link antibodies to liposomes using glutaraldehyde, dimethyl suberimidate [212,219,220], and carbodiimide [221-223]. These methods have the inherent disadvantages of producing homopolymers of vesicles, protein, or both. In general, the reported extent of antibody binding is small and not appreciably greater than the amount of nonspecific binding. The problem of cross-linking can be avoided, in principle, as shown by Jansons et al. [224], who protected amino groups of an antibody by citraconic anhydride during the cross-linking of carboxyl group of antibody with PEA by carbodiimide. Although some of these methods can be used to couple sufficient quantities of antibody to liposomes and site-specific liposome binding can be demonstrated, the potential damage to liposomal lipids, and more important, to the contents of the liposome due to the relatively harsh coupling conditions, remain as major shortcomings of these methods. Recently, liposomes have been coupled with monoclonal antibody through covalent linkage using assymetric or heterobifunctional cross-linking reagents that minimize the problems of homopolymerization seen with glutaraldehydes or suberimidate. These reagents include SPDP [123,225-230] and succinimidyl-4-(p-maleimidophenyl) butyrate [231-236], which are used to modify the amino groups of PE before liposome formation. The coupling method results in efficient binding of antibody to the liposomes without aggregation and without denaturation of the coupled antibody. Furthermore, liposome should not leak any encapsulated material as a consequence of the reaction.

The use of monovalent Fab′ fragments offers several advantages over other ligands in targeting liposomes. The Fab′ monomer is relatively small (50K) in comparison to intact IgG molecule (150K), $F(ab')_2$ fragments (100K), and many lectins. Bulky molecules or macromolecular complexes may inhibit cell fusion by sterically preventing close juxtaposition of vesicles and cell membranes [237]. Finally, unlike intact IgG molecules, Fab′ fragments lack an Fc region and are therefore incapable of complement activation in vivo. Complement activation by vesicle-born IgG is not an unlikely possibility, and should it occur, lysis of vesicles and/or target cell membrane could result. Martin et al. [238] and several other groups [239] described a novel method of coupling Fab′ fragments to the surface of lipid vesicles using SPDP. The coupling reaction is efficient, proceeds rapidly under mild conditions, and yields well-defined products. Each vesicle-linked Fab′ fragment retains its original antigenic specificity and full capacity to bind antigen.

V. PROBLEMS AND POSSIBLE SOLUTIONS

Although monoclonal antibodies have shown great promise in tumor therapy, their use as therapeutic agents suffers from a number of shortcomings generally associated with the native antibody molecules.

1. In most cases, monoclonal antibodies against human tumor cells are developed in mice by immunizing them with tumor cells or extracts therefrom. The antibodies, when applied either for drug delivery or antibody therapy, are recognized by the patients as foreign antigens and consequently a second antibody response is mediated. The second antibody can react with the therapeutic monoclonal antibody, rendering the therapy impossible. In about 50% of cases where tumors have been treated with native antibodies, the patients developed

an anti-antibody response, which in many cases correlated well with the further inefficacy of the therapy. The problem can be alleviated by using human monoclonal antibodies, which are less likely to generate an immune response. However, there are two major drawbacks associated with this approach. First, there are not many genetically marked human myeloma cell lines that can be used for fusion with the immunized cells; and second, it is difficult to obtain immunized cells from human sources [240,241]. Besides, preliminary results indicate that the human monoclonal antibodies have a low affinity and reduced half-life in vivo [242].

2. Immunoconjugates have to cross anatomical and physiological barriers to gain access to tumor cells [243], which limits the amount of drug conjugate reaching the tumor site. These barriers include vascular barriers, permeation through the tumor, and heterogeneity of the tumor itself. When antitumor antibodies conjugated with radioisotopes are delivered for tumor imaging, it takes several days for sufficient antibody to accumulate specifically in the tumor for a clear image to be visible. In another study with radiolabeled murine monoclonal antibodies, Epenestos et al. [244] have demonstrated that in mice only 1-25% of the injected radioactivity actually reaches the tumor site. Working with one of the best-characterized antigens, i.e., carcino embryonic antigen, Mach et al. have demonstrated that 0.1% of the injected ^{131}I-labeled antibody localized in the resected tumor [245].

3. For successful therapy, multiple administration of the drug-antibody conjugate is needed because the reticuloendothelial arms of the immune system must be avoided even during the first antibody administration. Such repeated administration might give rise to anaphylactic shock or development of human anti-mouse IgG antibody, which might remove the conjugate before it reaches the tumor. The problem may be circumvented by using hybrid antibodies (i.e., chimeric antibodies) composed of the antigen-combining site and a variable region of a mouse monoclonal antibody, attached to a human immunoglobulin constant region by genetic engineering [246,247]. However, it is not clear whether such antibodies will be free of the immunogenicity of mouse antibody or that they will have the required efficacy.

4. Binding of an antibody molecule (either native or conjugated) to a target site sometimes leads to an antigenic modulation where shedding or internalization of the antigen molecule may occur. This is particularly true for solid tumors, which are known to shed their antigens into the plasma. Internalization is necessary for the action of drug-antibody conjugates. If the antigens are not reexpressed, subsequent administration of the conjugates will not produce avid binding to tumor cells. Thus, the presence of a shed antigen in the plasma may lead to complexing of the immunoconjugate prior to its reacting to the target cells, thereby diluting specific cytotoxic capacity. Ideally, it is useful to define more

than one unique target antigen expressed on a cell so that when one antigen disappears, therapy can be directed at the second antigen. However, not all antibodies possess this particular property.

5. An immunoconjugate that is effective against one type of cancer in one patient may not be at all effective for a similar kind of cancer in another patient.

6. The use of antibodies in this form of therapy may sometimes give rise to enhancement of tumor growth [248], a problem that is difficult to obviate.

7. The immunoconjugate must be able to recognize a tumor-specific antigen that clearly distinguishes a tumor cell from a normal cell. However, with the notable exception of idiotypes on cell tumors, no such tumor-specific antigen has yet been found. Thus, it is possible that normal cells could be killed. Although this may not matter with drug-antibody conjugates, it could be quite serious with toxin-antibody conjugates due to their potency, as only one molecule is needed to kill a cell [249].

8. Tumor heterogeneity seems to be another serious problem in antibody-mediated drug delivery. In some instances a single antibody preparation may not be sufficient, but "cocktails" of antibodies that recognize different tumor-associated antigens may improve drug targeting. Combinations of monoclonal antibodies may also make it possible to employ conjugates with multiple cytotoxic agents. It has been shown that when monoclonal antibody 791T/36 and the anti-CEA monoclonal antibody C24 are injected, they localize in colon carcinoma xenographs [250]. This is just one example of how monoclonal antibody "cocktails" may be used for therapy in the future. But the development of hybridomas producing anticancer monoclonal antibodies is turning out to be far more difficult than was originally envisaged. In fact, the experience gained over the past few years indicates that, as in many scientific endeavors, perseverance and luck are two of the main characteristics for success in this field!

VI. REFERENCES

1. I. Zeidman, in *Cancer Biology*, Vol. 2 (J. Marchalonis, M. Hanna, and I. Fidler, eds.), Marcel Dekker, New York, 1981, p. 1.
2. T. Ghose and A. H. Blair, *J. Natl. Cancer Inst. 61*:657 (1978).
3. C. E. Newman, C. H. J. Ford, D. A. L. Davies, and G. J. O'Neil, *Lancet ii*: 163 (1977).
4. E. Arcamone, *Doxorubicin: Anticancer Antibiotics*, Academic Press, New York, 1981, p. 25.
5. P. Ehrlich, *Collected Studies on Immunity*, John Wiley, New York, 1906.
6. J. R. Robinson (ed.), *Sustained and Controlled Release Drug Delivery Systems*, Marcel Dekker, New York, 1978.
7. G. Gregoriadis (ed.), *Drug Carriers in Biology and Medicine*, Academic Press, New York, 1979.

8. P. Ehrlich, *Collected Papers of Paul Ehrlich*, Pergamon Press, London, 1956, p. 442.
9. G. Mathe, T. B. Loe, and J. Bernard, *C.R. Acad. Sci. Paris 246*:1626 (1958).
10. E. Calendi, G. Costanzi, F. Indiveri, G. Lotti, and C. Zini, *Boll. Chim. Farm. 108*:25 (1969).
11. T. Ghose, A. Guclu, and J. Tai, *J. Natl. Cancer Inst. 55*:1353 (1975).
12. T. Ghose, S. T. Norvell, A. Guclu, A. Bodurtha, and A. S. MacDonald, *J. Natl. Cancer Inst. 58*:845 (1977).
13. D. A. L. Davies and G. J. O'Neil, *Br. J. Cancer, Suppl. I, 28*:285 (1973).
14. D. A. L. Davies, *Cancer Res. 34*:3040 (1974).
15. D. A. L. Davies, A. J. Manstone, and S. Bucham, *Br. J. Cancer 30*:297 (1974).
16. I. Flechner, *Eur. J. Cancer 9*:741 (1973).
17. T. Ghose, S. T. Norvell, A. Guclu, D. Cameron, A. Bodurtha, and A. S. MacDonald, *Biomed. J. 3*:495 (1972).
18. T. Ghose, S. T. Norvell, A. Guclu, and A. S. MacDonald, *Eur. J. Cancer 11*: 321 (1975).
19. H. Isliker, J. C. Cerottini, J. C. Jaton, and G. Magne, *Chemotherapy of Cancer*, Elsevier, Amsterdam, 1969, p. 278.
20. F. L. Moolten, N. J. Capparel, S. H. Zajdel, and S. R. Cooperband, *J. Natl. Cancer Inst. 55*:473 (1975).
21. F. L. Moolten, N. J. Capparel, and S. R. Cooperband, *J. Natl. Cancer Inst. 49*:1057 (1972).
22. G. F. Rowland, G. J. O'Neil, and D. A. L. Davies, *Nature* (London) *255*:487 (1975).
23. W. C. J. Ross, P. E. Thorpe, A. J. Cumber, D. C. Edwards, C. A. Hinson, and A. J. S. Davies, *Eur. J. Biochem. 104*:381 (1980).
24. A. Nisonoff, F. C. Wissler, L. N. Lipman, and D. L. Woernley, *Arch. Biochem. Biophys. 89*:230 (1960).
25. J. M. Cassidy, and J. D. Douros (eds.), *Medicinal Chemistry*, Vol. 16, Academic Press, New York, 1980.
26. J. R. Bertino, *Cancer Res. 23*:1286 (1963).
27. J. Kanellos, G. A. Pietersz, and I. F. C. McKenzie, *J. Natl. Cancer Inst. 75*: 319 (1985).
28. M. J. Smyth, G. A. Pietersz, B. J. Classon, and I. F. C. McKenzie, *J. Natl. Cancer Inst. 76*:503 (1986).
29. R. W. Baldwin, in *Monoclonal Antibody Therapy of Human Cancer* (K. Foon, A. C. Morgan, eds.), Martinus Nijhoff Publishing, Boston, 1985, p. 23.
30. M. J. Smyth, G. A. Pietersz, and I. F. C. McKenzie, *Cancer Res. 47*(1): 62 (1987).
31. D. S. Zaharko, M. Przybylski, and V. T. Oliverio, *Methods Cancer Res. 16*:347 (1979).
32. A. Goldstein, L. Aronow, and S. M. Kalman, in *Principles of Drug Action* (A. Goldstein, L. Aronow, and S. Kalman, eds.), John Wiley, New York, 1974, p. 15.

33. R. Arnon and E. Hurwitz, in *Monoclonal Antibodies for Cancer Detection and Therapy* (R. W. Baldwin, and V. S. Byers, eds.), Academic Press, New York, 1985, p. 365.
34. G. Köhler and C. Milstein, *Nature* (London) *256*:494 (1975).
35. M. J. Embleton, B. Gunn, V. S. Byers, and R. W. Baldwin, *Br. J. Cancer 43*:582 (1981).
36. P. A. Farrands, A. C. Perkins, L. Sulley, J. S. Hopkins, M. V. Pimm, R. W. Baldwin, and J. D. Hardcastle, *J. Joint. Bone Surg. 65*:638 (1988).
37. P. A. Farrands, A. C. Perkins, M. V. Pimm, J. D. Hardy, B. J. Embleton, R. W. Baldwin, and J. D. Hardcastle, *Lancet 1*:397 (1982).
38. N. C. Armitage, M. V. Pimm, R. W. Baldwin, and J. D.Hardcastle, *Br. J. Surg. 70*:691 (1983).
39. F. C. Campbell, R. A. Robins, A. C. Perkins, J. G. Hardy, M. L. Wastie, and R. W. Blamey, *Br. J. Surg. 70*:681 (1983).
40. E. H. Blythman, P. Casellas, O. Gros, P. Gros, F. K. Jansen, F. Paolucci, B. Pau, and H. Vidal, *Nature* (London) *290*:145 (1981).
41. D. G. Gilliland, Z. Steplewski, R. J. Collier, K. F. Mitchell, T. H. Chang, and H. Koprawski, *Proc. Natl. Acad. Sci. USA 77*:4539 (1980).
42. K. A. Krolick, J. W. Uhr, and E. S. Vitetta, *Nature* (London) *295*:604 (1980).
43. I. S. Trowbridge and D. L. Domingo, *Nature* (London) *12*:171 (1981).
44. V. Raso, J. Ritz, M. Basela, and S. F. Schlossman, *Cancer Res. 42*:457 (1982).
45. M. V. Pimm, J. A. Jones, M. R. Price, J. G. Middle, M. J. Embleton, and R. W. Baldwin, *Cancer Immunol. Immunother. 12*:125 (1982).
46. M. Andreff, A. Bartal, C. Feif, and Y. Hirshaw, *Hydridoma 4*:277 (1985).
47. C. L. Reading, *J. Immunol. Methods 53*:261 (1982).
48. S. D. Wolpe, *Mammalian Cell Culture*, Plenum Press, New York, 1984, p. 103.
49. R. T. Taggart and I. M. Samloff, *Science 219*:1228 (1983).
50. K. A. Krolick, C. Villemez, P. C. Isakson, E. S. Vitetta, and J. W. Uhr, *Proc. Natl. Acad. Sci. USA 77*:5419 (1980).
51. R. Youle, and D. M. Neville, *J. Biol. Chem. 257*:1598 (1982).
52. T. R. Mosmann, M. Gallalin, and B. M. Longnecker, *J. Immunol. 125*: 1152 (1980).
53. M. J. Embleton and M. C. Garnett, in *Monoclonal Antibodies for Cancer Detection and Therapy* (R. W. Baldwin and V. S. Byers, eds.), Academic Press, New York, 1985, p. 317.
54. G. F. Rowland, R. G. Simmonds, J. R. F. Corvalan, R. W. Baldwin, J. P. Brown, M. J. Embleton, C. H. J. Ford, K. E. Hellström, J. T. Hellström, C. E. Newman, and C. S. Woodhouse, *Protides Biol. Fluids 30*:375 (1982).
55. W. P. Drake and M. R. Mardiney, Jr., *J. Immunol. 114*:1052 (1975).
56. R. D. Wochmer, W. Strober, and T. A. Waldmann, *J. Exp. Med. 126*:207 (1967).
57. H. L. Spiegelberg and W. O. Weigle, *J. Exp. Med. 121*:323 (1965).

58. D. Colcher, M. Zalutsky, W. Kaplan, D. Kufe, F. Austin, and J. Schlom, *Cancer Res. 43*:736 (1983).
59. F. Buchegger, C. M. Haskell, M. Schreyer, R. Bianca, R. Scazziga, S. Randin, S. Carrel, and J. P. Mach, *J. Exp. Med. 158*:202 (1983).
60. D. Herlyn, J. Power, A. Alavi, J. A. Mattis, M. Herlyn, C. Ernst, R. Vaum, and H. Koprowski, *Cancer Res. 43*:2731 (1983).
61. E. Hurwitz, R. Maron, R. Arnon, and M. Sela, *Cancer Biochem. Biophys. 1*:197 (1976).
62. R. Arnon and M. Sela, *Immunological Rev. 62*:5 (1982).
63. B. J. Taksacs and T. Staehelin, *J. Immunol. Methods 2*:27 (1981).
64. F. J. Primus and D. M. Goldenberg, *Cancer Res. 40*:2979 (1980).
65. J. W. Goding, *J. Immunol. Methods 13*:215 (1976).
66. M. Masquelier, R. Baurain, and A. Trouet, *J. Med. Chem. 23*:1166 (1980).
67. R. Duncan, J. B. Lloyd, and J. Kopecek, *Biochem. Biophys. Res. Commun. 94*:284 (1980).
68. R. Duncan, P. Rejmanova, J. Kopecek, and J. B. Lloyd, *Biochim. Biophys. Acta 678*:143 (1981).
69. M. Monsigny, C. Kieda, A. C. Roche, and F. Delmotte, *FEBS Lett. 119*: 181 (1980).
70. W. C. Shen, H. J. P. Ryser, *Biochem. Biophys. Res. Commun. 102*:1048 (1981).
71. A. Trouet, M. Masquelier, R. Baurain, and D. Deprezde Campeneere, *Proc. Natl. Acad. Sci. USA 79*:626 (1982).
72. G. E. Abraham and P. K. Grover, in *Principles of Competitive Protein-Binding Assay* (W. D. Odell and W. H. Daughaday, eds.), J. B. Lippincott, Philadelphia, 1971, p. 140.
73. F. A. Fitzpatrick and G. L. Bundy, *Proc. Natl. Acad. Sci. USA 75*:2689 (1978).
74. B. F. Erlanger, F. Borek, S. M. Beiser, and S. Liberman, *J. Biol. Chem. 234*: 1090 (1959).
75. B. Africa and E. Haber, *Immunochemistry 8*:479 (1971).
76. S. Spector and C. W. Parker, *Science 168*:1347 (1970).
77. N. Weliky and H. H. Weetall, *Immunochemistry 2*:293 (1965).
78. P. T. Tsui, K. A. Kelly, M. M. Ponpipon, and A. H. Sehon, *Can. J. Biochem. 52*:252 (1974).
79. A. Levy, K. Kawashima, and S. Spector, *Life Sci. 19*:1421 (1976).
80. K. Landsteiner, *The Specificity of Serological Reactions*, Harvard Univ. Press, Cambridge, Mass., 1945.
81. W. F. Goebel, *J. Exp. Med. 64*:29 (1936).
82. W. F. Goebel, *J. Exp. Med. 68*:469 (1938).
83. W. F. Goebel and R. D. Hotchkiss, *J. Exp. Med. 66*:191 (1937).
84. M. H. Karol and S. W. Tanenbaum, *Proc. Natl. Acad. Sci. USA 57*:713 (1967).
85. M. Sela and H. Ungar-Waron, *Fed. Proc. 24*:1438 (1965).
86. M. Sela, H. Ungar-Waron, and Y. Schechter, *Proc. Natl. Acad. Sci. USA 52*: 285 (1964).

87. E. Hurwitz, R. Levy, R. Maron, M. Wilchek, R. Arnon, and M. Sela, *Cancer Res. 35*:1175 (1975).
88. G. A. Pietersz, J. Kanellos, M. J. Smith, J. Zalcberg, and I. F. C. McKenzie, *Immunol. Cell Biol. 65*:111 (1987).
89. J. R. Vaughan, Jr., and R. L. Ostao, *J. Am. Chem. Soc. 73*:5553 (1951).
90. J. R. Vaughan, Jr., and R. L. Ostao, *J. Am. Chem. Soc. 74*:676 (1952).
91. B. F. Erlanger, S. M. Beiser, F. Borek, F. Edel, and S. Lieberman, *Methods Immunol. Immunochem. 1*:144 (1967).
92. B. M. Jaffe, J. W. Smith, W. T. Newton, and C. W. Parker, *Science 171*:494 (1971).
93. S. Comoglio and F. Celada, *J. Immunol. Methods 10*:161 (1976).
94. B. G. Joyce, G. F. Read, and D. R. Fahmy, *Steroids 29*:761 (1977).
95. J. J. Pestka, P. K. Gaur, and F. S. Chu, *Appl. Environ. Microbiol. 40*:1027 (1980).
96. B. F. Erlanger, *Pharmacol. Rev. 25*:271 (1973).
97. S. Burstein and R. Knapp, *J. Med. Chem. 20*:950 (1977).
98. P. N. Kulkarni, A. H. Blair, and T. I. Chose, *Cancer Res. 41*:2700 (1981).
99. Z. A. Latif, B. B. Lozzio, C. J. Wust, S. Kraus, M. C. Aggio, and C. B. Lozzio, *Cancer 45*:1325 (1980).
100. V. R. Mattox and A. N. Nelson, *J. Steroid Biochem. 110*:167 (1979).
101. G. C. Oliver, Jr., B. M. Parker, D. L. Brasfield, and C. W. Parker, *J. Clin. Invest. 47*:1035 (1968).
102. A. N. Miekle, J. A. Weed, and F. H. J. Tyler, *J. Clin. Endocrinol. Metab. 41*:717 (1975).
103. P. K. Nakane, J. SriRam, and G. B. Pierce, *J. Histochem. Cytochem. 14*: 789 (1966).
104. T. L. Goodfriend, L. Levine, and G. Fasman, *Science 143*:1344 (1964).
105. D. Clyne, S. Norris, R. R. Modesto, and A. J. Pesce, *J. Histochem. Cytochem. 21*:233 (1973).
106. T. Ghose, A. H. Blair, K. Vaughan, and P. Kulkarni, in *Targeted Drugs* (E. P. Goldberg, ed.), John Wiley, New York, 1983, p. 1.
107. R. N. Hamburger, *Science 152*:203 (1966).
108. H. J. Creech, *Cancer Res. 12*:557 (1952).
109. M. J. Hunter and M. L. Ludwig, *J. Am. Chem. Soc. 84*:3491 (1962).
110. A. Broughton, J. E. Strong, L. K. Pickering, and G. P. Brodey, *Antimicrob. Agents Chemother. 10*:652 (1976).
111. H. G. Kopp, A. Eberle, P. Vitius, W. Lichensteiger, and R. Schwyzer, *Eur. J. Biochem. 75*:417 (1977).
112. Y. H. Chien and L. Levine, *Immunochemistry 12*:291 (1975).
113. B. Peskar and S. Spector, *Science 179*:1340 (1973).
114. F. Bartos, D. Bartos, A. M. Dolney, D. P. Grettie, and R. A. Campbell, *Res. Commun. Chem. Pathol. Pharmacol. 19*:295 (1978).
115. D. J. Brunswick, B. Needleman, and J. Mendels, *Life Sci. 22*:137 (1978).
116. E. Haber, L. B. Page, and G. A. Jacoby, *Biochemistry 4*:693 (1965).
117. L. A. Frohman, H. Reichlin, and J. E. Sokal, *Endocrinology 87*:1055 (1970).

118. B. A. Peskar, B. M. Peskar, and L. Levine, *Eur. J. Biochem. 26*:191 (1972).
119. M. Reichlin, J. J. Schnure, and V. K. Vance, *Proc. Soc. Exp. Biol. Med. 128*:347 (1968).
120. E. Ishikawa, *J. Immunoassay 4*:209 (1983).
121. J. E. Leonard, R. Taetlle, D. To, and K. Rhyner, *Blood 65*:1149 (1985).
122. P. J. Lachman, A. Vyakarnam, and K. Sikora, *Immunology 42*:329 (1981).
123. L. D. Leserman, J. Barket, F. Kourilsky, and J. N. Weinstein, *Nature 288*: 602 (1980).
124. M. Richards and J. R. Knowles, *J. Mol. Biol. 37*:231 (1968).
125. S. Avrameas, T. Ternynck, and J. L. Guesdon, *Scand. J. Immunol. 8*(Suppl. 7):7 (1978).
126. T. Ghose and S. P. Nijam, *Cancer Res. 29*:1398 (1972).
127. T. Ghose, S. T. Norvell, A. Guclu, D. Cameron, A. Bodurtha, and A. S. MacDonald, *Br. Med. J. 3*:495 (1972).
128. J. H. Linford, G. Froesse, I. Broczi, and L. G. Israels, *J. Natl. Cancer Inst. 52*:1665 (1974).
129. R. D. Rubens and R. Dubecco, *Nature 248*:81 (1974).
130. M. Szekerke, R. Wade, and M. E. Whisson, *Neoplasma 19*:211 (1972).
131. J. H. Linford, *Can. J. Biochem. Biophys. 41*:931 (1963).
132. D. Blakeslee, M. Chen, and J. C. Kennedy, *Br. J. Cancer 34*:882 (1974).
133. M. Segerling, S. M. Ohaninam, and T. Borsos, *J. Natl. Cancer Inst. 53*: 1141 (1974).
134. D. A. L. Davies and G. J. O'Neil, *Proc. XI Int. Cancer Cong.*, Vol. 1, Exerpta Medica, Amsterdam, pp. 218–221.
135. W. C. Shen and H. J. P. Ryser, *Proc. Natl. Acad. Sci. USA 75*:1872 (1978).
136. B. C. F. Chu and J. M. Whiteley, *Mol. Pharmacol. 13*:80 (1977).
137. M. Szekerke, R. Wade, and M. E. Whisson, *Neoplasma 19*:211 (1972).
138. J. M. Varga, N. Asato, S. Lande, and A. B. Lerner, *Nature* (London) *267*: 56 (1977).
139. B. C. F. Chu and J. M. Whiteley, *Mol. Pharmacol. 17*:382 (1980).
140. T. Kitao and K. Hattori, *Nature* (London), *265*:81 (1977).
141. A. Bernstein, E. Hurwitz, R. Maron, R. Arnon, M. Sela, and M. Wilchek, *J. Natl. Cancer Inst. 60*:379 (1978).
142. G. F. Rowland, *Eur. J. Cancer 13*:593 (1977).
143. F. Zunino, F. Giuliani, G. Savi, T. Dasdia, and R. Gambetta, *Int. J. Cancer 30*:465 (1982).
144. E. Hurwitz, M. Wilchek, and J. Pitha, *J. Appl. Biochem. 2*:25 (1980).
145. M. C. Garnett, M. Embleton, E. Jacobs, and R. W. Baldwin, *Int. J. Cancer 31*:661 (1983).
146. E. Hurwitz, R. Maron, A. Bernstein, M. Wilchek, M. Sela, and R. Arnon, *Int. J. Cancer 21*:747 (1978).
147. Y. J. Schneider, J. Abarca, E. Aboud-Pirak, R. Baurain, F. Ceulemans, D. Deprez-De-Campeneere, B. Lesur, M. Masquelier, C. Ohe-Slachmuylder, D. Rolin-van Swieten, and A. Trouet, in *Receptor-Mediated Targeting of Drugs* (G. Gregoriadis, G. Posta and A. Trouet, eds.), Plenum Press, New York, 1984, p. 1.

148. R. Levy, E. Hurwitz, R. Maron, R. Arnon, and M. Sela, *Cancer Res. 35*: 1181 (1975).
149. T. Ghose, R. Ramakrishnan, P. Kulkarni, A. H. Blair, K. Vaughan, H. Nolido, S. T. Norvell, and P. Belitsky, *Transplant. Proc. 13*:1970 (1981).
150. N. Shigeru, T. Nobuko, H. Nozomu, and H. Hidematsu, *Gan. tokagaku Ryoho 11*(8):1591 (1984).
151. M. K. Ghosh, D. O. Kildsig, and A. K. Mitra, *Drug Design and Delivery 4*: 13 (1989).
152. J. Tai, A. H. Blair, and T. Ghose, *Eur. J. Cancer 15*:1357 (1979).
153. M. J. S. Warzynski, K. W. Cochran, and W. W. Ackerman, *J. Immunol. Methods 35*:157 (1980).
154. B. T. Kaufman, in *Methods in Enzymology* (W. B. Jakoby and M. Wilchek, eds.), Academic Press, New York, 1970, p. 272.
155. B. A. Hurn and S. M. Chantler, in *Methods in Enzymology* (H. V. Vanakis and John J. Langone, eds.), Academic Press, New York, 1980, p. 104.
156. J. H. Linford and G. Froese, *J. Natl. Cancer Inst. 60*:307 (1978).
157. G. E. Abraham, R. Swerdloff, D. Tulchinsky, and W. D. Odell, *J. Clin. Endocrinol. Metab. 32*:619 (1971).
158. R. C. Lauer, P. Soloman, K. Nakanishi, and B. F. Erlanger, *Fed. Proc. 32*:500 (1971).
159. R. C. Lauer, P. H. Soloman, K. Nakanishi, and B. F. Erlanger, *Experientia 30*:558 (1974).
160. R. C. Lauer, P. H. Soloman, K. Nakanishi, and B. F. Erlanger, *Experientia 30*:560 (1974).
161. F. Sanger, *Biochem. J. 45*:563 (1949).
162. C. R.Parish and I. F. C. McKenzie, *J. Immunol. Methods 20*:173 (1978).
163. Y. Tsukada, W. K. D. Kischof, N. Hibi, H. Hirai, E. Hurwitz, and M. Sela, *Proc. Natl. Acad. Sci. USA 79*:621 (1982).
164. R. A. Robins, R. R. Laxton, M. Garnett, M. R. Price, and R. W. Baldwin, *J. Immunol. Methods 90*:165 (1986).
165. H. Ichihashi, T. Kondo, S. Sakakibara, S. Akiyama, and T. Watanabe, *Oncology 41*:88 (1984).
166. D. Mew, V. Lum, C. K. Wat, G. H. N. Towers, C. H. C. Sun, R. J. Walter, W. Wright, M. W. Berns, and J. G. Levy, *Cancer Res. 45*:4380 (1985).
167. J. Epstein, R. W. Rosenthal, and R. J. Ess, *Anal. Chem. 27*:1435 (1955).
168. I. D. Goldman, N. S. Lichtenstein, and U. T. Oliverio, *J. Biol. Chem. 243*: 5007 (1968).
169. J. E. Byfield and P. M. Calabro-Jones, *Nature* (London) *294*:281 (1981).
170. W. C. Shen and H. J. P. Ryser, *Biochem. Biophys. Res. Commun. 102*: 1048 (1981).
171. G. J. O'Neil, *Br. J. Cancer 41*:839 (1980).
172. G. J. O'Neil, B. A. Pearson, and D. A. L. Davies, *Immunology 28*:323 (1975).
173. H. Hirschberg, G. Rowland, and E. Thorsby, *Transplantation 26*:292 (1978).
174. M. J. Smyth, G. A. Pietersz, and I. F. C. McKenzie, *Immunol. Cell Biol. 65*:189 (1987).

175. P. Uadea, A. H. Blair, T. Ghose, and S. Ferrone, *J. Natl. Cancer Inst. 74*: 29 (1985).
176. M. C. Garnett, E. Jacobs, M. J. Embleton, and R. W. Baldwin, *Biochem. Soc. Trans. 12*:1035 (1984).
177. M. C. Willingham, J. A. Hanover, R. B. Dickson, and I. H. Pastan, *Proc. Natl. Acad. Sci. USA 81*:175 (1984).
178. R. W. Baldwin, M. J. Embleton, M. C. Garnett, and M. V. Pimm, *Natl. Cancer Inst. 3*:96 (1987).
179. J. Heller and R. W. Baker, *Controlled Release of Bioactive Materials*, Academic Press, New York, 1980, p. 1.
180. E. P. Goldberg, *Targeted Drugs*, John Wiley, New York, 1982.
181. D. S. T. Heish, R. Langer, and J. Folkman, *Proc. Natl. Acad. Sci. USA 78*: 1863 (1981).
182. G. Gregoriadis, *Pharmacol. Ther. 10*:103 (1980).
183. U. Zimmerman, in *Targeted Drugs* (E. P. Goldberg, ed.), John Wiley, New York, 1982.
184. I. S. Joholm and P. Edman, *J. Pharmacol. Exp. Ther. 211*:656 (1979).
185. G. Poste, *Biol. Cell 47*:19 (1983).
186. A. Huang, L. Huang, and J. J. Kennel, *J. Biol. Chem. 255*:8015 (1980).
187. K. S. Bragman, T. D. Heath, and D. Papahadjopoulus, *Biochim. Biophys. Acta 730*:187 (1983).
188. P. S. Wu, G. W. Tin, and J. D. Baldeschwieler, *Proc. Natl. Acad. Sci. USA 78*:2033 (1981).
189. S. Kochwa, M. Brownell, R. E. Rosenfield, and L. R. Wasserman, *J. Immunol. 5*:981 (1967).
190. B. W. Morrissey and C. C. Han, *J. Colloid Interface Sci. 65*:423 (1977).
191. F. Milgram and R. Goldstein, *Vox. Sang. 7*:86 (1962).
192. P. Cutrecasas, *J. Biol. Chem. 245*:3059 (1970).
193. E. Bergman, W. T. Tsatsos, and R. F. Fischer, *J. Polymer Sci. 3*:3685 (1965).
194. C. F. Vanoss, C. F. Gillman, and A. W. Newman, *Phagocytic Engulfment and Cell Adhesiveness*, Marcel Dekker, New York, 1975, p. 1.
195. L. Illum, P. D. E. Jones, J. Kreuter, R. W. Baldwin, and S. S. Davis, *Int. J. Pharm. 17*:65 (1983).
196. J. L. Brash and V. J. Davidson, *Thromb. Res. 9*:249 (1976).
197. L. Illum, P. D. E. Jones, and S. S. Davis, in *Microspheres and Drug Therapy* (S. S. Davis, L. Illum, J. G. McVice, and E. Tomlinson, eds.), Elsevier North-Holland, Biomedical Press, Amsterdam, 1984, p. 353.
198. G. Loper-Berestein, V. Fainstein, R. Hofer, K. Mehta, M. P. Sulivan, M. Keating, M. G. Rosenblum, R. Mehta, M. Luna, E. M. Hersh, J. Reuben, R. L. Juliano, and G. P. Bodey, *J. Infect. Dis. 151*:704 (1985).
199. I. R. McDougal, J. K. Dunnick, M. C. McNamee, and J. B. Kriss, *Proc. Natl. Acad. Sci. USA 71*:3487 (1974).
200. R. L. Juliano and D. Stamp, *Biochem. Biophys. Res. Commun. 63*:651 (1975).

201. C. R. Alving, *Pharmacol. Ther. 22*;407 (1983).
202. C. R. Alving, E. A. Steck, and W. L. Chapman, Jr., *Proc. Natl. Acad. Sci. USA 75*:295 (1978).
203. W. L. Chapman, W. L. Hanson, C. R. Alving, and L. D. Hendricks, *Am. J. Vet. Res. 45*:1028 (1984).
204. C. D. V. Black, G. J. Watson, and R. J. Ward, *Trans. Roy. Soc. Trop. Med. Hyg. 71*:550 (1984).
205. R. R. C. New, M. L. Chance, S. C. Thomas, and W. Peters, *Nature 272*:55 (1978).
206. G. Gregoriadis and E. D. Neerjunjun, *Biochem. Biophys. Res. Commun. 65*:537 (1975).
207. M. Schieren, G. Weissman, M. Seligman, and P. Coleman, *Biochem. Biophys. Res. Commun. 82*:1160 (1978).
208. M. C. Finkelstein, S. H. Kuhn, H. Schieren, G. Weissman, and S. Hoffstein, *Biochim. Biophys. Acta 673*:286 (1981).
209. L. B. Margolis and N. A. Dorfman, *Bull. Exp. Biol. Med. 83*:53 (1977).
210. L. Huang and S. J. Kennel, *Biochemistry 18*:1702 (1979).
211. D. L. Urdal and S. Hakomori, *J. Biol. Chem. 255*:10509 (1980).
212. V. P. Torchilin, V. S. Goldmacher, and V. N. Smirnov, *Biochem. Biophys. Res. Commun. 85*:983 (1978).
213. D. Sinha and F. Karush, *Biochem. Biophys. Res. Commun. 90*:534 (1979).
214. V. P. Torchilin, A. L. Klibanov, A. I. Mikhailnov, V. I. Goldanskii, and V. N. Smirnov, *Biochim. Biophys. Acta 602*:511 (1980).
215. A. Huang, S. J. Kennell, and L. Huang, *J. Immunol. Methods 46*:141 (1981).
216. A. Huang, Y. S. Tsao, S. J. Kennell, and L. Huang, *Biochim. Biophys. Acta 716*:140 (1982).
217. D. F. Shen, A. Huang, and L. Huang, *Biochim. Biophys. Acta 689*:31 (1982).
218. M. Harsch, P. Walther, H. G. Weder, and H. Hengartner, *Biochem. Biophys. Res. Commun. 103*:1069 (1981).
219. V. P. Torchilin, B. D. Khaio, V. N. Smirnov, and E. Haber, *Biochem. Biophys. Res. Commun. 89*:114 (1979).
220. V. P. Torchilin, V. A. Berdichevsky, A. A. Barsukov, and V. N. Smirnov, *FEBS Lett. 111*:184 (1980).
221. H. Endoh, Y. Suzuki, and Y. Hashimoto, *J. Immunol. Methods 44*:79 (1981).
222. J. K. Dunnick, R. S. Badger, Y. Takeda, and J. P. Kriss, *J. Nucl. Med. 17*: 1073 (1976).
223. J. K. Dunnick, I. R. McDougall, S. Aragon, M. L. Goris, and J. P. Kriss, *J. Nucl. Med. 16*:483 (1955).
224. V. K. Jansons and P. L. Mallett, *Anal. Biochem. 111*:54 (1981).
225. Y. Ishimori, T. Yasude, T. Tsumita, M. Notshki, M. Koyama, and T. Tadakuma, *J. Immunol. Methods 75*:351 (1984).
226. J. Barbet, P. Machy, and L. D. Leserman, *J. Supramol. Struct. Cell Biochem. 16*:243 (1981).

227. W. Godfrey, B. Doe, E. E. Wallace, B. Bredt, and L. Wofsy, *Exp. Cell Res. 135*:137 (1981).
228. W. Godfrey, D. Doe, and L. Wofsay, *Proc. Natl. Acad. Sci. USA 80*:267 (1983).
229. B. Wolff and G. Gregoriadis, *Biochim. Biophys. Acta 802*:259 (1984).
230. A. Goundalkar, T. Ghose, and M. Mezei, *J. Pharm. Pharmacol. 36*:465 (1984).
231. F. J. Martin and D. Papahadjopoulos, *J. Biol. Chem. 257*:286 (1982).
232. Y. Hashimoto, M. Sugawara, and H. Endoh, *J. Immunol. Methods 62*: 155 (1983).
233. V. T. Kung, P. E. Maxim, R. W. Veltri, and F. J. Martin, *Biochim. Biophys. Acta 839*:105 (1985).
234. J. T. P. Derksen and G. Scherphof, *Biochim. Biophys. Acta 814*:151 (1985).
235. T. D. Heath, J. A. Montgomery, J. R. Piper, and D. Papahadjapoulos, *Proc. Natl. Acad. Sci. USA 80*:1377 (1983).
236. K. K. Matthay, T. D. Heath, and D. Papahadjopoulos, *Cancer Res. 44*: 1880 (1984).
237. F. J. Martin and R. C. MacDonald, *J. Cell Biol. 70*:515 (1976).
238. F. J. Martin, W. C. Hubell, and D. Papahadjopoulos, *Biochemistry 20*: 4229 (1981).
239. T. Heath, R. Fraley, and D. Papahadjopoulos, *Science 210*:539 (1980).
240. R. K. Oldham, *J. Clin. Oncol. 1*:582 (1983).
241. R. K. Oldham, *Cancer Treat. Rep. 68*:221 (1984).
242. K. Sikora, T. Alderson, J. Ellis, J. Phillips, and J. Watson, *Br. J. Cancer 47*:135 (1983).
243. M. J. Poznansky and R. J. Juliano, *Pharmacol. Rev. 36*:277 (1984).
244. A. A. Epenestos, D. Snook, H. Durbin, P. M. Johnson, and J. T. Papadimitriou, *Cancer Res. 46*:3183 (1986).
245. J. Mach, S. Carrel, M. Forni, J. Ritschard, A. Donath, and P. Alberto, *N. Engl. J. Med. 303*:5 (1980).
246. S. L. Morrison, M. J. Johnson, L. A. Herzenberg, and V. T. Oi, *Proc. Natl. Acad. Sci. USA 81*:6851 (1984).
247. M. S. Neuberger, G. T. Williams, E. B. Mitchell and T. H. Rabbits, *Nature 314*:268 (1985).
248. J. D. Everall, P. Dowd, D. A. L. Davies, G. J. O'Neill, G. F. Rowland, *Lancet i*:1105 (1977).
249. K. Eiklid, S. Olsnes, and A. Pihl. *Exp. Cell Res. 126*:321 (1980).
250. M. V. Pimm, A. C. Perkins, and R. W. Baldwin, *IRCS-Med. Sci. 13*:499 (1985).

8

Current Status of Antibody-Toxin Conjugates for Tumor Therapy

S. RAMAKRISHNAN*
Duke University Medical Center, Durham, North Carolina

I. INTRODUCTION

The concept of targeting toxic molecules to tumor cells using carriers such as antibodies was first proposed by Paul Ehrlich [1] almost 80 years ago. The rapid development in antibody-toxin conjugates, immunotoxins (IMTs), during the last decade is due mainly to the development of monoclonal antibodies by Kohler and Milstein [2]. IMTs contain three essential components. They are (1) a carrier molecule, which is often an antibody directed to a cell surface determinant, but growth factors and hormones are also used in some studies; (2) a toxic moiety; and (3) a chemical linkage that holds both these molecules together. Each of these components contributes to the ultimate efficacy of the IMT. The carrier molecule helps in binding and homing of IMT on the surface of tumor cells. From the surface, the conjugate is internalized. Once inside the cell, the toxin molecule acts catalytically and inhibits vital cellular functions such as translation. Since the toxin molecules inactivate the target cells enzymatically with a higher turnover rate, they are more effective than conventional chemotherapeutic drugs that act stoichiometrically. In this chapter, the latest developments in antibody-mediated delivery of toxin polypeptides is summarized; additional information can be found elsewhere [3,4].

**Current affiliation*: University of Minnesota Medical School, Minneapolis, Minnesota

Table 1 Potential Toxins/RIP for Immunotoxin Preparation

Toxin moiety	Source	Molecular weight	Carbohydrate	Biological activity
Bacterial toxins				
Diphtheria toxin	*Corynebacterium diphtheriae*	58,342	-	ADP-ribosylation of EF-2
Pseudomonas exotoxin A	*Pseudomonas aeruginosa*	66,000	-	ADP-ribosylation of EF-2
Plant toxins				
Ricin	*Ricinus communis*	62,057	+	Depurination of A4324 from 282 rRNA
Abrin	*Abrus precatorius*	65,000	+	60S ribosomal subunit
Modeccin	*Adenia digitata*	63,000	+	60S ribosomal subunit
Volkensin	*Adenia volkensii*	62,000	+	60S ribosomal subunit
Viscumin	*Viscum album*	60,000	+	60S ribosomal subunit
RIP				
PAP[a]	*Phytolacca americana*	28,000	-	Depurination of A4324 from 28S rRNA
PAP-II[a]	*Phytolacca americana*	30,000	-	Depurination of A4324 from 28S rRNA

PAP-S[a]	*Phytolacca americana*	31,000	-	Depurination of A4324 from 28S rRNA
Dodecandrin	*Phytolacca dodecandra*	30,000	-	60S ribosomal subunit
Gelonin[a]	*Gelonium multiflorm*	24,300	+	60S ribosomal subunit
Momordin[a]	*Momardica charantia*	24,300	+	60S ribosomal subunit
Saporin[a]	*Saponaria officinalis*	30,000	-	60S ribosomal subunit
Bryodin[a]	*Bayonia dioica*			60S ribosomal subunit
Luffin	*Luffa cylindrica*	26,000	-	60S ribosomal subunit
Dianthin 30	*Dianthus caryophylus*	29,500	+	60S ribosomal subunit
Dianthin 32	*Dianthus caryophylus*	31,700	+	60S ribosomal subunit
Analogous fungal toxins				
Alpha sarcin	*Aspergillus giganteus*	16,800	-	Large rRNA, EF-1 and EF-2
Mitogillin	*Aspergillus restrictus*	16,200	-	Large rRNA, EF-1 and EF-2
Restrictocin	*Aspergillus restrictus*	16,300	-	Large rRNA, EF-1 and EF-2
Enomycin	*Streptomyces mauvecolor*	11,000	-	60S ribosomal subunit
Phenomycin	*Streptomyces fervens* var. *phenomyceticus*	10,000	-	60S ribosomal subunit

[a]Hemitoxins thus far used in making active conjugates.

II. TOXIN MOLECULES USED IN IMT

A list of toxin molecules currently used in conjugation with antibodies is shown in Table 1. In general, two types of molecules are used successfully in making biologically active IMTs: (1) toxins derived from bacteria and (2) phytotoxins isolated from a variety of plants.

A. Bacterial Toxins

Two widely used and extensively studied bacterial toxins are Pseudomonas exotoxin E and diphtheria toxin (DT). Both these molecules have been genetically cloned and expressed in *E. coli*. The gene for DT is carried in a bacteriophage, beta, which is lysogenic to *Corynebacterium diphtheriae*. Production of this toxin is dependent on the iron content of the culture medium. Purified DT has a molecular weight of 58,342 and is released into the medium as a single polypeptide chain. There are two disulfide loops. The first disulfide loop is present in the amino-terminal part (Cys_{186}-Cys_{201}) and has an arginine-rich region that is highly sensitive to proteases [5,6]. Thus, cleavage of this site (nicked DT) results in two fragments held together by a disulfide bond. The A fragment of DT is made of 193 amino acid residues and the B fragment contains 342 residues. Upon reduction, the A fragment of DT (DTA; 21 kDa) can be purified.

DT binds to eukaryote cells via the B fragment. The nature of the cell surface receptor is not yet determined. Cells from certain animals are found to be more resistant to the action of DT than others. The binding site in DT is localized at the COOH-terminal end of the B fragment, since a truncated form of DT (CRM-45) lacking a C-terminal, 17,000-kDa fragment does not compete with native DT in competitive inhibition assays. Further analysis of point mutants reveals that several amino acid residues including Ser 508 and either Leu 325 or Ser 525 are involved in binding. Surface-bound DT is internalized by endocytosis. The endocytic vessicles have an acidic environment, which facilitates the translocation of toxin molecules to the cytoplasmic compartment. In fact, hydrophobic domains in the B fragment are exposed under low pH, which is suggested to help in insertion and translocation of DTA across the membrane.

DT inhibits protein synthesis of target cells by inactivating elongation factor-2 (EF-2) by ADP-ribosylation. The enzymatic activity is localized in fragment A of DT. Prokaryote elongation factor is not sensitive to ADP-ribosylation. The inactivation is dependent on NAD as a cofactor. The ADP-ribose moiety from NAD is transferred to a nitrogen atom on a unique histidine residue (diphthamide) present on EF-2. The histidine has been modified post-translationally with carboxy amidotrimethyl aminopropyl groups. Other experiments indicate that the glutamine residue at position 148 seems to be important for the catalytic activity of DT.

Unlike DT, the gene for Pseudomonas exotoxin A (PE) is carried in the bacterial chromosome. PE is a monomeric protein of 613 amino acid residues. X-ray crystallographic analysis reveal three structual/functional domains [7]. Domain I (residues 1-252 and 365-404) is involved in binding to cell surface receptors. Particularly, the lysine residue at position 57 seems to play a critical role in binding. Substitution of Lys 57 to Glu 57 by site-directed mutagenesis results in a 100-fold loss in binding/toxicity [8]. Domain II (253-364) is believed to facilitate translocation across the membrane, and Domain III (405-613) possesses the ADP-ribosylation activity. PE also inactivates EF-2 by a mechanism similar to DT. Although there is no distinct homology between DT and PE in the primary structure, aligning residues of DT involved in catalytic activity to Domain III of PE shows significant similarities. Further, PE does not undergo any proteolytic processing like DT. Nevertheless, the toxin becomes biologically active only after exposure to denaturing conditions. One of the problems anticipated in using DT-linked antibodies is the possible neutralization of the conjugate by circulating anti-DT antibodies in vaccinated people. The frequency of anti-PE antibody in humans is lower and hence is favored over DT in IMT preparation.

B. Toxins and Ribosomal Inhibitory Proteins (RIPs) Isolated from Plants

A variety of inhibitory molecules isolated from plants can be broadly categorized into two groups: (1) heterodimeric proteins such as ricin, abrin, and modeccin, containing a A chain and a B chain linked together by a disulfide bond, which are true toxins and are cytotoxic to eukaryotic cells at picomolar concentrations; and (2) single-chain polypeptides that are not toxic to intact cells but inhibit ribosomal function in cell-free systems. These proteins are collectively called hemitoxin or RIP. Toxin polypeptides used in the synthesis of IMT are summarized in Table 1.

Among the true toxins, detailed structural analysis is available for ricin. Ricin is a glycoprotein of about Mr 62,000, and the A chain is the functional moiety. The B chain of ricin has two binding sites for galactose residues, with which the toxin binds to the cell surface. Genetic cloning and expression studies reveal that ricin is produced as proricin with a leader sequence, and the A chain is joined to the B chain by a 12-amino-acid peptide (J peptide), which is removed post-translationally [9,10]. In the mature protein, the carboxy terminal of the A chain (Cys 259) is linked to the amino terminal of the B chain (Cys 4) by a disulfide bond [11]. Surface-bound ricin is internalized by endocytosis and either the A chain alone or the entire toxin molecule reaches the cytoplasm. A chain inactivates 60S ribosomes of eukaryotes in the absence of any cofactors. The mechanism of action of ricin has been recently elucidated [12,13]. Treatment of intact ribosomes leads to the removal of a purine base (adenine) from position 4324 of 28S ribosomal RNA by N-glycosidase activity. "Naked" 28S rRNA is less sensitive to the action of ricin. Depurination is suggested to be responsible for the reduction in EF-1-dependent interaction (GTPase activity and enzymatic binding of aminoacyl-tRNA) and the failure to form EF-2 GTP complex with the ribosomes. Inactivation of ribosomes is irreversible, and a single molecule of A chain could inhibit about 1500 ribosomes per minute. Thus, theoretically, the presence of even a single molecule could kill a cell.

The A chain of ricin is preferred in the construction of IMT, since conjugates made with intact toxin lack specificity. Ricin conjugates could be rendered specific by the addition of galactose in excess to competitively inhibit the direct binding of the toxin to cell surface glycoproteins. Alternatively, the galactose binding sites could be blocked by chemical modifications or altered by genetic engineering. These approaches are currently under investigation.

Ribosomal Inhibitory Proteins

Single-chain RIPs, on the other hand, do not have the binding moiety equivalent to ricin B chain and, therefore, is not toxic to cells. When linked to carrier molecules, RIP conjugates become highly cytotoxic. The majority of RIPs have a molecular weight of about 30,000. Pokeweed antiviral protein (PAP), dodecandrin, saporin, and Luffa ribosomal inhibitory protein (LRIP) are nonglycosylated. Gelonin, momordin, and dianthins contain variable amounts of carbohydrate (1.2–2.34%). In general, RIPs are highly basic proteins in contrast to the heterodimeric toxins. The partial amino acid sequence of the NH_2 terminus indicates that RIPs are structurally and evolutionarily divergent [14,15]. Immunochemical studies show no cross-reactivity between RIPs. Endo and Tsurugi [16] have shown that PAP also depurinates adenine 4324 from 28S rRNA.

In addition to plant toxins, some fungal toxins such as alpha sarcin is also considered for making antibody-toxin conjugates. Alpha sarcin inactivates eukaryote ribosomes (60S) by a different mechanism than does ricin toxin. Alpha sarcin cleaves a phosphodiester bond between residue G 4325 and A 4326 of 28S rRNA. This site is very close to the region at which ricin and PAP act. Fungal toxins also show exo- and endonuclease activity. It is interesting to note that this purine-rich region in rRNA is evolutionarily conserved among prokaryotes and eukaryotes. However, only the eukaryote ribosomes are sensitive to the toxin action.

III. CONJUGATION OF TOXIN MOLECULES TO ANTIBODIES

One of the most important considerations in chemical linkage of toxin polypeptides to antibodies is that the derivatization and coupling process should not interfere with the function of either of the molecules. A disulfide linkage between antibody and toxin molecule is highly preferred. Disulfide linkage mimics the native condition of dimeric toxins such as ricin and often forms biologically active conjugates. Purified RTA and DTA have a free thiol group that can be used in a thiol-exchange reaction to establish disulfide linkage. In this case, the antibody is derivatized with SPDP, which reacts to free amino groups. RIPs do not have free SH, and therefore thiol groups have to be introduced by derivatization. This can be achieved either by reacting the RIP with SPDP followed by reduction [17] or by using 2-iminothiolane [18]. The former method results in the loss of a positive charge on the amino group, whereas modification with 2-iminothiolane does not alter the charge density. Changes in charge could have subtle effects on the biological activity of the conjugate.

Alternatively, antibodies can be linked to toxin polypeptides by a nonreducible thioether bond. One of the way to make such a conjugate is to derivatize the antibody with MBS and react it with SH containing toxin polypeptide. Many of these procedures are described in detail elsewhere [19,20]. An important observation in using nonreducible linkage is the variations in biological activity of the IMT. Disulfide-linked conjugates of RTA and antibodies were at least two to three orders more inhibitory than the corresponding thioether-linked conjugates. The marked difference in cytotoxicity is attributed to the prerequisite that the linkage between antibody and toxin be broken before the toxin molecule can translocate into the cytoplasm.

In contrast, native toxins such as ricin and bacterial toxins, DT and PE, can be linked via thioether bonds and retain full biological activity. In ricin and DT conjugates this phenomenon can be explained by the existance of a natural disulfide bond between the A chain and the B chain, which could be reduced prior to translocation. However, PE is made of a single polyptide, and therefore reduction may not be the only prerequisite for transmembrane passage. It is possible that PE has endogenous signals for translocation or that PE may be proteolytically cleaved inside the endocytic vesicles before the translocation. Indeed, several proteolytically cleaved fragments of PE show significant ADP-ribosylation activity, alluding to such a possibility.

Conjugation of toxin molecules can sometimes result in steric hindrance. To obviate, inert or proteolytically sensitive stretches of amino acid residues can be introduced between the toxin and the antibody. The spacer arm method has been shown to be useful in making effective conjugates with genetically engineered CRM-45. The recombinant protein has an additional stretch of 19 amino acids terminating with a cysteine residue that can be used in linking to antibodies. Blattler et al. [21] synthesized acid-labile compounds that can be used in the covalent linkage of toxin molecules to antibodies. These conjugates would be unstable in the acid environment of endosomes. Another approach is to use a spacer arm derived from oxidized insulin B chain [22]. Conjugates made with the 29-residue insulin peptide interchelated between the antibody and toxin have shown a 10-fold increase in cytotoxicity when compared to the conventional conjugates. Other investigators have modified the carbohydrate moieties of antibodies to prepare IMT.

IV. ANTIBODY-TOXIN CONJUGATES SELECTIVE TO TUMOR CELLS

Currently, IMTs are being developed for three major applications: (1) for in vitro use in bone marrow purging prior to autologous or allogenic transplantation. This method is also used to eliminate mature alloreactive T-cells from bone marrow to prevent graft-versus-host disease; (2) an intracavitory approach wherein IMTs could be administered into a restricted anatomical space such as the peritoneal cavity; (3) systemic application for diffused and solid tumors. The first approach is the simplest procedure among the three methods. The tumor burden in bone marrow graft is small, and the IMT is easily accessible to all tumor cells. The other two methods are increasingly complicated because many pharmacological parameters play critical roles in determining therapeutic efficacy of systemically administered conjugates. A summary of immunotoxins made with various toxin polypeptides is shown in Tables 2 and 3.

Theoretically, any macromolecule that has selective binding to tumor/target cells can be used in conjugation to toxins. Monoclonal antibodies are used extensively, since experimentally one can select tumor-specific antibodies by hybridoma techniques. Other carriers used in targeting include hormones and growth factors. Receptors for some of these molecules are amplified in certain tumors. Except in certain isolated cases, this approach is intended solely for in vitro applications. Another concern about the use of hormones and growth factors is the possibility of competition with endogenous sources.

Many important criteria are used in selecting the ideal antibody for targeting: (1) Antibody should be directed to a cell surface determinant having higher affinity; (2) target antigen should be expressed in high numbers; (3) antibody should be highly selective, with minimal cross-reaction to normal cells; and (4) target antigen should be internalized with good efficiency. The first three parameters are relatively easy to achieve by intensive screening and selection of hybridomas. Internalization, however, cannot be assured for all antibodies. The screening process for selecting an ideal antibody with good internalization capacity has been laborious until recently. A novel strategy to simplify the screening was the use of secondary targeting [23,24]. In this approach, tumor-specific antibody is first allowed to bind target tumor cells at 4°C to prevent internalization. Subsequently, the F(ab) portion of a scond antibody directed to murine

Table 2 Immunotoxins Directed to Human Tumors: Partial List

Antibody	Toxin	References
Lymphoid tumors		
Anti-CD2, CD3, CD5, CD7, CD11	Ricin	28,29
Anti-CD5	RTA	30
Anti-CD7	PAP	31
Anti-TAC	RTA	32
	PE	33
Anti-Ig	RTA	34
Anti-CD19, CD22	RTA	35
Anti-CD19	PAP	36
Anti-CALLA	RTA	37
	PAP	38
Breast and ovarian tumors		
260F9	RTA	41
280D11	PE/RTA	43
245E7	RTA	43
520C9	RTA	41
OVB-3	PE	67
Melanoma		
9-2-27	DTA	46
Colorectal carcinoma		
1083-17-1A	RTA/DTA	44
Bladder carcinoma		
791T/36	RTA	45

IgG conjugated to a toxin molecule was added. Thus the cytotoxicity is elicited by an indirect method. Careful comparison between direct and indirect targeting has showed good correlation.

A. Lymphoid Antigens

Extensive studies on the differentiation of lymphoid cells have yielded a number of antibodies recognizing stage-specific antigens. Leukemia arising from various

Table 3 Hemitoxin Antibody Conjugates: In Vitro Cytotoxicity[a]

Antibody	Target antigen	Toxin	Target cell	TCID 50 (M)[b]
31E6	Thy 1.1	PAP	AKR-SL-3	1×10^{-11}
OX-7	Thy 1	PAP	AKR-SL-3	3.6×10^{-9}
3A-1	CD-7	PAP	HSB-2	5.6×10^{-10}
5E9-C11	HTFR	PAP	HSB-2	2.2×10^{-10}
	HFTR	PAP	CEM	2×10^{-10}
JL-1	HTFR	PAP-II	MCF-7	7×10^{-10}
B-43	CD19	PAP	B-ALL	1.1×10^{-8}
J5	CALLA	PAP-S	NAMALWA	8.4×10^{-10}
9.2.27	250-kd glycoprotein proteoglycan	PAP-II	FMX-Met-II	4.0×10^{-10} 6.6×10^{-10}
J5	CALLA	Gel	NALM-6	3×10^{-11}
I-2	Ia	Gel	NALM-6	1×10^{-11}
HB-5	C3d receptor	Gel	B-Cell	1×10^{-11}

HB-5	C3d receptor	Gel	Raji	$> 10^{-7}$
T32B11	Thy 1.1	Gel	BW5147	6×10^{-7}–1.5×10^{-6}
			AKR-T	2×10^{-7}–1.3×10^{-6}
			AKR-spleen cells	1×10^{-12}–4×10^{-10}
M549	Thy 1.1 and Thy 1.2	Gel	RADA	1×10^{-12}
38-13	Glycolipid (globotriaosyl ceramide)	Gel	L2210	1×10^{-9}
			Ramos	6×10^{-10}
9.2.27	250-kd glycoprotein proteoglycan	Gel	FMX-Met II	1.0×10^{-10}–3.0×10^{-11}
AT15E	Thy 1.2	Gel	EL-4	7×10^{-10}
OX-7	Thy 1	Sap	AKR-A	3×10^{-11}
			BW5147	3×10^{-12}
OX-7	Thy 1	Sap	BW5147	3×10^{-12}

[a]Abbreviations: Gel, gelonin. Sap, saporin; PAP, pokeweed antiviral proteins; HTFR, human transferrin receptor.
[b]TCID 50 is amount required to elicit 50% inhibition of target cell protein or DNA synthesis.

steps of the differentiation pathway are well characterized for their surface phenotype using these antibodies.

T-Cell-Directed IMT

In earlier work, murine model systems using antibodies to Thy-1 was studied by many investigators. Specific antibodies to variants of Thy-1 allele when conjugated to toxin molecules were highly selective in target cell cytotoxicity [17,25-27]. By comparing the effect of conjugates on nonlymphoid cells or antigen-negative phenotypes, the selectivity of these conjugates was determined. Most of the conjugates had a specificity index of over 1000, indicating that more than 3 orders higher concentrations of IMT were required to elicit comparative cytotoxicity on antigen-negative cell lines.

Antibodies to human T cells were of special significance. Some valuable information on IMT activity came out of these studies. Monoclonal antibodies directed to CD-2, CD-3, CD-5, CD-7, and CD-11 were used in making conjugates with either ricin A chain or intact ricin or PAP [28-31]. With the sole exception of CD-5 antibody, all conjugates were biologically active. CD-5 is a 67,000-dalton glycoprotein present in most T-cell malignancies and in some B-cell leukemia. Antibody T101, which recognizes CD5 antigen, was highly cytotoxic when linked to intact ricin; whereas substitution of ricin to A chain resulted in lower biological activity. Reasons for inactiveness and methods to overcome these problems are discussed in a following section.

Variations in cytotoxicity with RTA conjugates lead to the use of intact ricin conjugates by Vallera et al. [28-30] for bone marrow purging. In these studies, the effective elimination of almost all the leukemia cells was achieved by using a "cocktail" of three IMTs directed to independent antigens. Other IMTs used to remove T cells from bone marrow grafts include anti-CD-7 conjugates. Anti-CD-7-PAP conjugate was cytotoxic even at 10 ng/ml concentration [31]. During bone marrow purging, it is important to preserve the viability of stem cells. Binding of IMT to progenitor cells would result in the elimination of stem cells from the graft. This would significantly alter engraftment and prevent repopulation of the lymphoid system. Most of the anti-T-cell IMTs did not show any effect on the viability of the stem cells. Treatment of bone marrow with anti-transferrin receptor IMT, however, resulted in complete loss of multipotent stem cells [31]. Such conjugates are not suitable for removing contaminating leukemia cells from bone marrow.

Besides differentiation antigens and antibodies directed to a glycoprotein determinant present on T cells, Interleukin-2 receptor were also used. IL-2 receptors are expressed low in number in resting T cells. Upon mitogenic/antigenic activation, the density of IL-2 receptors increases. Moreover, some leukemia cells infected with HTLV-1 virus (adult T-cell leukemia) constitutively express a higher number of IL-2 receptors, thus offering a selective advantage in targeting. Anti-TAC antibody, recognizing human IL-2 receptor when linked to ricin A chain [32] or PE [33], formed highly cytotoxic conjugates. One such construct is currently under clinical evaluation.

Anti-B-Cell IMT

During the ontogeny of B cells, surface immunoglobulins are expressed in pre-B- and B-cell stages. These antibodies are specific for individual clones, and therefore B-cell leukemia from patients would have clonotypic idiotypes. Anti-idiotype antibodies could thus be linked to ricin A chain for selective cytotoxicity. Besides the idiotypic determinant, one can also prepare cytotoxic conjugates by using antibodies to surface immunoglobulin. Anti-IgM and IgD IMTs were effective in inhibiting B-cell leukemia cells in vitro [34]. In another study, B-43, an antibody recognizing pre-B and B cells (CD19) was used to eliminate B-ALL cells [36]. This conjugate was highly effective against "fresh" leukemia cells in clonogenic assays. Other antibodies used against B-ALL cells include anti-CALLA antibody, J-5 [37,38], and anti-CD-22 [35].

B cells have Class II MHC antigen on their surface, and in certain circumstances class II-bearing leukocytes are identified in pathogenesis. One example is the infiltration of leukocytes in Type I insulin-dependent diabetes mellitus (DM). DM is an autoimmune disease, and in an experimental model system, transplantation of insulin-secreting pancreatic islets are tried to reverse hyperglycemia. Islets are often rejected by the onset of host-versus-graft disease. Class II antigen-bearing lymphocytes from the donor seem to play a crucial role in the onset of rejection. Shizuru et al. [39] used an anti-Ia antibody linked to ricin A chain conjugates to overcome the lymphocyte reactivity to pancreatic transplants in vitro. Pretreatment of the graft with IMT prolonged the graft survival, with the return of normoglycemia in recipient rats (personal communication).

B. IMT to Nonlymphoid Antigens

Human epithelial tumors are difficult to treat, with poor prognosis. Monoclonal antibodies to epithelial tumors such as breast and ovarian cancer are being studied by many investigators. Eighty-five monoclonal antibodies reactive to breast cancer cells were developed by Frankel et al. [40]. Many of them showed higher selectivity and broader tumor range. These antibodies were linked to both native and recombinant ricin A chain. Only 25 of the antibodies formed cytotoxic conjugates. Antibodies reactive to 55-kd protein and 95-kd proteins were highly cytotoxic to many breast cancer cell lines. Some of the conjugates reactive to mucinous antigens were not effective [41]. Consistently, most active IMTs were directed to transferrin receptor molecules. Although valuabe information could be obtained by using this conjugate in model systems, its clinical application is limited due to nonspecific toxicity. The extent of toxicity could not be judged in the murine model system, since anti-human transferrin receptor (TR) antibodies do not cross-react with mouse TR. Therefore, it was important to investigate homologous conjugates in a murine system. Recent studies indicate that in spite of normal cell binding, anti-murine TR conjugates show antitumor effect in a syngenic tumor model system [42].

Many of the anti-breast cancer antibodies also reacted with ovarian carcinoma cells. Effective conjugates cytotoxic to ovarian cells were made with these antibodies using RTA and PE. Both types of conjugates were highly cytotoxic to breast cancer cells. However, parallel studies revealed that some antibodies formed inactive IMTs against ovarian carcinoma cells when linked to RTA. Same antibody (260F9), when conjugated to PE, however, was cytotoxic to both breast and ovarian cancer cells [43]. The underlying mechanism for such differences may be due to the efficiency in translocation of PE from the vesicular compartment. This notion is based on cytochemical studies. There was no significant difference in the intracellular level of both the conjugates, but biologically, only the PE IMT inhibited tumor cell protein synthesis.

Other toxin conjugates include antibodies directed to colorectal carcinoma [44], bladder carcinoma [45], melanoma [46], and oncofetal antigens. Overexpression of certain oncogene-related antigens on tumor cells have prompted the use of antibodies reactive to some of them. Some of the oncogene-coded proteins are located in the inner plasma membrane and are inaccessible to antibodies. Antibody 45-2D9, which reacts to 74-kd protein encoded by c-Ha-ras was cytotoxic to c-Ha-ras transfected cells when linked to RTA [47]. Further studies are necessary to determine the usefulness of these conjugates in clinical settings.

V. POTENTIATION OF IMT ACTIVITY

The first step in IMT-mediated cytotoxicity is binding. Extent of binding is dependent on the affinity of the antibody and the number of surface receptors per cell. Subsequent to binding, the conjugate is internalized by receptor-mediated endocytosis and reaches acidic compartments. Internalization is energy-dependent. The final step involves the escape of the toxin moiety from the vesicles to the cytoplasmic compartment. Individual toxin moieties play an important role in this process. Some of the toxins have special hydrophobic domains that help in translocation. Since IMTs pass through the endosomal/lysosomal compartment, compounds that alter the internal milieu of lysosomes could influence the biological activity. Lysosomes are cellular sinks involved in the breakdown of ingested material from cell surface. Proteolytic enzymes such as cathepsin reside in these acidic vesicles. External addition of amines such as chloroquine and ammonium chloride increase the internal pH of lysosomes, thus affecting the proteolytic degradation pathway. In this context, investigations were carried out to determine the effect of lysosomotropic compounds on the biological activity of IMT. Presence of lysosomotropic compounds increased the cytotoxic activity of many immunotoxins by 2 to 3 orders [48-50]. Among the various reagents studied, monovalent ionophores such as monensin showed dramatic effects on immunotoxin activity at relatively low concentrations. Calculations indicate that the low level at which monensin increases IMT activity could not be accounted for by a change in endosomal/lysosomal pH. Therefore, other indirect mechanisms may be involved in ionophore-mediated increase in IMT activity.

In another series of experiments, calcium channel blockers and attenuated adeno virus particles were found to augment cytotoxicity [51]. Potentiation is very useful in improving the cytotoxicity of relatively inactive conjugates. Anti-T cell IMT, T 101-RTA conjugate that was less cytotoxic could be made highly potent by the addition of 10 mM ammonium chloride to the incubation medium [52]. The latest studies show that in the case of T 101 conjugates, biologically active IMT could be prepared when the Fc portion of the IgG molecule was removed. Both divalent and monovalent F(ab) fragments linked to RTA molecule showed inhibition of tumor cell growth [53]. The lysosomotropic compounds seem to alter the lag period seen for cytotoxicity (time taken for internalization) and change the biphasic inhibition curve into a single-order process. Further, they also alter the intracellular fate of the conjugates by delaying and reducing the amount of IMT reaching lysosomes. The longer the conjugates remain in the endosomal vesicles, more is the probability of leakage/escape into the cytoplasm increases.

Another novel method to improve the cytotoxicity of RTA conjugates was developed by Vitetta et al. [54]. In this strategy, the translocation properties associated with the B chain of ricin were taken advantage of. RTA conjugates were preincubated with the target cells to allow binding, but under conditions that block internalization. Subsequently, addition of either free B chain or a second antibody directed to murine IgG linked to the B chain of ricin resulted in improved cytotoxicity. This method, although feasible in vitro, may not be applicable in vivo. However, these studies highlight the importance of B chain in facilitating RTA to reach the ribosomal site.

A. Effects of Chemotherapeutic Drugs and IMT

A different strategy was employed by Uckun et al. [55] to improve the effect of IMT on tumor cells. ASTA Z7557 is a stable derivative of 4-hydroperoxy cyclophosphamide (4-HC). Treatment with 4-HC has been shown to differentially inhibit leukemia cells with minimal damage to progenitor cells. 4-HC inhibits DNA synthesis, and the combined treatment with IMT eliminated tumor cells synergistically. One possible explanation for the increased cytotoxicity is that IMT and 4-HC act on different but vital cellular processes. Using 20 μg/ml of 4-HC and 10 μg/ml of B43-PAP conjugate, almost 6 logs of leukemia cells could be eliminated in the presence of excess bone marrow cells.

VI. IN VIVO STUDIES

Systemic administration of IMT is profoundly influenced by the pharmacological behavior of the conjugate in vivo. Some of the general parameters governing the success of in vivo application are represented schematically in Figure 1. Antibody-toxin conjugates from the vascular compartment have to permeate into the tumors that are present in the extravascular space. Therefore, the stability of the conjugate in blood and its ability to diffuse across the endothelial gap junctions play a critical role in the efficacy of IMT in vivo. Currently, all the parameters determining vascular permeability are under investigation. Studies indicate that there is an inverse correlation between molecular size and diffusion across the endothelial wall. Some investigations suggest that basic proteins cross the capillary bed more easily than acidic proteins. Thus, it may be important to compare the relative localization capacity of $F(ab')_2$ or $F(ab')$ fragments of antibodies linked to toxins having higher pI values with conventional conjugates.

A. Biological Stability of IMT

The therapeutic efficacy of antitumor agents is dependent on their biological stability and tissue distribution. Both of these factors, in turn, determine the toxicity associated with the drug. To increase the efficacy, therefore, one must understand and design therapeutic agents so as to reach the target tumors at optimal concentrations while bypassing vital organs. When antibody-toxin conjugates were injected into experimental animals, they were cleared from circulation in a biphasic manner. The immediate drop in the level of IMT in blood is attributed to the tissue distribution. The rate of decrease in circulation is used to quantitate half-life. As IMTs are not a single homogenous compound, studies on the pharmacokinetics are more complicated. Critical analysis of IMT by electrophoresis indicates that the conjugates dissociate in circulation [56]. With time, there was a preferential accumulation of IgG in the blood as the dissociated toxin molecules are rapidly cleared by filtration through the kidneys. Both glycosylated and nonglycosylated toxin polypeptides have much shorter half-lives. By 30 min the toxin was undetectable in blood. Free IgG, on the other hand, remains in blood over a longer period. The nature of chemical linkage may

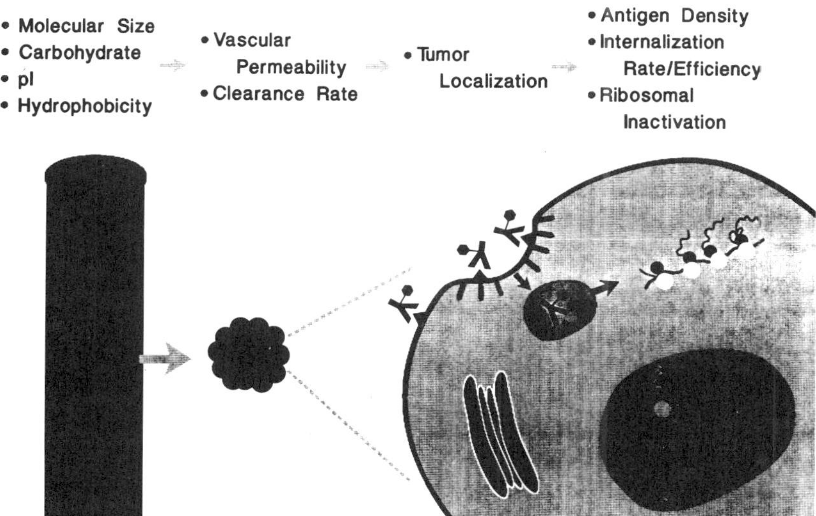

Figure 1 Critical factors determining the biological efficacy of immunotoxins in vivo. Efficacy of systemically administered antibody-toxin conjugates is determined by (1) vascular permeability, (2) clearance rate and (3) cellular processes such as endocytosis. Many physicochemical properties of immunotoxins influence the escape of conjugates from vascular compartment. The amount of immunotoxins localizing on tumor cells depends on the rate of permeability and biological half-life. Finally, cytotoxicity of the target cell is determined by the density of surface antigen, endocytosis and the ability of the conjugate to translocate.

contribute to the biological half-life of IMTs. Disulfide bonds, which are often used in the construction of conjugates, are subjected to reduction by thiol exchange, rendering the IMT inactive.

Antibodies linked to RTA [57], PAP [58], gelonin [59], saporin [59], and PE have been studied in vivo for clearance. All the IMTs had shorter half-lives than the respective free antibody. A majority of the injected RTA conjugates were sequestered by the reticulo-endothelial system of the animal [60]. Kupffer cells present in the liver have receptors for sugar residues by which they bind to glycosylated proteins and catabolize. To minimize liver uptake, one could use either nonglycosylated toxins or deglycosylated RTA. In fact, studies by Blakey et al. [56] suggest that chemical and enzymatic deglycosylation of RTA results in significant reduction in liver-mediated clearance of RTA. It is also possible to pretreat animals with a complex mannose homopolymer isolated from yeast to block the liver binding sites. Such methods reduced liver uptake and increased the relative amount of IMT in the blood.

IMTs are cleared from the blood in a biphasic manner: a rapid alpha phase followed by a slower beta phase. A majority of the injected conjugate is removed during the first phase. The amount of IMT cleared from blood during the alpha phase did not differ when the linkage was changed from disulfide to a non-reducible thioether bond. However, thioether conjugates showed a better stability during the beta phase in monkeys. The physiological significance of such differences is not clear. Since ultimately the amount of conjugate localized on tumor is more relevant than the quantity in circulation, it is important to study the tissue distribution of IMT in tumor-bearing animals. Investigations using radiolabeled conjugates indicate that less than 1% of the injected IMT reaches the tumor site. Improving localization of IMT is being investigated in many laboratories by optimizing circulatory half-life and vascular permeability.

B. Efficacy in Tumor Model Systems

Immunotoxins have been evaluated in multiple experimental systems. Initial studies were conducted on murine tumors in homologous models to determine whether the administration of conjugates could inhibit or delay the development of tumors. Later, some IMTs were tested in regression models wherein established malignancies were treated with conjugates. Regression models are closer to clinical situations. A general observation in all these investigations is the anamoly between in vitro cytotoxicity and efficacy in vivo. Many constructs that were highly potent in inhibiting tumor cell growth in vitro were not equally effective in vivo. A number of factors could contribute to this difference.

Ricin A Chain/RIP Conjugates

Injection of IMT prepared with RTA and RIP has been shown to be effective in many systems when the conjugates were administered within a short period after tumor transplantation. Antibodies to Thy 1.1 differentiation antigen inhibited the growth of T-lymphoma cells in congenic mice [61]. In this experimental system, host animals express a variant of Thy 1 allele (Thy 1.2), non-cross-reactive with the antigen present on the tumor cells (Thy 1.1). Thus, the normal T cells of the animal would not be affected by the injected conjugate. In another study, the relative efficacy of two conjugates made with two toxin polypeptides was evaluated. Saporin conjugated to anti-Thy 1.1 antigen was more inhibitory than the corresponding A chain conjugate. Interestingly, in vitro, the RTA conjugate was more cytotoxic than the saporin IMT [62]. Krolick et al. [63] used a murine B cell tumor (BCL1) that was highly tumoregenic. Even a few cells, when injected into the peritoneal cavity, produced tumors in nearly all the animals. Parallel to the clinical conditions, the tumor-bearing animals were first debulked by fractionated total lymphoid irradiation and spleenectomy prior to the administration of conjugates. In this study, IMT directed to surface immunoglobulins of B cells were found to be highly effective. Almost 99.9% of the residual tumor cells could be eliminated by injecting RTA IMT. Inhibition of tumor growth was specific, since injection of either unconjugated antibody or antibody admixed with RTA or an irrelevant IMT (control antibody or BSA conjugated to RTA) did not show any effect.

Intact Toxin Conjugates

As mentioned earlier, the use of intact toxin conjugates in vivo is limited by nonspecific binding to normal cells. However, these conjugates could be rendered specific by modifications of the sugar binding sites of ricin or by the presence of excess amounts of galactose in the medium. Weil-Hillman et al. [64] investigated this approach in vivo in a nude mouse model system using human xenografts. Subcutaneous injection of human T-lymphoblastic leukemia cells, CEM, results in the formation of solid tumors. Mice bearing established tumors were treated with anti-T-cell ricin conjugates in the presence of 100 mM lactose. Animals were also pretreated with lactose infusions to prevent/minimize toxicity by ricin. Under these conditions, no significant decrease in tumor size was noticed. Later, IMT was injected directly into the tumor. Intratumoral injection of ricin conjugates in the presence of lactose was effective in inducing complete regression. Parallel studies using nontarget cells (Burkitt's lymphoma, Daudi cells) indicated that the effect was indeed specific. While this model system is good for initial evaluation in animals, the extrapolation of these studies to clinical use is rather difficult.

PE and DT Conjugates

Pioneering work on DT conjugates for in vivo studies was carried out by Moolten et al. [65] using hapten-derivatized tumor cells and polyclonal antihapten antibodies linked to DT. Hamsters treated with IMT survived for longer times than untreated control animals. Other investigators have used the A fragment of DT. A conjugate reactive to proteoglycan present in human melanoma cells were effective in inhibiting the growth of human melanoma cell line M21 in athymic nude mice. The suppression of tumor growth is difficult to evaluate, however, since free antibody also elicited equal antitumor effects in this study [46]. Interestingly enough, the best antitumor effects were seen in mice injected with free DT. Even established tumors regressed completely after treatment with DT. Mice are resistant to DT toxicity, and human cells are highly sensitive to DT. It remains to be seen how a small amount of DT could bring about complete regression and prolonged remission in animals.

Human ovarian carcinoma model systems using nude mice are studied extensively to evaluate the efficacy of PE conjugates. OVCAR-3 cells, when injected into the peritoneum of nude mice, produce ascites with micro and macro metastasis resembling the clinical situations. These tumors grow very slowly. Antibodies directed to human transferrin receptor linked to PE were very effective against OVCAR-3 cells transplanted into mice. The conjugates were administered 4–5 days after the transplantation of tumor cells. Injection of 2 μg conjugates increased the median survival time to 90–100 days [66,67]. Control mice succumbed to the disease by 40–50 days. Similarly, an antibody specific for human ovarian tumors (OVB-3) linked to PE inhibited the growth of OVCAR-3 cells. The latter conjugate is currently under clinical investigation.

VII. PROBLEMS AND FUTURE DEVELOPMENTS

Remarkable progress has been made in conjugation chemistry and preparation of biologically active immunotoxins in the recent past. Although most of the conjugates are cytotoxic and specific in vitro, only a few conjugates have shown antitumor effect in vivo. Some of the issues concerning future developments are (1) stability of the conjugate, (2) vascular permeability and tumor penetration, (3) toxicity to liver and vital organs, (4) immune response against the conjugate, and (5) development of resistant clones.

A. Recombinant Toxins and Fusion Proteins

IMT are synthesized currently by random conjugation. At the toxin side, the reactive groups are localized while antibodies are often derivatized at random to facilitate conjugation. To produce defined conjugates with minimum steric hindrance, genetic engineering of the toxin and antibody molecules is favored. Ricin, DT, and PE have been cloned and expressed in *E. coli*. Bacteria-derived RTA is nonglycosylated and may have some potential advantage over native RTA. Cloning and expression of PE has rapidly improved our understanding about the structure/function relationship. Of interest is the elimination or substitution of critical amino acid residues involved in receptor binding while retaining the translocation domains. Recently, using genetic engineering techniques, a fusion protein of gp120 and PE was produced. Gp120 is expressed on HIV infected human T-cells. Incubation of the virus-infected T cells with this fusion protein resulted in specific inhibition [68]. The availability of recombinant toxin polypeptides would be very useful in making novel cytotoxic reagents.

Piatak et al. [69] have expressed RTA in soluble form. Recombinant ricin and the B chain of ricin are found to be difficult to produce in prokaryotic systems. Some investigators are successful in the genetic expression of the B chain of ricin in eukaryotic systems. Attempts are being made to delete or alter the contact residues involved in galactose binding sites in ricin. Such modifications could result in altered ricin molecules with little nonspecific toxicity but may retain the putative translocation domains. Further, genetic engineering would be useful in designing toxin molecules by transposing biologically active fragments to improve the tumor-inhibitory properties of IMTs. Recombinant fusion proteins containing growth factors/hormones and toxin fragments have been made during the past year. IL-2-DTA chimeric protein synthesized in this manner has been found to be very effective in inhibiting allogenic T-cell activation during tissue transplantation [70]. Studies in these directions would provide us the next generation of immunotoxins in the near future.

B. Antibody Response

Recent Phase I clinical trials [71] and animal data [61] suggest that antitoxin and antimurine IgG response in recipients would limit the use of immunotoxins for a prolonged period. Neutralizing antitoxin antibodies are unavoidable. However, the duration of treatment can be increased by switching toxin moieties that are immunologically non-cross-reactive. Antimurine IgG response in IMT-treated patients is directed mainly to the Fc region of the antibody. Chimeric antibodies having murine combining site [F(ab)] and human Fc regions could help in reducing the immune response against murine IgG. Recently, the variable region of the heavy chain of OVB-3 antibodies has been successfully cloned [72]. It is only a matter of time before genetically engineered fusion proteins become available for clinical application. An alternative approach to reducing immune response to IMTs is either to induce tolerance or to obliterate immunodominant regions by chemical modifications with polyethyleneglycol (PEG). These studies are currently in progress.

C. Resistance to IMT

Parallel to the development of multiple drug resistance, IMT-resistant tumor cells have been isolated. Most of the cells surviving after IMT treatment are variants expressing low antigen density or antigen-negative entities. Escape of antigen-negative cells has to be minimized for successful treatment. It is possible to use a mixture of IMT directed to independent determinants to prevent the growth of IMT-resistant clones. The probability of cells lacking a group of antigens would be much lower when compared to a single determinant. Moreover, multiple binding of IMT could also increase the efficiency of internalization and cytotoxicity. IMT resistance can also develop by a defect in the internalization pathway. Variants belonging to this category show comparable levels of surface antigens and to some extent cross-resistance to other IMTs as well. Further studies are necessary to understand the mechanism of resistance to IMTs, and methods have to be developed to overcome this problem.

VIII. ACKNOWLEDGMENTS

This work was supported by a grant from the National Cancer Institute (CA 48068-01).

IX. REFERENCES

1. P. Ehrlich, in *The Collected Papers of Paul Ehrlich*, Vol. I (F. Himmelweite, M. Marquardt, and H. Dale, eds.), Pergamon Press, London and New York, 1956, pp. 596–618.
2. G. Kohler and C. Milstein, *Nature 256*:495–497 (1975).
3. *Immunol. Rev. 62* (1982).
4. A. E. Frankel (ed.), *Immunotoxins*, Kluwer Academic Publishers, Boston, 1988.
5. R. Drazin, J. Kandel, and R. J. Collier, *J. Biol. Chem. 246*:1504–1510 (1971).
6. A. M. Pappenheimer, Jr. *Ann. Rev. Biochem. 46*:69–94 (1977).
7. V. S. Allured, R. J. Collier, S. F. Carroll, and D. B. McKay, *Proc. Natl. Acad. Sci. USA 83*:1320–1324 (1986).
8. J. Hwang, D. J. Fitzgerald, S. Adhya, and I. Pastan, *Cell 48*:129–136 (1987).
9. K. C. Halling, A. C. Halling, E. E. Murray, B. F. Ladin, L. L. Houston, and F. R. Weaver, *Nucleic Acids Res. 13*:8019–8033 (1985).
10. F. I. Lamb, L. M. Roberts, and J. M. Lord, *Eur. J. Biochem. 148*:265–270 (1985).
11. W. Montfort, J. E. Villafranca, A. F. Monzingo, S. Ernst, B. Katzin, E. Rutenber, N. H. Xuong, R. Hamlin, and J. D. Robertus, *J. Biol. Chem. 262*: 5398–5403 (1987).
12. Y. Endo, K. Mitsui, M. Motizuki, and K. Tsurugi, *J. Biol. Chem. 262*:5098–5912 (1987).
13. J. L. Bradley, H. M. Silva, and P. M. McGuire, *Biochem. Biophys. Res. Commun. 149*:588–593 (1987).
14. M. Ready, K. Wilson, M. Piatak, and J. D. Robertus, *J. Biol. Chem. 259*: 15252–15256 (1984).
15. L. L. Houston, S. Ramakrishnan, and M. A. Hermodson, *J. Biol. Chem. 258*: 9601–9604 (1983).
16. Y. Endo and K. Tsurugi, *J. Biol. Chem. 262*:8128–8130 (1987).
17. S. Ramakrishnan and L. L. Houston, *Cancer Res. 44*:201–208 (1984).
18. R. Jue, J. M. Lambert, L. R. Pierce, and R. R. Traut, *Biochemistry 17*: 5399–5406 (1978).
19. A. J. Cumber, J. A. Forrester, B. M. Foxwell, W. C. J. Ross, and P. E. Thorpe, *Methods Enzymol. 112*:207–225 (1985).
20. D. J. Fitzgerald, *Methods Enzymol. 151*:139–145 (1987).
21. W. A. Blattler, B. S. Kuenzi, and J. M. Lambert, and P. D. Senter, *Biochemistry 24*:1517–1524 (1985).

22. J. W. Marsh, K. Srinivasachar, and D. M. Neville, Jr., in *Immunotoxins* (A. E. Frankel, ed.), Kluwer Academic Publishers, Boston, 1988, pp. 213–237.
23. J. K. Weltman, P. Pedroso, S. A. Johnson, D. Davignon, L. D. Fast, and L. A. Leone, *Cancer Res. 47*:5552–5556 (1987).
24. M. Till, R. D. May, J. W. Uhr, P. E. Thorpe, and E. S. Vitetta, *Cancer Res. 48*:1119–1123 (1988).
25. R. J. Youle, and D. M. Neville, Jr., *Proc. Natl. Acad. Sci. USA 77*:5483–5486 (1980).
26. P. E. Thorpe, A. N. Brown, W. C. Ross, A. J. Cumber, S. I. Detre, D. C. Edwards, A. J. Davies, and F. Stirpe, *Eur. J. Biochem. 116*:447–454 (1981).
27. D. A. Vallera, R. J. Youle, D. M. Neville, Jr., and J. H. Kersey, *J. Exp. Med. 155*:949–954 (1982).
28. D. A. Vallera, R. C. Ash, E. D. Zanjani, J. H. Kersey, T. L. LeBien, P. C. Beverley, D. M. Neville, Jr., and R. J. Youle, *Science 22*:8512–8515 (1983).
29. R. C. Stong, F. M. Uckun, R. J. Youle, J. H. Kersey, and D. A. Vallera, *Blood 66*:627–635 (1985).
30. G. Laurent, J. Pris, J.-P. Farcet, P. Carayon, H. Blythman, P. Casellas, P. Poncelet, and F. K. Jansen, *Blood 67*:1690–1687 (1986).
31. S. Ramakrishnan and L. L. Houston, *Science 223*:58–61 (1984).
32. M. Kronke, J. M. Depper, W. J. Leonard, E. S. Vitetta, T. A. Waldmann, and W. C. Greene, *Blood 65*:1416–1421 (1985).
33. D. J. FitzGerald, T. A. Waldman, M. C. Willingham, and I. J. Pastan, *J. Clin. Invest. 74*:966–971 (1984).
34. M. Muirhead, P. J. Martin, B. Torok-Storb, J. W. Uhr, and E. S. Vitetta, *Blood 62*:327–332 (1983).
35. R. D. May, E. S. Vitetta, G. Moldenhauer, and B. Dorken, *Cancer Drug Deliv. 3*:261–272 (1987).
36. F. M. Uckun, S. Ramakrishnan, and L. L. Houston, *J. Immunol. 134*:2010–2016 (1985).
37. V. Raso, Ritz, M. Basala, and S. F. Schlossman, *Cancer Res. 42*:457–464 (1982).
38. J. M. Lambert, P. D. Senter, A. Yau-Young, W. A. Blattler, and V. S. Goldmacher, *J. Biol. Chem. 260*:12035–12041 (1985).
39. J. Shizuru, S. Ramakrishnan, T. Hunt, R. C. Merrell, and C. G. Fathman, *Transplantation 42*:660–666 (1986).
40. A. E. Frankel, D. B. Ring, F. Tringale, and S. T. Hsieh-Ma, *J. Biol. Resp. Modif. 4*:273–286 (1985).
41. M. J. Bjorn, D. Ring, and A. E. Frankel, *Cancer Res. 45*:1214–1221 (1985).
42. M. J. Bjorn and G. Groetsema, *Cancer Res. 47*:6639–6645 (1987).
43. R. Pirker, D. J. FitzGerald, T. C. Hamilton, R. F. Ozols, W. Laird, A. E. Frankel, M. C. Willingham, and I. Pastan, *J. Clin. Invest. 76*:1261–1267 (1985).
44. D. G. Gilliland, Z. Steplweski, R. J. Collier, K. F. Mitchell, T. H. Chang, and H. Koprowski, *Proc. Natl. Acad. Sci. USA 77*:4539–4543 (1980).

45. M. J. Embleton, V. S. Byers, H. M. Lee, P. Scannon, N. W. Blackhall, and R. W. Baldwin, *Cancer Res. 46*:5524–5528 (1986).
46. T. F. Bumol, Q. C. Wang, R. A. Reisfeld, and N. O. Kaplan, *Proc. Natl. Acad. Sci. USA 80*:529–533 (1983).
47. J. A. Roth, R. S. Ames, V. Byers, H. M. Lee, and P. J. Scannon, *J. Immunol. 136*:2305–2310 (1986).
48. P. Casellas, B. J. Bourrie, P. Gros, and F. K. Jansen, *J. Biol. Chem. 259*: 9359–9364 (1984).
49. V. Raso and J. Lawrence, *J. Exp. Med. 160*:1234–1240 (1984).
50. S. Ramakrishnan, M. J. Bjorn, and L. L. Houston, *Cancer Res. 49*:613–617 (1989).
51. P. Seth, D. J. FitzGerald, H. Ginsberg, M. Willingham, and I. Pastan, *Mol. Cell Biol. 4*:1528–1533 (1984).
52. P. Casellas, X. Canat, A. A. Fauser, O. Gros, G. Laurent, P. Poncelet, and F. K. Jansen, *Blood 65*:289–297 (1985).
53. J. M. Derocq, P. Casellas, G. Laurent, S. Ravel, H. Vidal, and F. K. Jansen, *J. Immunol. 141*:2837–2843 (1988).
54. E. S. Vitetta, R. J. Fulton, and J. W. Uhr, *J. Exp. Med. 160*:341–346 (1984).
55. F. M. Uckun, S. Ramakrishnan, and L. L. Houston, *Cancer Res. 45*:69–85 (1985).
56. D. C. Blakey, G. J. Watson, P. P. Knowles, and P. E. Thorpe, *Cancer Res. 47*:947–952 (1987).
57. B. J. Bourrie, P. Casellas, H. E. Blythman, and F. K. Jansen, *Eur. J. Biochem. 155*:1–10 (1986).
58. S. Ramakrishnan and L. L. Houston, *Cancer Res. 45*:2031–2036 (1985).
59. N. L. Letvin, V. S. Goldmacher, J. Ritz, J. M. Yetz, S. F. Schlossman, and J. M. Lambert, *J. Clin. Invest. 77*:977–984 (1986).
60. N. R. Worrell, A. J. Cumber, G. D. Parnell, W. C. J. Ross, and J. A. Forrester, *Biochem. Pharmacol. 35*:417–423 (1986).
61. S. Ramakrishnan and L. L. Houston, *Cancer Res. 44*:1398–1404 (1984).
62. P. Thorpe, A. Brown, J. Breamer, B. Foxwell, and F. Stirpe, *J. Natl. Cancer Inst. 75*:151–159 (1985).
63. K. A. Krolick, J. W. Uhr, S. Slavin, and E. S. Vitetta, *J. Exp. Med. 155*: 1797–1809 (1982).
64. G. Weil-Hillman, W. Runge, F. K. Jansen, and D. A. Vallera, *Cancer Res. 45*: 1328–1336 (1985).
65. F. L. Moolten, N. J. Caparell, and S. R. Cooperband, *J. Natl. Cancer Inst. 49*:1057–1062 (1972).
66. D. J. FitzGerald, M. Willingham, and I. Pastan, *Proc. Natl. Acad. Sci. USA 83*:6627–6630 (1986).
67. D. J. FitzGerald, M. J. Bjorn, R. J. Ferris, J. L. Winkelhake, A. E. Frankel, T. C. Hamilton, R. F. Ozols, M. C. Willingham, and I. Pastan, *Cancer Res. 47*:1407–1410 (1987).

68. V. K. Chaudhary, T. Mizukami, T. R. Fuerst, D. J. FitzGerald, B. Moss, I. Pastan, and E. A. Berger, *Nature 355*:369–371 (1988).
69. M. Piatak, J. Lane, W. Laird, M. Bjorn, A. Wang, and M. Williams, *J. Biol. Chem. 263*:4837–4843 (1988).
70. P. Bacha, D. D. Williams, C. Waters, J. M. Williams, J. R. Murphy, and T. B. Strom, *J. Exp. Med. 167*:612–622 (1987).
71. A. A. Hertler, D. M. Schlossman, M. J. Borowitz, G. Laurent, F. K. Jansen, C. Schmidt, and A. E. Frankel, *J. Biol. Resp. Modif. 7*:97–113 (1988).
72. M. Gallo, V. K. Chaudhary, D. J. FitzGerald, M. C. Willingham, and I. Pastan, *J. Immunol. 141*:1034–1040 (1988).

9

New Trends in the Use of Radioimmunoconjugates for the Therapy of Cancer

DONALD J. BUCHSBAUM and THEODORE S. LAWRENCE
The University of Michigan Medical Center, Ann Arbor, Michigan

I. INTRODUCTION

The in vivo use of radionuclide-labeled antibodies to tumor-associated antigens for the detection and treatment of cancer has been the goal of many investigators. For radioimmunodetection, the antibody moiety specifically binds to the tumor-associated antigen and external scintigraphic imaging methods are used to detect the radionuclide moiety. For radioimmunotherapy, the radionuclide is chosen to deliver potentially cytotoxic irradiation to the targeted tumor cell.

Radioimmunoconjugates offer a potential advantage over conventional diagnostic and therapeutic procedures by providing greater sensitivity and specificity as a result of antibody-specific binding to tumor cells and detection of the radionuclide emissions. Despite these evident potential advantages, many obstacles have to be overcome before radioimmunoconjugates become valuable reagents in the diagnosis and therapy of cancer. The purpose of this review is to summarize the work that has been done with radioimmunoconjugates for the localization and treatment of human lymphoma and colorectal cancer in both experimental animal models and humans, and to discuss new trends in the use of radioimmunoconjugates for the treatment of human cancer.

Pressman and Bale and their colleagues pioneered in the use of radioiodine-labeled polyclonal antibodies against fibrin for the localization and therapy of transplantable animal tumors [1-7] and human tumors [8-10]. Goldenberg used radioiodine-labeled polyclonal antibodies against carcinoembryonic antigen (CEA) for the localization of xenografts of human colon tumors in the cheek pouch of hamsters [11,12], and Mach used radioiodine-labeled polyclonal antibodies against CEA for the localization of xenografts of human colon cancer in athymic nude mice [13]. These investigators performed clinical localization studies in patients with colorectal cancer using ^{131}I-labeled polyclonal antibodies against CEA [14,15]. In 1975, Kohler and Milstein [16] developed the hybridoma technology that eventually led to the production of many murine and human monoclonal antibodies (MoAbs) against tumor-associated antigens from most types of human cancer. This development greatly stimulated attempts to conjugate radionuclides to MoAbs for the detection and treatment of cancer. A description of the most widely used of these MoAbs raised against human lymphomas and colorectal cancers is presented in the next two sections.

A. Murine MoAbs Against Human Lymphoma

Over 80% of adult cases of non-Hodgkin's lymphoma have membrane characteristics of B lymphocytes. A large number of MoAbs have been produced against tumor-associated antigens expressed by these malignancies, several of which have been radiolabeled and used in localization and therapy studies (Table 1). The murine MoAb BA-1, of the IgM subclass, recognizes the CD24 antigen and binds to most cases of non-T acute lymphoblastic leukemia, most cases of B chronic lymphocytic leukemia, and occasional cases of hairy cell leukemia, but not myeloma. Most surface immunoglobulin-positive B-cell lymphomas bind BA-1 [17-20]. With regard to normal tissues, BA-1 binds to a small percentage of peripheral blood lymphocytes and weakly to granulocytes, but not to T lymphocytes, monocytes, RBC, hematopoietic precursor cells, or platelets. Anti-B1 has a pan-B distribution, reacting with virtually all immunoglobulin-bearing malignancies as well as many non-T acute lymphoblastic leukemias [21]. Murine MoAb anti-B1 is an IgG2a antibody reactive with a 35,000-MW phosphorylated cell surface molecule (CD20). This antibody reacts with greater than 85% of peripheral blood B cells. The anti-B1 antibody also reacts with B cells in normal lymphoid tissues and bone marrow [22]. The anti-J5 murine MoAb [23] is an IgG2a antibody that reacts with a 100,000-MW common acute lymphoblastic leukemia antigen (CD10) found on cells from most patients with non-T acute lymphoblastic leukemia and follicular lymphomas. This antibody also reacts with granulocytes and renal tubular and glomerular cells. The Lym-1 murine MoAb [24] is an IgG2a that recognizes the HLA-Dr antigen or a polymorphic variant of the HLA-Dr antigen and binds to 60% of human lymphomas. While

Table 1 MoAbs Against Human Leukemia/Lymphoma That Have Been Used in Radioimmunodetection and Radioimmunotherapy Studies

MoAb (subclass)	Cluster designation of antigen	Investigators	References
B cell			
BA-1 (IgM)	CD24	Buchsbaum et al.	17–20,84
Anti-B1 (IgG2a)	CD20	Buchsbaum et al.	21,22,91, 133
Anti-J5 (IgG2a)	CD10	Buchsbaum et al.	23,91,133
Lym-1 (IgG2a)	HLA-Dr	DeNardo et al.	24,25,92
MB-1 (IgG1)	CD37	Press et al.	26
B43 (IgG1)	CD19	Buchsbaum and Uckun	27,95,134
T cell			
Anti-T1 (IgG2a)	CD5	Buchsbaum et al.	28,91
T101 (IgG2a)	CD5	Carrasquillo et al.	29,98,100–102

Lym-1 reacts with some normal B cells, the intensity of reactivity is less than that for malignant B cells [25]. Another MoAb recognizing an antigen (CD37) expressed by mature human normal and malignant B cells is the MB-1 IgG1 antibody [26]. The B43 murine MoAb [27] is of the IgG1 subclass and recognizes an antigen of the CD19 cluster that is expressed on neoplastic cells from malignant lymphoma patients and leukemic B-lineage lymphoid progenitor cells freshly obtained from B-lineage acute lymphoblastic leukemia patients but not on normal lymph nodes, spleens, normal myeloid, erythroid, megakaryocytic, or multilineage bone marrow progenitor cells, or nonlymphohematopoietic tissues.

Two anti-T-cell MoAbs, anti-T1 and T101, have been well characterized and used in localization and therapy studies. These are murine IgG2a antibodies that react with a 65,000-MW antigen (CD5) present on normal human T cells, and most T-cell leukemias and lymphomas [28,29].

B. Murine MoAbs Against Human Colorectal Cancer

A number of MoAbs against human colorectal cancer have been developed [30]. Many of these MoAbs have been radiolabeled and used in animal and clinical localization and therapy studies. Those that have been most widely used are presented in Table 2 and will be described below. The murine MoAbs, 35 and 115,

Table 2 MoAbs Against Human Colorectal Cancer That Have Been Used in Radioimmunodetection and Radioimmunotherapy Studies

MoAb (subclass)	MW of antigen (kd)	Investigators	References
35, 115, 202 (IgG1)	180 CEA glycoprotein	Mach, Buchegger, Buchsbaum	31,32,62, 103,109, 118,120, 145,167
NP-4 (IgG1)	180 CEA glycoprotein	Goldenberg et al.	33,146, 147
17-1A (IgG2a)	30-41 glycoprotein	Koprowski et al.	34,35,50, 69,109, 122,123
B72.3 (IgG1)	$> 10^3$ mucin-like glycoprotein	Colcher and Schlom	36,37, 106–109, 125,126, 143
791T/36 (IgG2b)	72 glycoprotein	Pimm and Baldwin	38–40,97, 127,128

are IgG1 antibodies that recognize the same or a closely related epitope of CEA [31]. MoAbs 35 and 115 do not react with either granulocytes or purified granulocyte glycoproteins NCA-55 or NCA-95 [31]. The affinities of these antibodies for CEA were $5.8 \times 10^9\ M^{-1}$ and $1.4 \times 10^9\ M^{-1}$, respectively [32]. NP-1 is a murine MoAb of the IgG1 subclass that is reactive with an epitope of CEA [33]. These anti-CEA monoclonal antibodies retain minimal reactivity with normal colon. Murine MoAb 17-1A is an IgG2a antibody reactive with a 30,000- to 41,000-MW gastrointestinal cancer-associated antigen confined to the cell surface [34,35], with an affinity of $0.7 \times 10^8\ M^{-1}$. The antibody was reported to bind to colon, gastric, and pancreatic carcinoma but did not bind to melanoma, fibrosarcoma, astrocytoma, myeloma, lymphomas, and normal epithelium in many human tissues. MoAb B72.3 is a murine IgG1 antibody reactive with the high-MW TAG-72 glycoprotein antigen expressed on colorectal, breast, lung, gastric, endometrial, and ovarian carcinoma [36,37]. The TAG-72 antigen has been shown to be distinct from CEA and the antigen recognized by MoAb 17-1A. MoAb 791T/36 is a murine IgG2b that was raised against osteogenic sarcoma cells [38] and recognizes a 72,000-MW glycoprotein and binds to most colon cancers [30], and has been shown to localize in primary osteogenic sarcoma [39] and ovarian and colonic cancer [40].

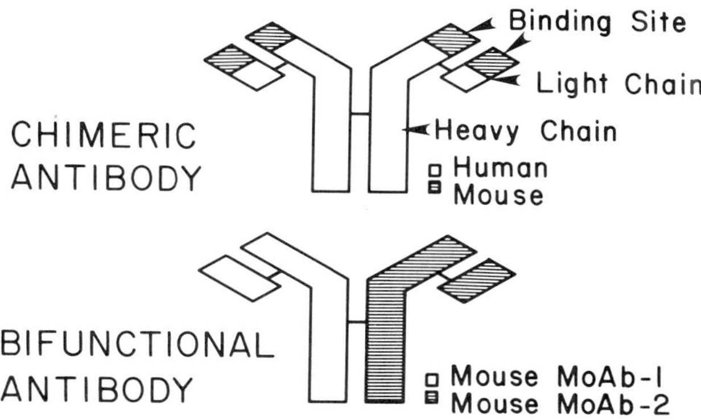

Figure 1 Structure of chimeric and bifunctional antibodies.

C. New-Generation MoAbs

Shortly after their introduction, it was recognized that the immunogenicity of mouse MoAbs may limit their use in humans for diagnosis and therapy. The human anti-mouse antibody (HAMA) would be a major deterrent to fractionated radioimmunotherapy with murine MoAb immunoconjugates because subsequent injections are cleared from the blood more rapidly [41]. One way of reducing MoAb immunogenicity is through the use of human, rather than murine, MoAbs. Some human MoAbs against human colorectal cancer have been produced, radiolabeled, and used for radioimmunodetection in nude mice bearing human colon cancer xenografts, and clinically [42,43]. Human MoAbs 16-88 and 28A32 were developed from the peripheral blood lymphocytes of colon carcinoma patients. Antibody 28A32 is an IgM antibody that is the product of a human-mouse heterohybridoma and recognizes an antigen on the surface and in the cytoplasm of colon carcinoma cells. Antibody 16-88 is an IgM produced by a spontaneously transformed human lymphoblastoid cell line. It recognizes a predominantly cytoplasmic antigen and has an affinity of 5×10^8 M^{-1}. The affinity of 28A32 is lower. These human MoAbs were reactive with most colon cancers and had weak reactivity with normal gastrointestinal tissues. There has been little or no antibody produced by patients against the administered human MoAbs [43,44], in contrast to the high levels of HAMA produced by colon cancer patients against administered mouse MoAbs [45–47].

A second method of reducing the HAMA response has been to use mouse-human chimeric MoAbs [48,49], which contain the mouse antigen-binding end bound to the human F_C region of the antibody (Figure 1). Because these antibodies are largely human in composition, they should be less immunogenic than

Table 3 Factors Affecting the Accumulation of Radioimmunoconjugates in Tumor and Normal Tissue

Physiological and pathological factors
Vascularity and blood flow
Vascular permeability
Concentration gradient
Metabolism and excretion
Route of injection
Site of tumor/normal tissue
Tumor mass
Degree of infiltration and necrosis in the tumor mass
Immunological factors
Number of cells expressing reactive antigen
Number of reactive antigens on cell surface
Affinity, avidity, and specificity of MoAb to reactive antigen
Immunoreactivity, affinity, and avidity of radioimmunoconjugate
Immunoglobulin subclass
Intact immunoglobulin or antibody fragment
Antigen modulation: internalization or shedding
Level of circulating reactive antigen in the blood or body fluids
Dose of radioimmunoconjugate
Ability of MoAb to be conjugated with radionuclide, drug, or toxin
Choice of radionuclide, drug, toxin, and conjugation techniques
Specific activity of radioimmunoconjugate
Nonspecific binding
Development of human anti-mouse antibody (HAMA) or anti-human antibody (HAHA) response

mouse MoAbs and hence more suitable for in vivo application in patients. Four subclasses of mouse-human chimeric 17-1A MoAb have been produced [49]. We have been investigating whether the binding in vitro of these chimeric antibodies is the same as the parental murine 17-1A MoAb, and the relative biodistribution of the radiolabeled chimeric MoAbs and parental MoAb in nude mice bearing xenografts of human colon cancer [50]. The results of our preliminary studies have indicated a longer blood half-life of the IgG1 chimeric MoAb compared to the IgG2 and IgG4 chimeric MoAbs [50]. Interestingly, Steplewski et al. [51] found that chimeric IgG1 was superior in antitumor activity in antibody-dependent cell-mediated cytotoxicity (ADCC) assays. The chimeric IgG1 and IgG4 antibodies were nearly as effective as the parental 17-1A antibody in inhibiting tumor growth in nude mice [51]. Some work has been done to produce class-switch variants of MoAbs [52], with a particular immunoglobulin subclass that is more effective in ADCC, and these have been studied in nude mouse xenograft models with colon cancer. We have some preliminary data on the improved localization of the IgG1 subclass of an anti-CEA MoAb C03D6 obtained from Dr. Steplewski at The Wistar Institute as compared to the IgG2a class-switch variant. Other investigators have reported new-generation MoAbs against the TAG-72 antigen produced by immunization with the purified antigen [53] and CEA [54], which ultimately will be investigated in radioimmunodetection and radioimmunotherapy trials.

D. Factors Affecting Uptake of Radioimmunoconjugates by Tumor and Normal Tissues

Table 3 gives a listing of the factors that affect uptake of radioimmunoconjugates by tumor and normal tissues. There have been several review articles published that discuss these factors [55-65]. Many of these physiological, pathological, and immunological factors will influence the success of tumor imaging and therapy. Examples will be given below.

II. CURRENT STATUS OF TUMOR IMAGING: EXPERIMENTAL MODELS AND CLINICAL APPLICATIONS

A. Radiolabeling of MoAbs

The radionuclides that are most widely used for radiolabeling MoAbs for experimental animal and clinical localization and imaging studies have been ^{125}I, ^{131}I, ^{123}I, ^{99m}Tc, and ^{111}In. Iodine can be bound covalently to the MoAb, usually to a tyrosine residue. The chemistry of ^{111}In and ^{99m}Tc do not allow direct covalent binding to a MoAb. These radionuclides must be attached to a MoAb through the use of an intermediate molecule or chelate, which is covalently bound to the MoAb and which binds (chelates) the radionuclide. The most commonly used chelate is diethylenetriaminepentaacetic acid (DTPA). The most important issues of radiolabeling concern the optimal choice of labeling technique, radionuclide, and conditions under which the radioimmunoconjugate is used.

There has been little work comparing the effect of radioiodine labeling techniques and on the subsequent tumor localization and imaging. Pettit et al. [66] investigated several iodination methods and sources of iodine radionuclides, and provided a summary of the results for incorporation of iodine radionuclides into normal goat IgG or goat anti-CEA. Saha [67] reviewed the techniques for radioiodination of antibodies and the importance of the immunoreactivity of the radioiodinated antibodies and the factors that affect it. The effect of labeling damage to the immunoreactivity of antibodies with the chloramine-T method due to exposure to oxidizing and reducing agents is well known and contrasts with the iodine monochloride method [68]. The iodogen method of radioiodination has also been reported to produce less damage than the chloramine-T method [69].

A number of reports on the localization of experimental animal tumors have claimed a superiority of ^{111}In-labeled monoclonal antibodies over ^{125}I- or ^{131}I-labeled MoAbs in terms of (1) better physical properties necessary for imaging, (2) higher concentrations in tumor, and (3) a more stable radiopharmaceutical in contrast to the in vivo dehalogenation that occurs with radioiodine-labeled MoAbs [70-75]. However, the published results have shown a high accumulation of ^{111}In, presumably attached to the chelate, in the liver, kidney, or spleen, whereas catabolism of localized or circulating antibody causes accumulation of radioiodine in the stomach and bladder [71-75]. There is no question that ^{111}In has more favorable physical characteristics (2.8-day half-life, 173- and 247-keV photons, no beta emission) than ^{131}I (8-day half-life, 364-keV photon, 608-keV beta emission) or ^{123}I (13-h half-life, 159-keV photon, no beta emission) for imaging purposes. In addition, ^{123}I cannot be used effectively for obtaining images beyond 24 h because of its short half-life. One argument put forward for the use of strong chelating groups, which can be covalently attached to proteins and then radiolabeled with metallic radionuclides, is that this results in greater stability in vivo than ^{131}I [72]. Furthermore, if the chelate becomes unconjugated from the antibody during metabolism, the radionuclide attached to DTPA is rapidly excreted by the kidneys [76], unlike the iodine radionuclides, which accumulate in the stomach and bladder [77]. Another advantage of ^{111}In-labeled antibodies over radioiodinated antibodies is that they are not eluted out of the tumor once localized [73]. However, ^{111}In-labeled MoAbs have prolonged retention in some normal tissues [78]. Larson and Carrasquillo have summarized the advantages of radioiodine over ^{111}In-labeled MoAbs for imaging solid tumors clinically [78]. These include the labeling of Fab fragments and the imaging of tumors in the liver.

The proportion of the radiolabeled MoAb molecules able to bind to antigen, the immunoreactivity, and the tightness of those molecules that are immunologically active, the avidity, are often impaired during the antibody production, purification, and storage process or by the radiolabeling procedure [79,80]. This has been shown by several investigators to result in decreased tumor uptake in animal tumor models [79-83]. The presence of labeled antibody molecules that do not have antigen-binding activity could increase background radioactivity when MoAbs are used for imaging and contribute an increased dose to normal tissues when radiolabeled MoAbs are used for therapy [80]. Pimm and Baldwin [81] decreased the immunoreactive fraction of an anti-CEA MoAb by adding increasing proportions of a control immunoglobulin. Following radioiodination, the immunoreactive fractions of the preparations were determined and their localization in human gastric carcinoma xenografts in nude mice was assessed. There was a progressive decline in tumor localization that correlated with a decrease in immunoreactive fraction of anti-CEA MoAb preparations.

B. Localization and Imaging of Radiolabeled MoAbs in Nude Mice Bearing Human Lymphoma Xenografts

Buchsbaum et al. [84] investigated the biodistribution of ^{131}I- and ^{111}In-labeled BA-1, and IgM MoAb reactive with human B-cell lymphomas, in groups of five normal Balb/c mice. The % ID/g ± SE for ^{131}I-labeled antibody 24 h after injection in blood, spleen, liver, and kidney were 2.10 ± 0.16, 0.48 ± 0.06, 0.31 ± 0.02, and 0.51 ± 0.05, respectively. The corresponding values for ^{111}In-labeled antibody were 2.81 ± 0.20, 1.21 ± 0.2, 1.71 ± 0.21, and 3.82 ± 0.85. The binding of the labeled antibody preparations to human leukemia cells in vitro were similar. Thus, it was concluded that the high uptake of ^{111}In-labeled BA-1 antibody in liver, kidney, and spleen would limit the usefulness of this radionuclide for tumor localization in mice bearing human B-cell lymphoma xenografts that are reactive with this antibody, and perhaps in patients with B-cell lymphoma as well. Halpern et al. [85] also observed rapid and marked uptake of ^{111}In-labeled IgM MoAbs by the liver. There is disagreement as to whether antibodies of the IgM subclass will be useful for localization and therapy of cancer [42,43,85-90]. Our unpublished studies show that ^{125}I-labeled fragments of BA-1 MoAb resemble intact BA-1 antibody, in that they localize poorly to human lymphoma xenografts in nude mice.

Buchsbaum et al. [91] reported localization and imaging results for ^{125}I- and ^{131}I-labeled anti-J5, anti-B1, and anti-T1 MoAbs in nude mice bearing human lymphoma xenografts. Namalwa (B-cell) and MOLT-4 (T-cell) tumors were grown subcutaneously in irradiated nude mice. The highest tissue concentration of ^{125}I-labeled anti-J5 in Namalwa-bearing mice was in blood and tumor. The tumor/blood ratio ranged from 0.7 to 1.2, with the highest ratio 4 days after injection. The area under the curve for tumor was two- to fivefold higher than the area under the curve for liver, kidney, skin, and muscle. The peak tissue levels of ^{125}I-labeled anti-B1 in Namalwa-bearing mice were again in blood and tumor and 6 days following injection more than fivefold greater activity was found in tumor than in normal tissues other than blood. The tumor/blood ratio was 1.2 and 0.7 at 4 and 6 days after injection. ^{125}I-labeled anti-J5 and anti-B1 showed minimal uptake in antigen-negative MOLT-4 tumors, and ^{125}I-labeled anti-T1 showed little uptake in Namalwa tumors. Scintigraphic images were obtained following the injection of ^{131}I-labeled anti-J5 and anti-B1 in nude mice bearing Namalwa tumors [91].

Epstein et al. [92] imaged nude mice bearing Raji Burkitt's lymphoma xenografts using ^{131}I-labeled Lym-1 antibody up to 7 days after injection, at which time tissues were dissected and biodistribution determined by well-type gamma counting. Tumors were well visualized 5-7 days after injection; a constant 8% of the injected dose in tumor combined with decreasing whole-body activity resulted in increasing tumor/whole-body ratios over time. The tissues showing the highest % ID/g at 7 days after injection were tumor, blood, and spleen, with values of 3.9, 1.4, and 0.7, respectively. Meares et al. [93,94] have been investigating the use of enzymatically metabolizable bifunctional chelates for conjugating ^{111}In to MoAbs with the hope of diminishing the normal tissue radioactivity relative to tumor. Circulating background activity due to enzymatic cleavage of the chelates into small radionuclide-containing fragments would be rapidly excreted by the kidneys. This would result in an increased tumor/whole-body ratio if cleavage in the tumor occurs more slowly than cleavage in the liver and other normal organs. This has obvious importance not only for tumor imaging, but also for radioimmunotherapy. The preliminary findings in mice bearing mouse lymphoma showed that enzyme-metabolizable bifunctional chelates had the advantage of lower blood and liver background, shorter initial biological half-life, and a higher tumor/blood ratio, but with a small decrease in absolute tumor uptake. Meares et al. [94] investigated several ^{111}In-labeled chelates conjugated to the Lym-1 MoAb in normal mice and found that a disulfide chelate cleared most rapidly from the liver and the whole body. Although the disulfide linker cleared from the body too rapidly to be useful, linkers related to the disulfide might be useful.

Preliminary studies by Buchsbaum and Uckun [27,95] have been conducted on the localization and imaging of radioiodine-labeled B43 MoAb. The B43 MoAb is a B-cell-specific antibody that recognizes an antigen of the CD19 cluster, that is expressed on neoplastic cells from 88% of B-lineage lymphoma patients, and is the most reliable B-cell-lineage surface marker. In a study of the relative tissue biodistribution of ^{125}I-labeled B43 IgG in antigen-positive Namalwa tumor-bearing mice [95], the highest concentration of B43 IgG was in blood and tumor at all time points after I.P. injection. Thus % ID/g in tumor was 6.6 and 3.0 at 2 and 4 days after injection, respectively. The tumor/blood ratio was about 0.5. At 7 days after I.P. injection, the % ID/g in Namalwa tumors was 3.7, while the % ID/g in negative control MOLT-4 tumors was 1.0. The tumor/blood ratio in Namalwa tumors was 0.6 at 7 days after injection, while the tumor/blood ratio in MOLT-4 tumors was 0.2. In nude mice bearing EB-3 tumors, the

% ID/g in tumor was 6.8 and 4.3 at 4 and 6 days after injection, respectively. The tumor/blood ratio was about 0.4. The highest concentrations of ^{125}I-labeled B43 $F(ab')_2$ in Namalwa tumor-bearing mice were in blood, tumor, and kidney. The % ID/g in tumor was 0.2 at 4 days after injection, considerably lower than B43 IgG. The tumor/blood ratio was 1.6, while the kidney/blood ratio was 1.2. Scintigraphic images of tumor localization were obtained with ^{131}I-labeled B43 IgG and $F(ab')_2$. In contrast, control anti-CD5 antibodies did not localize and image Namalwa tumors. These preliminary findings provide new evidence that B43/CD19 MoAb and its $F(ab')_2$ fragment can be effectively radiolabeled to yield highly specific probes for in vivo detection of $CD19^+$ malignant cells from B-lineage lymphoma patients. Our data do not indicate any obvious advantage for using the $F(ab')_2$ fragment of B43 MoAb rather than the intact IgG, particularly with regard to therapy trials.

C. Localization and Imaging of Radiolabeled MoAbs in Patients with Lymphoma

Localization and imaging studies have been performed in patients using radionuclides conjugated with T101 and Lym-1. For all the work described below, imaging was performed using conventional gamma cameras with or without image enhancement. Although the use of single-photon emission-computed tomography (SPECT) is a promising approach, only limited clinical data currently exist (for review see [96,97]).

A substantial amount of work has been performed in patients using the MoAb T101. As described above, this MoAb recognizes a glycoprotein of MW 65,000, found on normal circulating T cells and T-cell malignancies, but not on most B cells. In one trial, ^{111}In-labeled T101 was administered intravenously to a group of 11 patients with cutaneous T-cell lymphoma (CTCL) [98]. Escalating doses of unlabeled T101 up to 50 mg were given with 1 mg of labeled MoAb. The great majority of sites that were assessed as involved by clinical exam or CT scanning were imaged by T101. Forty-four sites positive by the T101 scan were clinically negative. Four of these sites were biopsied and all were histologically positive, suggesting subclinical disease. The liver and spleen showed significant nonspecific uptake. Although the dose of MoAb did not affect the number of lesions imaged, it did affect MoAb biodistribution. The data on biodistribution were consistent with the hypothesis that ^{111}In-labeled T101 binds to tumor cells that migrate to the involved regions. In addition, the MoAb appears to be internalized into target tumor cells. Evidence also suggests that, once internalized, ^{111}In is separated from the MoAb and binds tightly to intracellular sites [99]. If these factors turn out to be critical for imaging, it would suggest that MoAbs will be much less successful in solid tumors where neither migration nor internalization occurs to a significant degree in most cases. Localization has been obtained in another study using ^{131}I-labeled T101 [100], so it appears that the use of a radiometal is not an absolute requirement to produce positive tumor images.

Additional studies with T101 have investigated the importance of the radionuclide (^{131}I versus ^{111}In) and the route of administration. With regard to the radionuclide, a study was performed in which ^{131}I-labeled T101, prepared using the chloramine-T method, was administered to four patients with CTCL. Two of these patients subsequently received ^{111}In-labeled T101. Although the small number of patients precludes definitive conclusions, it appeared that imaging was superior in the patients who received ^{111}In-labeled T101. This difference was not secondary to degradation of MoAb immunoreactivity after labeling, but probably resulted from rapid de-iodination in the plasma and liver [101]. Other investigators have found that ^{131}I-labeled T101, also prepared by the chloramine-T method, was relatively stable, with approximately 90% of ^{131}I remaining associated with the MoAb [100]. The reason for these disparate results is not clear, but may simply reflect differences in the activity of serum and cellular dehalogenases among the small number of patients studied. Subcutaneous injection of ^{111}In-labeled T101 into the web spaces between the toes was attempted in 11 patients to increase the delivery of MoAb to the regional lymph nodes. Compared with intravenous administration, in which nodal uptake was minimal, the subcutaneous injected dose showed significant localization to nodal areas [102].

Lym-1, as described above, recognizes a cell membrane antigen with a MW of 31,000–35,000 that is, or is closely related to, the HLA-Dr antigen. Lym-1 has been radiolabeled with ^{123}I, using the chloramine-T method, and administered to patients with either breast cancer (four patients) or B-cell lymphoma (four patients). Only the patients with B-cell lymphoma showed MoAb localization [92].

D. Localization and Imaging of Radiolabeled MoAbs in Nude Mice Bearing Human Colon Cancer Xenografts

The tumor localization of radioiodinated $F(ab')_2$ and Fab fragments from MoAbs 35 and 115 reactive with the same or a closely related epitope of CEA was investigated by Buchegger et al. [103] in nude mice xenografted with C0112 human colon tumors. They found that the ratios of tumor concentration to normal organ antibody concentration increased dramatically with the use of fragments, although the highest absolute amount of radioactivity in tumor was obtained with intact antibody, and for longer periods of time than with fragments. A similar finding was reported by Douillard et al. [104] for intact 17-1A and its $F(ab')_2$ fragments. Munz et al. [105] reported an increase in tumor contrast when a mixture of two MoAb $F(ab')_2$ fragments against human gastrointestinal tract tumors labeled with ^{131}I were injected into nude mice bearing human colon carcinoma xenografts.

Preclinical studies using ^{125}I, ^{131}I, or ^{111}In-labeled B72.3 MoAb to target human colon carcinoma xenografts in athymic nude mice demonstrated that this MoAb specifically localized in tumor by external scintigraphic imaging and by direct analysis of tissue samples [106–108]. In paired-label studies with B72.3 and the new-generation higher-affinity MoAbs reactive with the TAG-72 antigen, the % ID/g of the new generation antibodies in LS174T human colon cancer xenografts was 1.4–5.4 times higher than that observed with the co-injected B72.3 IgG with higher tumor/normal tissue ratios [107]. With some of the new-generation antibodies, lower antibody concentrations were found in normal organs as compared to B72.3 IgG. One of the new-generation antibodies was labeled with ^{125}I and successfully used to image a LS174T tumor xenograft [107]. New approaches to ^{111}In chelation to MoAbs are being developed in an effort to produce more stable products and to lower the uptake in normal tissues, especially liver, spleen, and kidney [108]. The results of Esteban et al. [108] indicated a reduction of liver uptake by approximately 40% when the SCN-Bz-DTPA chelate and ^{111}In were conjugated to the B72.3 MoAb and administered to LS174T tumor-bearing mice as compared to the use of the bicyclic anhydride of DTPA.

In a direct comparison study, Buchsbaum et al. [109] studied the localization and imaging of four MoAbs (35, 115, 17-1A, and B72.3) reactive with human carcinoma surface antigens in athymic mice with LS174T, C0112, or SW948 colon carcinoma xenografts or negative control tumors. ^{125}I-labeled 115 showed the highest uptake (21.1 % ID/g at 4 days after injection) of any antibody in LS174T tumors. MoAbs 35 and B72.3 showed similar but lower levels of uptake at 4 days after injection in LS174T (12.2 and 9.9% ID/g, respectively) and C0112 tumors (5.8-9.7% ID/g), but B72.3 concentrated less in SW948 tumors (2.9% ID/g). 17-1A showed the highest degree of accumulation in SW948 tumor xenografts (8.4% ID/g at 4 days after injection). No specific uptake of the four anticarcinoma MoAbs was observed in human melanoma, lymphoma, or breast cancer xenografts. The specificity of the in vivo tumor localization of the four anticarcinoma MoAbs was confirmed by the low degree of accumulation of a control MoAb against influenza virus in LS174T tumors. Imaging studies with ^{131}I-labeled colorectal cancer MoAbs showed specific uptake and retention in LS174T tumors, with progressive clearance from the whole body. Differences in the immunohistochemical binding of the panel of four MoAbs against biopsies of colon adenocarcinoma were also found [109].

One possibility for increasing the amount of radiolabeled antibody uptake in tumors for radioimmunotherapy with less toxicity to the host would be to administer them regionally rather than systemically. Rowlinson et al. [110]

demonstrated a major advantage in tumor uptake when administering radiolabeled MoAbs I.P. for targeting intraperitoneal tumors. They compared the localization of ^{125}I- and ^{131}I-labeled AUA1 MoAb, raised by immunizing BALB/c mice with the colon carcinoma cell line LoVo, in athymic mice bearing intraperitoneal LoVo tumors. Tumor/tissue ratios 1 h after I.P. injection of radiolabeled antibody were about 50 times higher than after I.V. administration. The I.P./I.V. ratio fell to 4 by 8 h and to about 1 by 24 h. The % ID/g was almost 20%/g at 1-2 h after I.P. injection, then fell to approximately 10%/g by 24 h and remained at that level up to 5 days after administration. The tumor radioactivity 1 h after I.V. antibody administration was 1.4% ID/g, which was lower than normal organ levels of 5% ID/g, but increased to 10% ID/g at 24 h after administration. Following I.P. administration in mice, peak blood levels were reached by 2 h, whereas in clinical studies, peak blood levels were observed at 24-48 h [111]. This suggests that the regional pharmacological advantage of I.P.-administered radiolabeled MoAbs may be even higher in patients than in nude mice [110].

Because of its favorable photon energy (140 keV) properties for imaging, its low cost and general availability, ^{99m}Tc-labeled MoAbs have been investigated as possible imaging reagents. Zimmer et al. [112] compared the biodistribution and imaging of ^{99m}Tc- and ^{131}I-labeled MoAb fragments against CEA in nude mice bearing LS174T colon carcinoma xenografts. Although the immunoreactivities of the labeled antibodies were similar in vitro, there were marked biodistribution differences in vivo. Predominant kidney uptake and low tumor uptake were observed with ^{99m}Tc-labeled anti-CEA $F(ab')_2$ fragments. Higher tumor uptake and a higher blood concentration were obtained with ^{131}I-labeled $F(ab')_2$ fragments. Overall, ^{131}I-labeled fragments imaged tumors better than ^{99m}Tc-labeled fragments.

Besides antibody fragments, another approach to reducing blood pool radioactivity caused by circulating free, radiolabeled, intact MoAb has been the administration of a second antibody directed against the radiolabeled primary antibody. The resulting complexes are removed by the reticuloendothelial system [113-115]. This method may have applicability in radioimmunotherapy studies to reduce the whole-body dose and toxicity.

E. Localization and Imaging of Radiolabeled MoAbs in Patients with Colorectal Cancer

Early investigations demonstrated that a number of MoAbs against colorectal tumor-associated antigens, described above, could produce some evidence of tumor localization in patients [116,117]. Some of these early studies were conducted with MoAbs with only low affinity or cross-reactivity with granulocytes. Here we shall focus on some of the more recent work in order to suggest areas of potential development.

Several of these recent studies used MoAbs that react with CEA. One such MoAb is ZCE-025 (also called MoAb 35). This MoAb was chosen because of its high affinity for CEA (5.8×10^9 M^{-1}) and the lack of cross-reactivity with granulocytes [103]. Recent investigations have focused on the use of fragments to improve MoAb imaging, as well as the use of this MoAb in relatively large clinical studies, in order to better assess its ultimate applicability. In one study [118], 31 patients received a mixture of ^{123}I and ^{125}I prior to undergoing planned surgery, in order to correlate gamma images with radioactivity counted in the surgical specimen. The investigators incorporated several features into their clinical protocol to attempt to maximize sensitivity. First, they used ^{123}I, which, as described above, has almost ideal imaging properties. Second, they used $F(ab')_2$ and Fab fragments, to decrease nonspecific uptake by the reticuloendothelial system. In addition, they used SPECT instead of planar scanning. Under these conditions, they obtained 86% visualization (38 of 44 tumors) compared to either CT or surgical findings. The percentage of the injected dose that localized to tumor remained low (< 1%). These authors suggested that although the use of MoAb fragments may be optimal for imaging (i.e., low background), they may not be optimal for the delivery of the greatest dose of radiation for therapy. Similar results were obtained by Halpern and colleagues [119]. They found that 34% of known colorectal tumor lesions were imaged by ^{111}In-labeled ZCE-025 intact MoAb. The use of $F(ab')_2$ fragments increased the sensitivity to 85% [119]. In another study, a fixed quantity of DTPA-conjugated ^{111}In-labeled MoAb (1 mg) was administered with increasing amounts of cold MoAb to investigate the sensitivity of planar gamma imaging, compared to CT scans, in detecting metastatic deposits [120]. Up to 77% of lesions were imaged at some MoAb doses. This study identified several variables that were critical in obtaining optimal imaging, including quantity of cold MoAb and location of the metastasis. These results are qualitatively similar to those reported in a smaller study using an anti-CEA MoAb developed by Goldenberg and colleagues [121]. In none of the above studies did the presence of circulating CEA inhibit imaging.

Localization studies have also been performed in patients using MoAb 17-1A, which, as described above, recognizes a gastrointestinal cancer associated antigen with a MW of 30,000-41,000 that is confined to the cell surface [34,35]. This MoAb, or the $F(ab')_2$ fragment, was radiolabeled using the chloramine-T method and administered to 52 colorectal cancer patients with 63 tumor sites, as well as 15 patients with other types of cancer [122]. The tumor was imaged in 61% of the cases of those who received fragments, and 51% of those who received intact MoAb. The tumor/background ratio increased up to 10 days after administration. In three patients who underwent surgery, the ratio of counts measured in the tumor exceeded the surrounding normal tissue by a ratio of 2 to 7. As described above, however, tumor-associated radioactivity was in the nCi/g range, and therefore, accounted for only a small fraction of the injected dose. ^{125}I-labeled 17-1A has also been used to detect colorectal cancers intraoperatively using a small, hand-held gamma probe [123], although it is too early to determine the utility of this approach. Another MoAb developed by the same group of investigators is MoAb 19-9. Although this MoAb has been used chiefly to detect the antigen in the serum as a potential tumor marker (for review see [124]), radiolabeled 19-9 has also been employed in imaging studies [69]. Although scans performed after administration of ^{131}I-labeled 19-9 alone demonstrated the same sensitivity as 17-1A for the detection of colorectal metastasis (approximately 60%), the simultaneous administration of both of these radiolabeled MoAbs revealed 10 of 13 metastatic sites. This suggests that the use of such mixtures or "cocktails" may increase the sensitivity of immunodetection.

B72.3 is a murine IgG1 that has been extensively studied at the National Cancer Institute. It reacts with a high-MW glycoprotein called TAG (tumor-associated glycoprotein) 72 found on 85% of colon cancers, as well as 70% of breast cancers and 95% of ovarian cancers [125]. These investigators recently reported on the results of administration of increasing doses of ^{131}I-labeled B72.3 followed by gamma imaging and surgery [126]. Forty-six percent (16 of 35 patients) had positive scans. It should be noted that, unlike most studies, ^{99m}Tc background subtraction was not used. There was no relationship between the quantity of injected MoAb and scan sensitivity. The optimal time for imaging was approximately 1 week. The sensitivity of scanning was significantly less than the 70–85% result predicted from biopsy data. It is uncertain why the antigen does not bind radiolabeled MoAb in vivo.

In an attempt to improve sensitivity of scanning for patients with predominantly intraperitoneal tumors, Colcher and colleagues administered ^{131}I-labeled B72.3 intraperitoneally prior to gamma scanning and surgery [37]. Seven of 10 scans were positive in this setting. The ratio of counts in the tumor compared to normal tissue varied from 1.6 to 66, with 85% having ratios over 3. In a group of four patients who received ^{131}I-labeled B72.3 intraperitoneally and ^{125}I-labeled B72.3 intravenously, the intraperitoneal route localized tumor up to twice as well. These data suggest that in the presence of intraperitoneal disease, local delivery of MoAb may prove more beneficial for therapeutic purposes.

An IgG2b mouse MoAb, 791T/36, has also been used in patients to localize colorectal tumors. ^{131}I-labeled 791T/36 could be imaged by gamma scanning in 13/15 patient tumors prior to surgery [127]. Both immunohistochemistry and autoradiography were used to demonstrate that MoAb binding had occurred in the patient. The ratio of counts in the tumor compared to normal tissue was 2.5 to 1. Similar results were obtained with ^{111}In-labeled 791T/36 [128]. High sensitivity of detection of metastatic deposits in patients has also been obtained with external gamma scanning using intravenously administered ^{131}I-labeled 250-30.6, a murine IgG2b [129].

F. Future Developments

Positron emission tomography (PET) is a method for quantitative imaging of regional function and chemical reactions within various organs of the human body [130]. The dual annihilation photons, which are the final products of positron decay, allow coincidence detection. Systems with high resolution and high sensitivity are possible [131]. Very little work has been done on conjugating positron-emitting radionuclides to MoAbs. This is certainly a promising area that needs to be explored in the future. Another area that is likely to receive greater attention is SPECT imaging in clinical localization studies. Research will continue on the chemistry of labeling MoAbs with ^{123}I, ^{99m}Tc, ^{111}In, and other radiometals and positron emitters to take advantage of the greater resolution capability of SPECT and PET imaging devices, which has the promise of increasing the sensitivity of radioimmunodiagnosis by lowering the limit of tumor size detection below the current level of 1.0–1.5 cm. New approaches to ^{111}In and ^{99m}Tc labeling chemistry can be applied to the beta-emitting radionuclides ^{90}Y, ^{186}Re, and ^{188}Re for radioimmunotherapy, in an effort to reduce liver, kidney, and bone marrow toxicity. Another approach that was recently developed to minimize the high background in blood and liver utilizes the preadministration of an unlabeled bifunctional antibody (Figure 1) in which one binding site recognizes a tumor-associated antigen and the other a chelate. At the time of maximum tumor concentration of the pretargeted unlabeled antibody, the blood is cleared of excess circulating unlabeled antibody by administering a nonradioactive chase consisting of antigen bound to human transferrin. Then a radiolabeled chelate is administered, and imaging is done 1–3 h later [132]. With this approach, short-lived radionuclides could be used for imaging, and the method may be useful for radioimmunotherapy because there would be rapid elimination of radiolabeled hapten not bound to bifunctional antibody in the tumor. It is expected that further work will be done using this promising approach.

III. CURRENT STATUS OF RADIOIMMUNOTHERAPY: EXPERIMENTAL MODELS AND CLINICAL APPLICATIONS

A. Radioimmunotherapy of Lymphomas and Leukemias in Animal Models

MoAbs labeled with ^{131}I have been used for radioimmunotherapy of leukemia and lymphoma in experimental animal models [133-138]. A feature of many lymphomas is that they are radiosensitive, chemosensitive, and have a better supply of blood than solid tumors [139,140]. Buchsbaum et al. [133] have studied the therapeutic effects of ^{131}I- and ^{90}Y-labeled anti-J5 in athymic mice bearing Namalwa tumors. A group of 12 mice bearing S.C. Namalwa tumors of approximately 10 mm in maximal dimension were injected I.P. with a single 300-μCi dose of ^{131}I-labeled anti-J5. The second group of four mice were injected I.P. with a single dose of 175 μCi of ^{90}Y-labeled anti-J5. A third group of four mice were injected with 175 μCi of ^{90}Y-labeled anti-B1. The fourth group of mice did not receive any antibody. There was significant tumor growth inhibition in the mice injected with ^{131}I-labeled anti-J5, ^{90}Y-labeled anti-J5, or ^{90}Y-labeled anti-B1 as compared to the untreated controls. Tumors from the untreated control mice experienced rapid growth, and were approximately twice as large as the tumors from the mice injected with radiolabeled MoAbs at 25 days after injection. Six of the tumors in mice injected with radiolabeled B-cell antibodies regressed, and three were eradicated completely [133]. Adams et al. [137] successfully treated Raji human B-cell lymphoma xenografts in athymic nude mice with a single injection of 310-393 μCi ^{131}I-labeled Lym-1 (about 30 μg) monoclonal antibody specific for this tumor. In three mice treated with ^{131}I-labeled Lym-1, 4 of 6 tumors regressed completely and did not recur.

B. Radioimmunotherapy of Lymphomas in Patients

Although there is an evident dearth of data that demonstrate efficacy in animal models, a few clinical radioimmunotherapy trials have been initiated. Rosen and colleagues treated six patients with CTCL with up to 150 mCi of ^{131}I-labeled T101 in a single administration [100]. Two patients achieved a partial response lasting 2 months. All patients had the acute toxicities of fever and pruritis, while three of the five patients evaluable had the chronic, dose-limiting toxicities of myelosuppression and thrombocytopenia. Sufficient data were not presented to determine whether these responses could be due to MoAb alone or the total body irradiation alone. Indeed, in vivo binding of T101 could not be demonstrated. All patients developed a HAMA response, which was predicted based on a small previous experience with T101 administered to patients with CTCL [141]. Since it was suspected that the HAMA response would cause subsequent administrations of MoAb to lose efficacy, plasmapheresis was used prior to attempts to treat patients for a second time with T101 [142]. Although plasmapheresis did lower HAMA levels by 28–61%, these retreated patients still cleared the administered T101 MoAb faster after the second administration compared to the first injection. Of the three patients undergoing retreatment, one partial response lasting 1 month was observed.

In addition to T101, clinical trials have commenced with ^{131}I Lym-1 [25]. Three of the first five patients (three with lymphoma, one with chronic lymphocytic leukemia, and one with Richter's syndrome) showed evidence of tumor shrinkage after intravenous injection of 20-60 mCi of ^{131}I Lym-1. Patients received multiple injections. Only one patient developed a HAMA response. This was attributed to the fact the patients with B-cell malignancies manifest decreased ability to generate antibodies to foreign proteins of all types. It is surprising that responses were observed, considering the small dose actually received by the tumor, and the fact that not all tumors were even visualized. These findings may reflect the inherent radiosensitivity of these tumors and emphasize the difficulties in dosimetry that exist.

C. Radioimmunotherapy of Colon Cancer in Animal Models

Radioimmunotherapy studies in nude mice bearing human colon carcinoma xenografts have been conducted with ^{131}I-labeled murine MoAbs [50,143-147]. In our preliminary radioimmunotherapy study in nude mice bearing LS174T colon cancer xenografts, a single injection of 300 μCi ^{131}I-labeled murine 17-1A MoAb slowed the growth of established tumor xenografts as compared to control animals injected with unlabeled murine 17-1A MoAb or no antibody [50]. Similar tumor growth inhibition results were obtained by Esteban et al. [143] using 300 μCi ^{131}I-labeled B72.3 administered to nude mice bearing LS174T tumor xenografts. They found no visible toxic effect in the mice with 300 μCi of ^{131}I-labeled B72.3, although 500 μCi of ^{131}I-labeled B72.3 showed greater inhibition of tumor growth and produced toxic effects in the mice, including early death. Zalcberg et al. [144] found that 1 mCi of ^{131}I-labeled 250-30.6 MoAb directed against an antigen present on human colonic secretory epithelium inhibited the growth of COLO 205 colon carcinoma xenografts in nude mice, whereas a similar quantity of ^{131}I-labeled control MoAb or unlabeled specific antibody did not. Buchegger et al. [145] employed a mixture of ^{131}I-labeled anti-CEA MoAbs and F(ab$'$)$_2$ fragments for radioimmunotherapy of colon cancer xenografts in nude mice. After injection of a mixture of 200 μCi of intact MoAb and 400 μCi of F(ab$'$)$_2$ fragments, the T380 human colon carcinoma xenografts increased in size up to 6-10 days and then regressed for 4-12 weeks. A control group of mice injected with the same amount of ^{131}I-labeled normal mouse IgG1 and its F(ab$'$)$_2$ fragments showed retarded tumor progression for 1-3 weeks as compared to untreated controls, but no tumor regression was observed.

The use of F(ab$'$)$_2$ fragments may increase the tumor/normal tissue ratio because fragments are cleared from the circulation more rapidly than intact antibody [50,57,58,62,95,103-105]. However, the absolute amount of F(ab$'$)$_2$ fragments accumulating in tumors has been less than with intact antibody, and F(ab$'$)$_2$ fragments accumulate to a significant extent in the kidneys [50,95,103-105], which could result in renal toxicity.

In our preliminary radioimmunotherapy study with three injections of about 300 μCi of ^{131}I-labeled murine 17-1A MoAb administered on days 10, 17, and

29 after tumor cell injection to nude mice bearing LS174T tumor xenografts, prolonged tumor growth inhibition was observed as compared to unlabeled murine 17-1A MoAb treatment or no treatment at all [50]. The ^{131}I-labeled chimeric IgG1 17-1A MoAb, following a single I.P. injection of 300 μCi, produced tumor growth inhibition comparable to that of multiple doses of ^{131}I-labeled murine 17-1A.

One of the major factors that would interfere with the successful clinical use of radiolabeled MoAbs for tumor radioimmunotherapy would be the production of the HAMA response [148-153]. Fractionated therapy doses of radiolabeled murine MoAbs would be cleared from the blood more rapidly, resulting in lower tumor doses and possibly higher bone marrow and normal tissue doses. Order et al. [154] used radiolabeled polyclonal antibodies from several species of animals in order to avoid the problems of the host response to the administered radiolabeled antibody. Approaches that have been taken with MoAbs reactive with colon carcinoma to avoid the host response have been the production of human MoAbs [42,43] and human-mouse chimeric MoAbs [48,49,51]. LoBuglio et al. [155] reported that multiple injections of chimeric IgG1 17-1A MoAb in humans did not elicit a host response against the administered antibody. The ADCC activity of the chimeric IgG1 MoAb [51,156] and the fact that it did not elicit an immune response by patients makes it an interesting MoAb for unlabeled immunotherapy trials. The results of our study suggest that chimeric IgG1 17-1A MoAb might be promising in clinical localization and radioimmunotherapy studies [50].

D. Radioimmunotherapy of Colon Cancer in Patients

Very little information is available concerning the use of radiolabeled MoAb therapy for the treatment of colorectal cancer in patients. Several trials are in progress using ^{125}I-labeled 17-1A as part of a multimodality approach to metastatic colorectal cancer [157], but the contribution of the radiolabeled MoAb to the results is unclear.

E. Dosimetry of Radioimmunoconjugates

Although a detailed discussion of the dosimetry of radioimmunonconjugates is beyond the scope of this chapter, it is important to mention some of the critical concerns in this area. One important question is how to estimate the actual dose delivered to the tumor after administration of a radioimmunoconjugate. Most investigators use a Medical Internal Radiation Dose type of calculation. This calculation is based on the assumption that the sources are homogeneously distributed throughout the tumor. The experimental animal and patient studies clearly demonstrate that this is not the case. The dose delivered to volumes within a tumor that do not bind radiolabeled MoAb is probably critical to the ultimate success of therapy [158].

Another important issue concerns the choice of radionuclide for therapy. Wessels and Rogus [159] have listed many of the important variables that must be considered in choosing a radionuclide, including the physical and chemical properties, the biodistribution and biological half-life of the radiolabeled MoAb, and the binding of MoAb to its actual target compared to normal tissues. They point out that the high-specific-activity radiolabel needed to produce a dose rate similar to that used in traditional brachytherapy (0.3-1 Gy/h) necessitates the use of a radionuclide with a short half-life. This requirement conflicts, however, with the biodistribution data for intact MoAbs showing that the maximum tumor/normal tissue ratios, described above, are not obtained before 3-7 days.

Attempts have been made to estimate the actual dose delivered by ^{131}I- and ^{90}Y-labeled anti-ferritin in experimental therapy for hepatoma [160,161]. Estimates were in the range of 0.05 Gy/h for ^{131}I and 0.15 Gy/h for ^{90}Y. However, the hepatoma system is unique, in that up to 46% of the MoAb is deposited in the liver. Specific uptake in the tumor is considerably less. For tumors outside the liver, specific uptake accounts for less than 1% of the injected dose per gram of tumor when the MoAb is administered intravenously [57,162,163].

IV. FUTURE PERSPECTIVES ON THE EXPERIMENTAL AND CLINICAL USE OF RADIOIMMUNOCONJUGATES FOR CANCER THERAPY

Animal and human studies have demonstrated a lower and less prolonged tumor content of radioiodinated antibodies as compared to radiometal-labeled antibodies [59,61,70-75]. This finding has resulted in the investigation of several radiometals and chelation techniques to eventually be used in experimental radioimmunotherapy studies. Deshpande et al. [164] have investigated the radiolabeling of Lym-1 MoAb with ^{67}Cu and its biodistribution in nude mice bearing Raji human B-cell tumor xenografts. ^{67}Cu has a 61.5-h physical half-life, which is similar to the residence time of an antibody in tumor; and it releases abundant beta particles, which would be useful to treat solid tumors, and less abundant gamma emissions, which would be suitable for imaging and pharmacokinetic estimation of radiation dosimetry from pretherapy doses of ^{67}Cu-labeled antibody. The ^{67}Cu was stably chelated to the Lym-1 antibody with the macrocyclic chelating agent TETA [164-166]. Nude mice bearing S.C. Raji tumors of 0.1-0.4 g size were given an I.V. injection of 155-165 μCi Lym-1 antibody (20 μg), and biodistribution and imaging studies were performed. Tumor uptake increased from 8.2% ID/g at 1 day after injection to 14.7%/g at 3 days after injection, with a small decline to 13.2%/g at 5 days after injection.

Table 4 Selected Nuclides for Tumor Radioimmunotherapy

Nuclide	Half-life	References
Beta-particle emitters		
^{131}I	8.04 d	4,7,9,10,25,50,111,133–138,143–146,154
^{90}Y	64 h	147,158–160,167–174
^{67}Cu	62 h	164
^{186}Re	91 h	175
^{188}Re	17 h	176
^{99}Au	74 h	177
^{105}Rh	36 h	178
Alpha-particle emitters		
^{212}Bi	1 h	179–181
^{211}At	7.2h	182–186
Auger-electron emitters		
^{125}I	60.2 d	187–190
^{77}Br	57 h	191
^{80m}Br	4.4 h	192
^{123}I	13 h	
Fission nuclide		
^{10}B		193,194

The normal tissues with the highest concentrations of ^{67}Cu-labeled antibody were blood, liver, and spleen. The blood % ID/g values were 13.8, 8.7, and 5.8 at 1, 3, and 5 days after injection, respectively. The corresponding values for liver were 10.9, 9.4, and 4.8, while the respective concentrations in spleen were 9.1, 8.5, and 3.2% ID/g. It is important to note that the radioactivity cleared from liver and other normal tissues, in contrast to the findings with other radiometal-chelated antibodies [59,61,71,75,101,119]. These promising biodistribution results, along with the fact that the tumor could be imaged, suggest that future radioimmunotherapy studies will be conducted with ^{67}Cu-labeled antibodies. Deshpande et al. [164] plan to proceed with imaging of ^{67}Cu-labeled Lym-1 to determine dosimetry in patients. If this is successful, they plan to conduct a clinical radioimmunotherapy trial.

There are a number of potentially useful radionuclides that could be attached to antibodies and tested for therapeutic efficacy. These are listed in Table 4. The factors that influence the choice of a radionuclide for radioimmunotherapy have been reviewed elsewhere [158,159]. Most work to date has been with ^{131}I-labeled MoAbs as summarized previously. However, the other radionuclides in Table 4 are currently being investigated. A potential advantage of some of these radionuclides would be a higher tumor/whole-body dose, resulting in less toxicity to normal tissue, particularly bone marrow. In addition, other cytotoxic agents such as toxins and chemotherapeutic drugs coupled to MoAbs have been widely investigated as potential immunoconjugates for tumor therapy [195–198]. Buchsbaum and Vallera [171,199] have been investigating the use of radiolabeled immunotoxins (radioimmunotoxins) for tumor therapy. Toxins and radionuclides are cytotoxic by different mechanisms when conjugated to antibody. The killing of cells by radiolabeled MoAbs likely results from the direct or indirect effects of radiation on DNA or the plasma membrane, culminating in interphase or reproductive death [171]. The cytotoxic activity of immunotoxins occurs after internalization of toxin into the cell. Once inside the cytoplasm, the toxin enzymatically inactivates cellular ribosomes, which terminates subsequent protein synthesis and induces cell death [171,198]. We have demonstrated that T101/CD5 ricin immunotoxins and ricin A-chain immunotoxins can be labeled with ^{125}I, ^{131}I, or ^{90}Y and retain antibody binding specificity in vitro, inhibit protein synthesis of CEM tumor cells in vitro, inhibit clonogenic replication of CEM tumor cells in vitro, and localize and inhibit CEM tumor growth in athymic nude mice [171,199]. The potential advantages of these dual-labeled reagents are that many toxins are potent catalytic inhibitors of protein synthesis, so if they are bound to a tumor cell by the MoAb portion of the immunoconjugate and internalized, one molecule of toxin reaching the cytosol is capable of killing the cell. The advantage of the radionuclide would be that a beta emitter would be capable of killing neighboring tumor cells that did not bind the immunoconjugate. Because of the heterogeneity of binding of MoAbs in solid tumors, such dual-labeled immunoconjugates may be more effective than singly labeled immunoconjugates. Another approach that Buchsbaum and Sinkule [200,201] developed was dual-labeled immunoconjugates consisting of a drug and radionuclide attached to the same MoAb. In preliminary studies, we demonstrated that the immunoconjugate consisting of doxorubicin conjugated to the 115 anti-CEA MoAb could be radiolabeled with ^{125}I or ^{131}I and shown to localize in LS174T colon cancer xenografts to an equivalent extent as radioiodine-labeled 115 MoAb [200]. It is conceivable that such chemo-radio-immunoconjugates may be useful for in vivo radioimmunotherapy.

Another potential area for future investigation is the combined use of radiosensitizers and radiolabeled MoAbs to enhance the lethality of ionizing radiation, especially in the radiation therapy of hypoxic tumors. Little work has been reported in this area. The low dose rate produced by radionuclides is typically less than 10 rad/h. This low rate produces markedly less killing per rad than high-dose-rate radiation (see review [202]). Therefore, it may be necessary to increase the tumor cytotoxic effect of radioimmunoconjugates by the use of radiosensitizers. Borlinghaus et al. [203] conjugated the radiosensitizer misonidazole to the 19-9 MoAb reactive with a colorectal carcinoma tumor-associated antigen, with retention of in vitro tumor cell-binding activity and stability of the product. This is a potentially important approach toward radiosensitization because site-specific delivery of the radiosensitizer may produce less normal tissue toxicity than systemic administration of nonconjugated radiosensitizer. Morstyn et al. [204] found that the cell kill produced by ^{131}I-labeled polyclonal antibody in vitro was enhanced in the presence of the radiosensitizer bromodeoxyuridine.

Other areas of active research include methods to decrease the HAMA response, genetic and chemical changes of MoAbs to increase their potential for radiolabeling and tumor localization [205], the use of cocktails of MoAbs [105, 122], and the use of multimodality strategies including the use of radiolabeled MoAbs in combination with chemotherapy, lymphokines, or other biological-response modifiers [206].

V. SUMMARY

Since 1980, there has been intense research in the area of radiolabeled immunoconjugates for the localization and therapy of cancer in both experimental animal models and in humans. This is based on the potential of MoAbs reactive with tumor-associated antigens as specific carriers of radionuclides, toxins, and chemotherapeutic drugs. Even though there has been much progress, a great deal of work and innovation will be required before the use of radiolabeled immunoconjugates becomes an established method for clinical tumor detection and therapy. The greatest problem is the relatively low tumor uptake of radiolabeled MoAbs in humans. This results in relatively high background levels of radioactivity in normal tissues and insufficient radioactivity in the tumor for the detection of small lesions and for therapeutic efficacy. In the future, one can expect the development of new MoAbs with greater specificity for particular tumor types, improvements in radiolabeling procedures, and the use of more effective nuclides than ^{131}I. Biological, physiological, pathological, and dosimetric considerations need to be addressed in detail. With this additional information, it is hoped that the full potential for radiolabeled immunoconjugates will be realized.

VI. ACKNOWLEDGMENTS

This work was supported in part by National Institutes of Health Grants R01 CA43368, R01 CA44173, and P01 CA42768, and by a Special Grant awarded by the Medical Research Foundation, Inc., Atlanta, Georgia. The authors wish to acknowledge the helpful editorial comments of Valeri Terry and would like to thank Dorothy Meade for typing the manuscript.

VII. REFERENCES

1. D. Pressman and L. Korngold, The *in vivo* localization of anti-Wagner-osteogenic sarcoma antibodies. *Cancer 6*:619–623 (1953).
2. L. Korngold and D. Pressman, The localization of antilymphosarcoma antibodies in the Murphy lymphosarcoma of the rat. *Cancer Res. 14*:96–99 (1954).
3. W. F. Bale, I. L. Spar, R. L. Goodland, and D. E. Wolfe, *In vivo* and *in vitro* studies of labeled antibodies against rat kidney and Walker carcinoma. *Proc. Soc. Exp. Biol. Med. 89*:564–568 (1955).
4. W. F. Bale and I. L. Spar, Studies directed toward the use of antibodies as carriers of radioactivity for therapy. *Adv. Biol. Med. Physics 5*:285–356 (1957).
5. E. D. Day, J. A. Planinsek, and D. Pressman, Localization *in vivo* of radioiodinated anti-rat-fibrin antibodies and radioiodinated rat fibrinogen in the Murphy rat lymphosarcoma and in other transplantable rat tumors. *J. Natl. Cancer Inst. 22*:413–426 (1959).
6. I. L. Spar, R. L. Goodland, and W. F. Bale, Localization of I^{131} labeled antibody to rat fibrin in transplantable rat lymphosarcoma. *Proc. Soc. Exp. Biol. Med. 100*:259–262 (1959).
7. W. F. Bale, I. L. Spar, and R. L. Goodland, Experimental radiation therapy of tumors with I^{131} carrying antibodies to fibrin. *Cancer Res. 20*:1488–1494 (1960).
8. W. C. Dewey, W. F. Bale, R. G. Rose, and D. Marrack, Localization of antifibrin antibodies in human tumors. *Acta Unio. Int. Contra Cancer 19*:185–196 (1963).
9. R. J. McCardle, P. V. Harper, I. L. Spar, W. F. Bale, G. Andros, and F. Jiminez, Studies with iodine-131-labeled antibody to human fibrinogen for diagnosis and therapy of tumors. *J. Nucl. Med. 7*:837–847 (1966).
10. I. L. Spar, W. F. Bale, D. Marrack, W. C. Dewey, R. J. McCardle, and P. V. Harper, ^{131}I-labeled antibodies to human fibrinogen. Diagnostic studies and therapeutic trials. *Cancer 20*:865–870 (1967).
11. D. M. Goldenberg, D. F. Preston, F. J. Primus, and H. J. Hansen, Photoscan localization of GW-39 tumors in hamsters using radiolabeled anticarcinoembryonic antigen immunoglobulin G. *Cancer Res. 34*:1–9 (1974).
12. F. J. Primus, R. MacDonald, D. M. Goldenberg, and H. J. Hansen, Localization of GW-39 human tumors in hamsters by affinity-purified antibody to carcinoembryonic antigen. *Cancer Res. 37*:1544–1547 (1977).
13. J.-P. Mach, S. Carrel, C. Merenda, B. Sordat, and J.-C. Cerottini, *In vivo* localisation of radiolabelled antibodies to carcinoembryonic antigens in human colon carcinoma grafted into nude mice. *Nature* (London) *248*:704–706 (1974).

14. D. M. Goldenberg, F. DeLand, E. Kim, S. Bennett, F. J. Primus, J. R. van Nagell, N. Estes, P. DeSimone, and P. Rayburn, Use of radiolabeled antibodies to carcinoembryonic antigen for the detection and localization of diverse cancers by photoscanning. *N. Engl. J. Med. 298*:1384–1388 (1978).
15. J.-P. Mach, S. Carrel, M. Forni, J. Ritschard, A. Donath, and P. Alberto, Tumor localization of radiolabeled antibodies against carcinoembryonic antigen in patients with carcinoma. *N. Engl. J. Med. 303*:5–10 (1980).
16. G. Kohler and C. Milstein, Continuous cultures of fused cells secreting antibody of predefined specificity. *Nature* (London) *256*:495–497 (1975).
17. C. S. Abramson, J. H. Kersey, and T. W. LeBien, A monoclonal antibody (BA-1) reactive with cells of human B lymphocyte lineage. *J. Immunol. 126*:83–88 (1981).
18. J. Jansen, R. C. Ash, E. D. Zanjani, T. W. LeBien, and J. H. Kersey, Monoclonal antibody BA-1 does not bind to hematopoietic precursor cells. *Blood 59*:1029–1035 (1982).
19. K. J. Gajl-Peczalska, C. D. Bloomfield, G. Frizzera, J. H. Kersey, and T. W. LeBien, Diversity of phenotypes of non-Hodgkin's malignant lymphoma. In *B and T Cell Tumors: Biological and Clinical Aspects* (E. Vitetta, ed.), Academic Press, New York, 1982, pp. 63–67.
20. T. W. LeBien, D. R. Boue, J. G. Bradley, and J. H. Kersey, Antibody affinity may influence antigenic modulation of the common acute lymphoblastic leukemia antigen *in vitro*. *J. Immunol. 129*:2287–2292 (1982).
21. L. M. Nadler, P. Stashenko, R. Hardy, A. van Agthoven, C. Terhorst, and S. F. Schlossman, Characterization of a human B cell specific antigen (B2) distinct from B1. *J. Immunol. 126*:1941–1947 (1981).
22. P. Stashenko, L. M. Nadler, R. Hardy, and S. F. Schlossman, Characterization of a human B lymphocyte-specific antigen. *J. Immunol. 125*:1678–1685 (1980).
23. J. Ritz, J. M. Pesando, J. Notis-McConarty, H. Lazarus, and S. F. Schlossman, A monoclonal antibody to human acute lymphoblastic leukaemia antigen. *Nature* (London) *283*:583–585 (1980).
24. A. L. Epstein, R. J. Marder, J. N. Winter, E. Stathopoulos, F.-M. Chen, J. W. Parker, and C. R. Taylor, Two new monoclonal antibodies, Lym-1 and Lym-2, reactive with human B-lymphocytes and derived tumors, with immunodiagnostic and immunotherapeutic potential. *Cancer Res. 47*:830–840 (1987).
25. S. J. DeNardo, G. L. DeNardo, L. F. O'Grady, N. B. Levy, S. L. Mills, D. J. Macey, J. P. McGahan, C. H. Miller, and A. L. Epstein, Pilot studies of radioimmunotherapy of B cell lymphoma and leukemia using I-131 Lym-1 monoclonal antibody. *Antibody, Immunoconjugates, and Radiopharm. 1*: 17–33 (1988).
26. M. P. Link, J. Bindl, T. C. Meeker, C. Carswell, C. A. Doss, R. A. Warnke, and R. Levy, A unique antigen on mature B cells defined by a monoclonal antibody. *J. Immunol. 137*:3013–3018 (1986).

27. F. M. Uckun, W. Jaszcz, J. L. Ambrus, A. S. Fauci, K. Gajl-Peczalska, C. W. Song, M. R. Wick, D. E. Myers, K. Waddick, and J. A. Ledbetter, Detailed studies on expression and function of CD19 surface determinant by using B43 monoclonal antibody and the clinical potential of anti-CD19 immunotoxins. *Blood 71*:13–29 (1988).
28. E. L. Reinherz, P. C. Kung, G. Goldstein, and S. F. Schlossman, A monoclonal antibody with selective reactivity with functionally mature thymocytes and all peripheral human T cells. *J. Immunol. 123*:1312–1317 (1979).
29. I. Royston, J. A. Majda, S. M. Baird, B. L. Meserve, and J. C. Griffiths, Human T cell antigens defined by monoclonal antibodies: The 65,000-dalton antigen of T cells (T65) is also found on chronic lymphocytic leukemia cells bearing surface immunoglobulin. *J. Immunol. 125*:725–731 (1980).
30. C. M. Boyer, Y. Lidor, C. Lottich, and R. C. Bast, Jr., Antigenic cell surface markers in human solid tumors. *Antibody, Immunoconjugates, and Radiopharm. 1*:105–162 (1988).
31. F. Buchegger, M. Schreyer, S. Carrel, and J.-P. Mach, Monoclonal antibodies identify a CEA crossreacting antigen of 95 kD (NCA-95) distinct in antigenicity and tissue distribution from the previously described NCA of 55 kD. *Int. J. Cancer 33*:643–649 (1984).
32. C. M. Haskell, F. Buchegger, M. Schreyer, S. Carrel, and J.-P. Mach, Monoclonal antibodies to carcinoembryonic antigen: Ionic strength as a factor in the selection of antibodies for immunoscintigraphy. *Cancer Res. 43*: 3857–3864 (1983).
33. F. J. Primus, K. D. Newell, A. Blue, and D. Goldenberg, Immunological heterogeneity of carcinoembryonic antigen: Antigenic determinants on carcinoembryonic antigen distinguished by monoclonal antibodies. *Cancer Res. 43*:686–692 (1983).
34. M. Herlyn, Z. Steplewski, D. Herlyn, and H. Koprowski, Colorectal carcinoma-specific antigen: detection by means of monoclonal antibodies. *Proc. Natl. Acad. Sci. USA 76*:1438–1442 (1979).
35. H. Koprowski, Z. Steplewski, K. Mitchell, M. Herlyn, D. Herlyn, and P. Fuhrer, Colorectal carcinoma antigens detected by hybridoma antibodies. *Somatic Cell Genet. 5*:957–971 (1979).
36. D. Colcher, J. M. Esteban, J. A. Carrasquillo, P. Sugarbaker, J. C. Reynolds, G. Bryant, S. M. Larson, and J. Schlom, Quantitative analyses of selective radiolabeled monoclonal antibody localization in metastatic lesions of colorectal cancer patients. *Cancer Res. 47*:1185–1189 (1987).
37. D. Colcher, J. Esteban, J. A. Carrasquillo, P. Sugarbaker, J. C. Reynolds, G. Bryant, S. M. Larson, and J. Schlom, Complementation of intracavitary and intravenous administration of a monoclonal antibody (B72.3) in patients with carcinoma. *Cancer Res. 47*:4218–4224 (1987).
38. M. J. Embleton, B. Gunn, V. S. Byers, and R. W. Baldwin, Antitumour reactions of monoclonal antibody against a human osteogenic sarcoma cell line. *Br. J. Cancer 43*:582–587 (1981).

39. P. A. Farrands, A. C. Perkins, L. Sulley, J. S. Hopkins, M. V. Pimm, R. W. Baldwin, and J. D. Hardcastle, Localisation of human osteosarcoma by anti-tumour monoclonal antibody 791T/36. *J. Bone Joint Surg. 65*:638–640 (1983).
40. P. A. Farrands, A. C. Perkins, M. V. Pimm, J. D. Hardy, M. J. Embleton, R. W. Baldwin, and J. D. Hardcastle, Radioimmunodetection of human colorectal cancers by an anti-tumour monoclonal antibody. *Lancet 2*:397–400 (1982).
41. M. V. Pimm, A. C. Perkins, N. C. Armitage, and R. W. Baldwin, The characteristics of blood-borne radiolabels and the effect of anti-mouse IgG antibodies on localization of radiolabeled monoclonal antibody in cancer patients. *J. Nucl. Med. 26*:1011–1023 (1985).
42. R. P. McCabe, L. C. Peters, M. V. Haspel, N. Pomato, J. A. Carrasquillo, and M. G. Hanna, Jr., Preclinical studies on the pharmacokinetic properties of human monoclonal antibodies to colorectal cancer and their use for detection of tumors. *Cancer Res. 48*:4348–4353 (1988).
43. R. P. McCabe, L. C. Peters, M. V. Haspel, N. Pomato, and J. A. Carrasquillo, Development and characterization of human monoclonal antibodies and their application in the radioimmunodetection of colon carcinoma. In *Radiolabeled Monoclonal Antibodies for Imaging and Therapy* (S. C. Srivastava, ed.), Plenum Press, New York, 1988, pp. 75–94.
44. J. A. Carrasquillo, R. G. Steis, R. McCabe, J. C. Reynolds, M. Bookman, S. Del Vecchio, J. W. Smith, P. Perentesis, V. Dailey, E. Paris, M. Rotman, M. G. Hanna, Jr., M. V. Haspel, D. Longo, and S. M. Larson, Imaging of colon cancer with I-131 16-88 human monoclonal antibody. *J. Nucl. Med. 29*:833 (1988).
45. H. F. Sears, D. Herlyn, Z. Steplewski, and H. Koprowski, Phase II clinical trial of a murine monoclonal antibody cytotoxic for gastrointestinal adenocarcinoma. *Cancer Res. 45*:5910–5913 (1985).
46. U. C. Traub, R. L. DeJager, F. J. Primus, M. Losman, and D. M. Goldenberg, Antiidiotype antibodies in cancer patients receiving monoclonal antibody to carcinoembryonic antigen. *Cancer Res. 48*:4002–4006 (1988).
47. E. H. Ford, R. E. Lee, M. Benamor, R. M. Sharkey, J. A. Horowitz, T. C. Hall, and D. M. Goldenberg, Effect of human anti-mouse antibody (HAMA) on monoclonal antibody pharmacokinetics and imaging. *Proc. 2nd Conf. on Radioimmunodetection and Radioimmunotherapy of Cancer*. Princeton, N.J., September 8–10, 1988, p. 64.
48. S. L. Morrison, L. Wims, S. Wallick, L. Tan, and V. T. Oi, Genetically engineered antibody molecules and their application. *Ann. N.Y. Acad. Sci. 507*:187–196 (1987).
49. L. K. Sun, P. Curtis, E. Rakowicz-Szulczynska, J. Ghrayeb, N. Chang, S. L. Morrison, and H. Koprowski, Chimeric antibody with human constant regions and mouse variable regions directed against carcinoma-associated antigen 17-1A. *Proc. Natl. Acad. Sci. USA 84*:214–218 (1987).
50. D. Buchsbaum, P. Brubaker, D. Hanna, A. Glatfelter, V. Terry, D. M. Guilbault, and Z. Steplewski, Comparative binding and preclinical localization

and therapy studies with radiolabeled human chimeric and murine 17-1A monoclonal antibodies. *Cancer Res. Suppl.* 1989, in press.
51. Z. Steplewski, L. K. Sun, C. W. Shearman, J. Ghrayeb, P. Daddona, and H. Koprowski, Biological activity of human-mouse IgG1, IgG2, IgG3, and IgG4 chimeric monoclonal antibodies with antitumor specificity. *Proc. Natl. Acad. Sci. USA 85*:4852–4856 (1988).
52. D. Herlyn, M. Herlyn, Z. Steplewski, and H. Koprowski, Monoclonal anti-human tumor antibodies of six isotypes in cytotoxic reactions with human and murine effector cells. *Cell. Immunol. 92*:105–114 (1985).
53. R. Muraro, M. Kuroki, D. Wunderlich, D. J. Poole, D. Colcher, A. Thor, J. W. Greiner, J. F. Simpson, A. Molinolo, P. Noguchi, and J. Schlom, Generation and characterization of B72.3 second generation monoclonal antibodies reactive with the tumor-associated glycoprotein 72 antigen. *Cancer Res. 48*:4588–4596 (1988).
54. J.-P. Mach, Current status and future directions of radiolabeled monoclonal antibodies. Society of Nuclear Medicine Annual Meeting, San Francisco, Calif., June 13, 1988.
55. H. Sands, Radioimmunoconjugates: An overview of problems and promises. *Antibody, Immunoconjugates, and Radiopharm. 1*:213–226 (1988).
56. R. K. Jain, Determinants of tumor blood flow: A review. *Cancer Res. 48*: 2641–2658 (1988).
57. D. Goodwin, Pharmacokinetics and antibodies. *J. Nucl. Med. 28*:1358–1362 (1987).
58. D. G. Covell, J. Barbet, O. D. Holton, C. D. V. Black, R. J. Parker, and J. N. Weinstein, Pharmacokinetics of monoclonal immunoglobulin G_1, $F(ab')_2$ and Fab′ in mice. *Cancer Res. 46*:3969–3978 (1986).
59. S. E. Halpern and R. O. Dillman, Problems associated with radioimmunodetection and possibilities for future solutions. *J. Biol. Response. Mod. 6*: 235–262 (1987).
60. D. M. Goldenberg, Current status of cancer imaging with radiolabeled antibodies. *J. Cancer Res. Clin. Oncol. 113*:203–208 (1987).
61. S. E. Halpern, The advantages and limits of indium-111 labeling of antibodies: Experimental studies and clinical applications. *Nucl. Med. Biol. 13*: 195–201 (1986).
62. J.-P. Mach, F. Buchegger, J.-Ph. Grob, V. von Fliedner, S. Carrel, L. Barrelet, A. Bischof-Delaloye, and B. Delaloye, Improvement of colon carcinoma imaging: From polyclonal anti-CEA antibodies and static photoscanning to monoclonal Fab fragments and ECT. In *Monoclonal Antibodies for Cancer Detection and Therapy* (R. W. Baldwin and V. S. Byers, eds.), Academic Press, London, 1985, pp. 53–64.
63. A. H. Gobuty, E. E. Kim, and R. E. Weiner, Radiolabeled monoclonal antibodies: Radiochemical pharmacokinetic and clinical challenges. *J. Nucl. Med. 26*:546–547 (1985).
64. A. R. Bradwell, D. S. Fairweather, P. W. Dykes, A. Keeling, A. Vaughan, and J. Taylor, Limiting factors in the localization of tumours with radiolabelled antibodies. *Immunol. Today 6*:163–170 (1985).

65. W. F. Bale, M. A. Contreras, and E. D. Grady, Factors influencing localization of labeled antibodies in tumors. *Cancer Res. 40*:2965–2972 (1980).
66. W. A. Pettit, S. J. Bennett, F. DeLand, and D. M. Goldenberg, Iodination and acceptance testing of antibodies. In *Tumor Imaging: The Radiochemical Detection of Cancer* (S. W. Burchiel, B. A. Rhodes, and B. E. Friedmand, eds.), Masson Publishing USA, New York, 1982, pp. 99–109.
67. G. B. Saha, Radioiodination of antibodies for tumor imaging. In *Radioimmunoimaging and Radioimmunotherapy* (S. W. Burchiel and B. A. Rhodes, eds.), Elsevier, New York, 1983, pp. 171–184.
68. M. A. Contreras, W. F. Bale, and I. L. Spar, Iodine monochloride (ICl) iodination techniques. *Methods Enzymol. 92*:277–292 (1983).
69. J.-F. Chatal, J.-C. Saccavini, P. Fumoleau, J.-Y. Douillard, C. Curtet, M. Kremer, B. LeMevel, and H. Koprowski, Immunoscintigraphy of colon carcinoma. *J. Nucl. Med. 25*:307–314 (1984).
70. R. Rainsbury and J. Westwood, Tumour localisation with monoclonal antibody radioactively labelled with metal chelate rather than iodine. *Lancet 2*:1347–1348 (1982).
71. P. Stern, P. Hagan, S. Halpern, A. Chen, G. David, T. Adams, W. Desmond, K. Brautigam, and I. Royston, The effect of the radiolabel on the kinetics of monoclonal anti-CEA in a nude mouse-human colon tumor model. In *Hybridomas in Cancer Diagnosis and Treatment* (M. S. Mitchell and H. F. Oettgen, eds.), Raven Press, New York, 1982, p. 245–253.
72. D. J. Hnatowich, W. W. Layne, R. L. Childs, D. Lanteigne, and M. A. Davis, Radioactive labeling of antibody: a simple and efficient method. *Science 220*:613–615 (1983).
73. D. A. Scheinberg and M. Strand, Kinetic and catabolic considerations of monoclonal antibody targeting in erythroleukemic mice. *Cancer Res. 43*: 265–272 (1983).
74. M. I. Bernhard, K. M. Hwang, K. A. Foon, A. M. Keenan, R. M. Kessler, J. M. Frincke, D. J. Tallam, M. G. Hanna, Jr., L. Peters, and R. K. Oldham, Localization of ^{111}In- and ^{125}I-labeled monoclonal antibody in guinea pigs bearing line 10 hepatocarcinoma tumors. *Cancer Res. 43*:4429–4433 (1983).
75. S. E. Halpern, P. L. Hagan, P. R. Garver, J. A. Koziol, A. W. N. Chen, J. M. Frincke, R. M. Bartholomew, G. S. David, and T. H. Adams, Stability, characterization, and kinetics of ^{111}In-labeled monoclonal antitumor antibodies in normal animals and nude mouse-human tumor models. *Cancer Res. 43*:5347–5355 (1983).
76. J. G. McAfee, G. Gagne, H. Atkins, P. T. Kirchner, R. C. Reba, M. D. Blaufox, and E. M. Smith, Biological distribution and excretion of DTPA labeled with Tc-99m and In-111. *J. Nucl. Med. 20*:1273–1278 (1979).
77. D. A. Scheinberg, M. Strand, and O. A. Gansow, Tumor imaging with radioactive metal chelates conjugated to monoclonal antibodies. *Science 215*: 1511–1513 (1982).

78. S. M. Larson and J. A. Carrasquillo, Advantages of radioiodine over radioindium labeled monoclonal antibodies for imaging solid tumors, *Nucl. Med. Biol. 15*:231–233 (1988).
79. S. M. Larson, A tentative biological model for the localization of radiolabelled antibody in tumor: The importance of immunoreactivity. *Nucl. Med. Biol. 13*:393–399 (1986).
80. C. C. Badger, K. A. Krohn, and I. D. Bernstein, *In vitro* measurement of avidity of radioiodinated antibodies. *Nucl. Med. Biol. 14*:605–610 (1987).
81. M. V. Pimm and R. W. Baldwin, Comparative tumour localization properties of radiolabelled monoclonal antibody preparations of defined immunoreactivities. *Eur. J. Nucl. Med. 13*:348–352 (1987).
82. P. L. Beaumier, K. A. Krohn, J. A. Carrasquillo, J. Eary, I. Hellstrom, K. E. Hellstrom, W. B. Nelp, and S. M. Larson, Melanoma localization in nude mice with monoclonal Fab against p97. *J. Nucl. Med. 26*:1172–1179 (1985).
83. H. Sakahara, K. Endo, T. Nakashima, M. Koizumi, H. Ohta, K. Torizuka, T. Furukawa, Y. Ohmomo, A. Yokoyama, K.-i. Okada, O. Yoshida, and S. Nishi, Effect of DTPA conjugation on the antigen binding activity and biodistribution of monoclonal antibodies against α-fetoprotein. *J. Nucl. Med. 26*:750–755 (1985).
84. D. Buchsbaum, B. Randall, D. Hanna, R. Chandler, M. Loken, and E. Johnson, Comparison of the distribution and binding of monoclonal antibodies labeled with 131-iodine or 111-indium. *Eur. J. Nucl. Med. 10*:398–402 (1985).
85. S. E. Halpern, P. L. Hagan, A. Chen, C. R. Birdwell, R. M. Bartholomew, K. G. Burnett, G. S. David, K. Poggenburg, B. Merchant, and D. J. Carlo, Distribution of radiolabeled human and mouse monoclonal IgM antibodies in murine models. *J. Nucl. Med. 29*:1688–1696 (1988).
86. K. P. Ryan, R. O. Dillman, S. J. DeNardo, G. L. DeNardo, J. Beauregard, P. L. Hagan, D. G. Amox, M. L. Clutter, K. G. Burnett, C. M. Rulot, R. E. Sobol, I. Abramson, R. K. Bartholomew, J. M. Frincke, C. R. Birdwell, D. J. Carlo, L. F. O'Grady, and S. E. Halpern, Breast cancer imaging with In-111 human IgM monoclonal antibodies: Preliminary studies. *Radiology 167*:71–75 (1988).
87. G. Levine, B. Ballou, J. Reiland, D. Solter, L. Gumerman, and T. Hakala, Localization of I-131-labeled tumor-specific monoclonal antibody in the tumor-bearing BALB/c mouse. *J. Nucl. Med. 21*:570–573 (1980).
88. B. Ballou, J. Reiland, G. Levine, B. Knowles, and T. Hakala, Tumor location using $F(ab')_{2\mu}$ from a monoclonal IgM antibody: Pharmacokinetics. *J. Nucl. Med. 26*:283–292 (1985).
89. A. M. Zimmer, S. T. Rosen, S. M. Spies, M. R. Polovina, J. D. Minna, W. C. Spies, and E. A. Silverstein, Radioimmunoimaging of human small cell lung carcinoma with I-131 tumor specific monoclonal antibody. *Hybridoma 4*: 1–11 (1985).

90. M. V. Pimm and R. W. Baldwin, Distribution of IgM monoclonal antibody in mice with human tumour xenografts: Lack of tumour localization. *Eur. J. Cancer Clin. Oncol. 21*:765–768 (1985).
91. D. J. Buchsbaum, J. A. Sinkule, M. S. Stites, P. A. Fodor, D. E. Hanna, S. M. Ford, S. L. Warber-Matich, J. E. Juni, and K. A. Foon, Localization and imaging with radioiodine-labeled monoclonal antibodies in a xenogeneic tumor model for human B-cell lymphoma. *Cancer Res. 48*:2475–2482 (1988).
92. A. L. Epstein, A. M. Zimmer, S. M. Spies, D. Mills, G. DeNardo, and S. DeNardo, Radioimmunodetection of human B-cell lymphomas with a radiolabeled tumor-specific monoclonal antibody (Lym-1). In *Malignant Lymphomas and Hodgkin's Disease: Experimental and Therapeutic Advances*, (F. Cavalli, G. Bonadonna, and M. Rozencweig, eds.), Martinus Nijoff Publishing Co., Boston, 1985, pp. 569–577.
93. M. K. Haseman, D. A. Goodwin, C. F. Meares, M. S. Kaminski, T. G. Wensel, M. J. McCall, and R. Levy, Metabolizable ^{111}In chelate conjugated anti-idiotype monoclonal antibody for radioimmunodetection of lymphoma in mice. *Eur. J. Nucl. Med. 12*:455–460 (1986).
94. C. F. Meares, M. J. McCall, S. V. Deshpande, S. J. DeNardo, and D. A. Goodwin, Chelate radiochemistry: Cleavable linkers lead to altered levels of radioactivity in the liver. *Int. J. Cancer 2*(Suppl.):99–102 (1988).
95. D. J. Buchsbaum and F. M. Uckun, Localization and imaging with radioiodine-labeled B43 monoclonal antibody and fragments in nude mice bearing human B-cell lymphoma xenografts. *Antibodies, Immunoconjugates, and Radiopharm.*, Submitted, 1988.
96. S. M. Spies, A. M. Zimmer, W. G. Spies, S. Rosen, and E. A. Silverstein, Considerations for tomographic imaging of monoclonal antibodies. *Sem. Nucl. Med. 17*:267–272 (1987).
97. A. C. Perkins, D. R. Whalley, K. C. Ballantyne, and M. V. Pimm, Gamma camera emission tomography using radiolabelled antibodies. *Eur. J. Nucl. Med. 14*:45–49 (1988).
98. J. A. Carrasquillo, P. A. Bunn, Jr., A. M. Keenan, J. C. Reynolds, R. W. Schroff, K. A. Foon, S. Ming-Hsu, A. F. Gazdar, J. L. Mulshine, R. K. Oldham, P. Perentesis, M. Horowitz, J. Eddy, P. James, and S. M. Larson, Radioimmunodetection of cutaneous T-cell lymphoma with ^{111}In-labeled T101 monoclonal antibody. *N. Engl. J. Med. 315*:673–680 (1986).
99. M. L. Thakur, A. W. Segal, W. L. Louis, M. J. Welch, J. Hopkins, and T. J. Peters, Indium-111-labeled cellular blood components: Mechanism of labeling and intracellular location in human neutrophils. *J. Nucl. Med. 18*:1022–1026 (1977).
100. S. T. Rosen, A. M. Zimmer, R. Goldman-Leikin, L. I. Gordon, J. M. Kazikiewicz, E. H. Kaplan, D. Variakojis, R. J. Marder, M. S. Dykewicz, A. Piergies, E. A. Silverstein, H. H. Roenigk, Jr., and S. M. Spies, Radioimmunodetection and radioimmunotherapy of cutaneous T cell lymphomas using an ^{131}I-labeled monoclonal antibody: An Illinois Cancer Council study. *J. Clin. Oncol. 5*:562–573 (1987).

101. J. A. Carrasquillo, J. L. Mulshine, P. A. Bunn, Jr., J. C. Reynolds, K. A. Foon, R. W. Schroff, P. Perentesis, R. G. Steis, A. M. Keenan, M. Horowitz, and S. M. Larson, Indium-111 T101 monoclonal antibody is superior to iodine-131 T101 in imaging of cutaneous T-cell lymphoma. *J. Nucl. Med.* *28*:281–287 (1987).
102. A. M. Keenan, J. N. Weinstein, J. A. Carrasquillo, P. A. Bunn, Jr., J. C. Reynolds, K. A. Foon, N. C. Smarte, B. Ghosh, R. M. Fejka, S. M. Larson, and J. L. Mulshine, Immunolymphoscintigraphy and the dose dependence of ^{111}In-labeled T101 monoclonal antibody in patients with cutaneous T-cell lymphoma. *Cancer Res.* *47*:6093–6099 (1987).
103. F. Buchegger, C. M. Haskell, M. Schreyer, B. R. Scazziga, S. Randin, S. Carrel, and J.-P. Mach, Radiolabeled fragments of monoclonal antibodies against carcinoembryonic antigen for localization of human colon carcinoma grafted into nude mice. *J. Exp. Med.* *158*:413–427 (1983).
104. J.-Y. Douillard, J.-F. Chatal, J. C. Saccavini, C. Curtet, M. Kremer, P. Peuvrel, and H. Koprowski, Pharmacokinetic study of radiolabeled anti-colorectal carcinoma monoclonal antibodies in tumor-bearing nude mice. *Eur. J. Nucl. Med.* *11*:107–113 (1985).
105. D. L. Munz, A. Alavi, H. Koprowski, and D. Herlyn, Improved radioimmunoimaging of human tumor xenografts by a mixture of monoclonal antibody $F(ab')_2$ fragments. *J. Nucl. Med.* *27*:1739–1745 (1986).
106. D. Colcher, A. M. Keenan, S. M. Larson, and J. Schlom, Prolonged binding of a radiolabeled monoclonal antibody (B72.3) used for the *in situ* radioimmunodetection of human colon carcinoma xenografts. *Cancer Res.* *44*:5744–5751 (1984).
107. D. Colcher, M. F. Minelli, M. Roselli, R. Muraro, D. Simpson-Milenic, and J. Schlom, Radioimmunolocalization of human carcinoma xenografts with B72.3 second generation monoclonal antibodies. *Cancer Res.* *48*:4597–4603 (1988).
108. J. M. Esteban, J. Schlom, O. A. Gansow, R. W. Atcher, M. W. Brechbiel, D. E. Simpson, and D. Colcher, New method for the chelation of indium-111 to monoclonal antibodies: Biodistribution and imaging of athymic mice bearing human colon carcinoma xenografts. *J. Nucl. Med.* *28*:861–870 (1987).
109. D. Buchsbaum, R. Lloyd, J. Juni, I. Wollner, P. Brubaker, D. Hanna, J. Spicker, F. Burns, Z. Steplewski, D. Colcher, J. Schlom, F. Buchegger, and J.-P. Mach, Localization and imaging of radiolabeled monoclonal antibodies against colorectal carcinoma in tumor-bearing nude mice. *Cancer Res.* *48*:4324–4333 (1988).
110. G. Rowlinson, D. Snook, A. Busza, and A. A. Epenetos, Antibody-guided localization of intraperitoneal tumors following intraperitoneal or intravenous antibody administration. *Cancer Res.* *47*:6528–6531 (1987).
111. A. A. Epenetos, A. J. Munro, S. Stewart, R. Rampling, H. E. Lambert, C. G. McKenzie, P. Soutter, A. Rahemtulla, G. Hooker, G. B. Sivolapenko, D. Snook, N. Courtenay-Luck, B. Dhokia, T. Krausz, J. Taylor-

Papadimitriou, H. Durbin, and W. F. Bodmer, Antibody-guided irradiation of advanced ovarian cancer with intraperitoneally administered radiolabeled monoclonal antibodies. *J. Clin. Oncol. 5*:1890–1899 (1987).

112. A. M. Zimmer, J. M. Kazikiewicz, S. T. Rosen, and S. M. Spies, Pharmacokinetics of ^{99m}Tc (Sn)- and ^{131}I-labeled anti-carcinoembryonic antigen monoclonal antibody fragments in nude mice. *Cancer Res. 47*:1691–1694 (1987).
113. A. R. Bradwell, A. Vaughan, D. S. Fairweather, and P. W. Dykes, Improved radioimmunodetection of tumours using second antibody. *Lancet 1*:247 (1983).
114. D. Goodwin, C. Meares, C. Diamanti, Use of specific antibody for rapid clearance of circulating blood background from radiolabeled tumor imaging proteins. *Eur. J. Nucl. Med. 9*:209–215 (1984).
115. D. M. Goldenberg, R. Sharkey, and E. Ford, Anti-antibody enhancement of iodine-131 anti-CEA radioimmunodetection in experimental and clinical studies. *J. Nucl. Med. 28*:1604–1610 (1987).
116. J.-P. Mach, F. Buchegger, M. Forni, J. Ritschard, C. Berche, J.-D. Lumbroso, M. Schreyer, C. Girardet, R. S. Accolla, and S. Carrel, Use of radiolabelled monoclonal anti-CEA antibodies for the detection of human carcinomas by external photoscanning and tomoscintigraphy. *Immunol. Today 2*:239–249 (1981).
117. R. O. Dillman, J. C. Beauregard, R. E. Sobol, I. Royston, R. M. Bartholomew, P. S. Hagan, and S. E. Halpern, Lack of radioimmunodetection and complications associated with monoclonal anticarcinoembryonic antigen antibody cross-reactivity with an antigen on circulating cells. *Cancer Res. 44*:2213–2218 (1984).
118. B. Delaloye, A. Bischof-Delaloye, F. Buchegger, V. von Fliedner, J.-P. Grob, J.-C. Volant, J. Pettavel, and J.-P. Mach, Detection of colorectal carcinoma by emission-computerized tomography after injection of ^{123}I-labeled Fab or F(ab$'$)$_2$ fragments from monoclonal anti-carcinoembryonic antigen antibodies. *J. Clin. Invest. 77*:301–311 (1986).
119. S. E. Halpern, W. Haindl, J. Beauregard, P. Hagan, M. Clutter, D. Amox, B. Merchant, M. Unger, C. Mongovi, R. Bartholomew, R. Jue, D. Carlo, and R. Dillman, Scintigraphy with In-111-labeled monoclonal antitumor antibodies: kinetics, biodistribution, and tumor detection. *Radiology 168*: 529–536 (1988).
120. Y. Z. Patt, L. M. Lamki, T. P. Haynie, M. W. Unger, M. G. Rosenblum, A. Shirkhoda, and J. L. Murray, Improved tumor localization with increasing dose of indium-111-labeled anti-carcinoembryonic antigen monoclonal antibody ZCE-025 in metastatic colorectal cancer. *J. Clin. Oncol. 6*:1220–1230 (1988).
121. F. H. DeLand, A. Lieber, M. D. Ram, and D. M. Goldenberg, Preliminary findings in the evaluation of hepatic malignancies by radioimmunodetection, X-ray computed tomography, and magnetic resonance imaging. *Eur. J. Nucl. Med. 12*:429–435 (1986).

122. J.-P. Mach, J.-F. Chatal, J.-D. Lumbroso, F. Buchegger, M. Forni, J. Rtischard, C. Berche, J.-Y. Douillard, S. Carrel, M. Herlyn, Z. Steplewski, and H. Koprowski, Tumor localization in patients by radiolabeled monoclonal antibodies against colon carcinomas. *Cancer Res. 43*:5593–5600 (1983).
123. E. W. Martin, Jr., S. E. Tuttle, M. Rousseau, C. M. Mojzisik, P. J. O'-Dwyer, G. H. Hinkle, E. A. Miller, R. A. Goodwin, O. A. Oredipe, R. F. Barth, J. O. Olsen, D. Houchens, S. D. Jewell, D. M. Bucci, D. Adams, Z. Steplewski, and M. O. Thurston, Radioimmunoguided surgery: Intraoperative use of monoclonal antibody 17-1A in colorectal cancer. *Hybridoma 5*(Suppl. 1):S97–S108 (1986).
124. E. P. Mitchell and J. Schlom, Monoclonal antibodies in gastrointestinal cancers. *Sem. in Oncol. 15*:170–180 (1988).
125. S. M. Larson, Lymphoma, melanoma, colon cancer: Diagnosis and treatment with radiolabeled monoclonal antibodies. *Radiology 165*:297–304 (1987).
126. J. A. Carrasquillo, P. Sugarbaker, D. Colcher, J. C. Reynolds, J. Esteban, G. Bryant, A. M. Keenan, P. Perentesis, K. Yokoyama, D. E. Simpson, P. Ferroni, R. Farkas, J. Schlom, and S. M. Larson, Radioimmunoscintigraphy of colon cancer with iodine-131-labeled B72.3 monoclonal antibody. *J. Nucl. Med. 29*:1022–1030 (1988).
127. N. C. Armitage, A. C. Perkins, M. V. Pimm, R. W. Farrands, R. W. Baldwin, and J. D. Hardcastle, The localization of an anti-tumour monoclonal antibody (791T/36) in gastrointestinal tumours. *Br. J. Surg. 71*:407–412 (1984).
128. N. C. Armitage, A. C. Perkins, M. V. Pimm, M. L. Wastie, R. W. Baldwin, and J. D. Hardcastle, Imaging of primary and metastatic colorectal cancer using an ^{111}In-labelled antitumour monoclonal antibody (791T/36). *Nucl. Med. Commun. 6*:623–631 (1985).
129. M. J. Leyden, C. H. Thompson, M. Lichtenstein, J. T. Andrews, J. R. Sullivan, J. R. Zalcberg, and I. F. C. McKenzie, Visualization of metastases from colon carcinoma using an iodine 131-radiolabeled monoclonal antibody. *Cancer 57*:1135–1139 (1986).
130. ACNP/SNM Task Force on Clinical PET, Positron emission tomography: Clinical status in the United States in 1987. *J. Nucl. Med. 29*:1136–1143 (1988).
131. R. M. Kessler, C. L. Partain, R. R. Price, and A. E. James, Jr., Positron emission tomography: Prospects for clinical utility. *Invest. Radiol. 22*:529–537 (1987).
132. D. A. Goodwin, C. F. Meares, M. J. McCall, and W. Chaovapong, Pretargeted immunoscintigraphy of murine tumors with indium-111-labeled bifunctional haptens. *J. Nucl. Med. 29*:226–234 (1988).
133. D. J. Buchsbaum, D. E. Hanna, M. S. Stites, P. A. Fodor, M. S. Kaminski, and K. A. Foon, Localization and therapy with radiolabeled anti-B1 and anti-J5 monoclonal antibodies in nude mice bearing human B-cell lymphoma xenografts. *Cancer Res.* Submitted, 1989.

134. D. Buchsbaum, D. Hanna, V. Terry, D. Guilbault, D. Myers, K. Waddick, M. Chandan, and F. Uckun, Localization, imaging and therapy with radio-labeled B43 monoclonal antibody and fragments in nude mice bearing human B-cell lymphoma xenografts. *4th Int. Conf. Monoclonal Antibody Immunoconjugates for Cancer*, San Diego, March 30–April 1, 1989.
135. C. C. Badger, K. A. Krohn, A. V. Peterson, H. Shulman, and I. D. Bernstein, Experimental radiotherapy of murine lymphoma with ^{131}I-labeled anti-Thy 1.1 monoclonal antibody. *Cancer Res. 45*:1536–1544 (1985).
136. C. C. Badger, K. A. Krohn, H. Shulman, N. Flournoy, and I. D. Bernstein, Experimental radioimmunotherapy of murine lymphoma with ^{131}I-labeled anti-T-cell antibodies. *Cancer Res. 46*:6223–6228 (1986).
137. D. A. Adams, G. L. DeNardo, S. J. DeNardo, G. B. Matson, A. L. Epstein, and E. M. Bradbury, Radioimmunotherapy of human lymphoma in athymic, nude mice as monitored by ^{31}P nuclear magnetic resonance. *Biochem. Biophys. Res. Commun. 131*:1020–1027 (1985).
138. W. R. Redwood, T. D. Tom, and M. Strand, Specificity, efficacy, and toxicity of radioimmunotherapy in erythroleukemic mice. *Cancer Res. 44*:5681–5687 (1984).
139. M. Mantyla, J. Kuikka, and A. Rekonen, Regional blood flow in human tumours with special reference to the effect of radiotherapy. *Br. J. Radiol. 49*:335–338 (1976).
140. V. T. DeVita, Jr., S. Hellman, and S. A. Rosenberg (eds.), *Cancer, Principles and Practice of Oncology*, 2nd ed. J. B. Lippincott, Philadelphia, 1985.
141. R. W. Schroff, K. A. Foon, S. M. Beatty, R. K. Oldham, and A. C. Morgan, Jr., Human anti-murine immunoglobulin responses in patients receiving monoclonal antibody therapy. *Cancer Res. 45*:879–885 (1985).
142. A. M. Zimmer, S. T. Rosen, S. M. Spies, R. Goldman-Leikin, J. M. Kazikiewicz, E. A. Silverstein, and E. H. Kaplan, Radioimmunotherapy of patients with cutaneous T-cell lymphoma using an iodine-131-labeled monoclonal antibody: Analysis of retreatment following plasmapheresis. *J. Nucl. Med. 29*:174–180 (1988).
143. J. M. Esteban, J. Schlom, F. Mornex, and D. Colcher, Radioimmunotherapy of athymic mice bearing human colon carcinomas with monoclonal antibody B72.3: Histological and autoradiographic study of effects on tumors and normal organs. *Eur. J. Cancer Clin. Oncol. 23*:643–655 (1987).
144. J. R. Zalcberg, C. H. Thompson, M. Lichtenstein, and I. F. C. McKenzie, Tumor immunotherapy in the mouse with the use of ^{131}I-labeled monoclonal antibodies. *J. Natl. Cancer Inst. 72*:697–704 (1984).
145. F. Buchegger, A. Vacca, S. Carrel, M. Schreyer, and J.-P. Mach, Radioimmunotherapy of human colon carcinoma by ^{131}I-labelled monoclonal anti-CEA antibodies in a nude mouse model. *Int. J. Cancer 41*:127–134 (1988).

146. R. M. Sharkey, M. J. Pykett, J. A. Siegel, E. A. Alger, F. J. Primus, and D. M. Goldenberg, Radioimmunotherapy of the GW-39 human colonic tumor xenograft with ^{131}I-labeled murine monoclonal antibody to carcinoembryonic antigen. *Cancer Res. 47*:5672–5677 (1987).
147. R. M. Sharkey, F. A. Kaltovich, L. B. Shih, I. Fand, G. Govelitz, and D. M. Goldenberg, Radioimmunotherapy of human colonic cancer xenografts with ^{90}Y-labeled monoclonal antibodies to carcinoembryonic antigen. *Cancer Res. 48*:3270–3275 (1988).
148. H. F. Sears, D. Herlyn, Z. Steplewski, and H. Koprowski, Effects of monoclonal antibody immunotherapy on patients with gastrointestinal adenocarcinoma. *J. Biol. Response Modif. 3*:138–150 (1984).
149. H. F. Sears, D. Herlyn, Z. Steplewski, and H. Koprowski, Phase II clinical trial of a murine monoclonal antibody cytotoxic for gastrointestinal adenocarcinoma. *Cancer Res. 45*:5910–5913 (1985).
150. H. F. Sears, D. Herlyn, Z. Steplewski, and H. Koprowski, Initial trial use of murine monoclonal antibodies as immunotherapeutic agents for gastrointestinal adenocarcinoma. *Hybridoma 5*(Suppl. 1):S109–S115 (1986).
151. J.-E. Frodin, P. Biberfeld, B. Christensson, P. Philstedt, S. Sundelius, M. Sylven, B. Wahren, H. Koprowski, and H. Mellstedt, Treatment of patients with metastasizing colo-rectal carcinoma with mouse monoclonal antibodies (MoAb 17-1A): A progress report. *Hybridoma 5*(Suppl. 1):S151–S161 (1986).
152. A. F. LoBuglio, M. Saleh, L. Peterson, R. Wheeler, R. Carrano, W. Huster, and M. B. Khazaeli, Phase I clinical trial of C017-1A monoclonal antibody. *Hybridoma 5*(Suppl. 1):S117–S123 (1986).
153. H. M. Blottiere, C. Maurel, and J.-Y. Douillard, Immune function of patients with gastrointestinal carcinoma after treatment with multiple infusions of monoclonal antibody 17.1A. *Cancer Res. 47*:5238–5241 (1987).
154. S. E. Order, G. B. Stillwagon, J. L. Klein, P. K. Leichner, S. S. Siegelman, E. K. Fishman, D. S. Ettinger, T. Haulk, K. Kopher, K. Finney, M. Surdyke, S. Self, and S. Leibel, Iodine 131 antiferritin, a new treatment modality in hepatoma: A Radiation Therapy Oncology Group study, *J. Clin. Oncol. 3*:1573–1582 (1985).
155. A. F. LoBuglio, R. Wheeler, R. D. Leavitt, E. B. Harvey, H. T. Holden, A. Haynes, K. Rogers, J. Lee, and M. Khazaeli, Pharmacokinetics and immune response to chimeric mouse/human monoclonal antibody (Ch-17-1A) in man. *Proc. 1st Int. Symp. Immunotoxins*, Durham, N.C., June 9–11, 1988, p. 59.
156. D. R. Shaw, G. Harrison, L. K. Sun, C. Shearman, J. Ghrayeb, S. McKinney, P. E. Daddona, and A. F. LoBuglio, Human lymphocyte and monocyte lysis of tumor cells mediated by a mouse/human IgG1 chimeric monoclonal antibody. *J. Biol. Response Modif. 7*:204–211 (1988).
157. L. W. Brady, D. V. Woo, A. M. Markoe, H. Koprowski, and C. T. Miyamoto, Monoclonal antibodies in the diagnosis and treatment of cancer.

In *Radiobiology in Radiotherapy* (N. M. Bleehen, ed.), Springer-Verlag, London, England, 1988, pp. 233–242.

158. J. L. Humm, Dosimetric aspects of radiolabeled antibodies for tumor therapy. *J. Nucl. Med. 27*:1490–1497 (1986).
159. B. W. Wessels and R. D. Rogus, Radionuclide selection and model absorbed dose calculations for radiolabeled tumor associated antibodies. *Med. Phys. 11*:638–645 (1984).
160. P. K. Leichner, Y. Nai-Chuen, T. L. Frenkel, D. M. Loudenslager, W. G. Hawkins, J. L. Klein, and S. E. Order, Dosimetry and treatment planning for ^{90}Y-labeled antiferritin in hepatoma. *Int. J. Radiat. Oncol. Biol. Phys. 14*:1033–1042 (1988).
161. P. K. Leichner, J. L. Klein, J. B. Garrison, R. E. Jenkins, E. L. Nickoloff, D. S. Ettinger, and S. E. Order, Dosimetry of ^{131}I-labeled anti-ferritin in hepatoma: A model for radioimmunoglobulin dosimetry. *Int. J. Radiat. Oncol. Biol. Phys. 7*:323–333 (1981).
162. A. A. Epenetos, D. Snook, H. Durbin, P. M. Johnson, and J. Taylor-Papadimitriou, Limitations of radiolabeled monoclonal antibodies for localization of human neoplasms. *Cancer Res. 46*:3183–3191 (1986).
163. A. T. M. Vaughan, A. R. Bradwell, P. W. Dykes, and P. Anderson, Illusions of tumour killing using radiolabelled antibodies. *Lancet 1*:1492–1493 (1986).
164. S. V. Deshpande, S. J. DeNardo, C. F. Meares, M. J. McCall, G. P. Adams, M. K. Moi, and G. L. DeNardo, Copper-67-labeled monoclonal antibody Lym-1, a potential radiopharmaceutical for cancer therapy: Labeling and biodistribution in RAJI tumored mice. *J. Nucl. Med. 29*:217–225 (1988).
165. W. C. Cole, S. J. DeNardo, C. F. Meares, M. J. McCall, G. L. DeNardo, A. L. Epstein, H. A. O'Brien, and M. K. Moi, Comparative serum stability of radiochelates for antibody radiopharmaceuticals. *J. Nucl. Med. 28*:83–90 (1987).
166. M. K. Moi, C. F. Meares, M. J. McCall, W. C. Cole, and S. J. DeNardo, Copper chelates as probes in biological systems: Stable copper complexes with macrocyclic bifunctional chelating agents. *Anal. Biochem. 148*:249–253 (1985).
167. D. J. Buchsbaum, D. E. Hanna, B. C. Randall, F. Buchegger, and J.-P. Mach, Radiolabeling of monoclonal antibody against carcinoembryonic antigen with ^{88}Y and biodistribution studies. *Int. J. Nucl. Med. Biol. 12*: 79–82 (1985).
168. D. J. Hnatowich, F. Virzi, and P. W. Doherty, DTPA-coupled antibodies labeled with yttrium-90. *J. Nucl. Med. 26*:503–509 (1985).
169. S. E. Order, J. L. Klein, P. K. Leichner, J. Frincke, C. Lollo, and D. J. Carlo, ^{90}Yttrium antiferritin–A new therapeutic radiolabeled antibody. *Int. J. Radiat. Oncol. Biol. Phys. 12*:277–281 (1986).
170. W. T. Anderson-Berg, R. A. Squire, and M. Strand, Specific radioimmunotherapy using ^{90}Y-labeled monoclonal antibody in erythroleukemic mice. *Cancer Res. 47*:1905–1912 (1987).

171. D. J. Buchsbaum, L. A. Nelson, D. E. Hanna, and D. A. Vallera, Human leukemia cell binding and killing by anti-CD5 radioimmunotoxins. *Int. J. Radiat. Oncol. Biol. Phys. 13*:1701–1712 (1987).
172. M. Chinol and D. J. Hnatowich, Generator-produced yttrium-90 for radio-immunotherapy. *J. Nucl. Med. 28*:1465–1470 (1987).
173. D. J. Hnatowich, M. Chinol, D. A. Siebecker, M. Gionet, T. Griffin, P. W. Doherty, R. Hunter, and K. R. Kase, Patient biodistribution of intraperitoneally administered yttrium-90-labeled antibody. *J. Nucl. Med. 29*: 1428–1434 (1988).
174. S. E. Order, H. M. Vriesendorp, J. L. Klein, and P. K. Leichner, A phase I study of ^{90}yttrium antiferritin: Dose escalation and tumor dose. *Antibody, Immunoconjugates, and Radiopharm. 1*:163–168 (1988).
175. S. M. Quadri, and B. W. Wessels, Radiolabeled biomolecules with ^{186}Re: Potential for radioimmunotherapy. *Nucl. Med. Biol. 13*:447–451 (1986).
176. K. A. Foon, Monoclonal antibodies in the diagnosis and treatment of cancer. NeoRx Corporation, 1988, pp. 1–27.
177. P. Anderson, A. T. M. Vaughan, and N. R. Varley, Antibodies labeled with ^{199}Au: Potential of ^{199}Au for radioimmunotherapy. *Nucl. Med. Biol. 15*: 293–297 (1988).
178. B. Grazman and D. E. Troutner, ^{105}Rh as a potential radiotherapeutic agent. *Appl. Radiat. Isot. 39*:257–260 (1988).
179. R. W. Kozak, R. W. Atcher, O. A. Gansow, A. M. Friedman, J. J. Hines, and T. A. Waldmann, Bismuth-212-labeled anti-Tac monoclonal antibody: α-particle-emitting radionuclides as modalities for radioimmunotherapy. *Proc. Natl. Acad. Sci. USA 83*:474–478 (1986).
180. R. M. Macklis, B. M. Kinsey, A. I. Kassis, J. L. M. Ferrara, R. W. Atcher, J. J. Hines, C. N. Coleman, S. J. Adelstein, and S. J. Burakoff, Radioimmunotherapy with alpha-particle-emitting immunoconjugates. *Science 240*:1024–1026 (1988).
181. S. H. Kurtzman, A. Russo, J. B. Mitchell, W. DeGraff, W. F. Sindelar, M. W. Brechbiel, O. A. Gansow, A. M. Friedman, J. J. Hines, J. Gamson, and R. W. Atcher, ^{212}Bismuth linked to an antipancreatic carcinoma antibody: Model for alpha-particle-emitter radioimmunotherapy. *J. Natl. Cancer Inst. 80*:449–452 (1988).
182. A. T. M. Vaughan, W. J. Bateman, and D. R. Fisher, The *in vivo* fate of ^{211}At labelled monoclonal antibody with known specificity in a murine system. *Int. J. Radiat. Oncol. Biol. Phys. 8*:1943–1946 (1982).
183. A. Harrison and L. Royle, Preparation of a ^{211}At-IgG conjugate which is stable *in vivo*. *Int. J. Appl. Radiat. Isot. 35*:1005–1008 (1984).
184. I. Brown, Astatine-211: Its possible applications in cancer therapy, *Appl. Radiat. Isot. 37*:789–798 (1986).
185. J. L. Humm, A microdosimetric model of astatine-211 labeled antibodies for radioimmunotherapy. *Int. J. Radiat. Oncol. Biol. Phys. 13*:1767–1773 (1987).

186. M. R. Zalutsky and A. S. Narula, Astatination of proteins using an N-succinimidyl tri-n-butylstannyl benzoate intermediate. *Appl. Radiat. Isot. 39*:227–232 (1988).
187. T. Lindmo, E. Boven, J. B. Mitchell, G. Morstyn, and P. A. Bunn, Jr., Specific killing of human melanoma cells by ^{125}I-labeled 9.2.27 monoclonal antibody. *Cancer Res. 45*:5080–5087 (1985).
188. E. Boven, T. Lindmo, J. B. Mitchell, and P. A. Bunn, Jr., Selective cytotoxicity of ^{125}I-labeled monoclonal antibody T101 in human malignant T cell lines. *Blood 67*:429–435 (1986).
189. A. I. Kassis, F. Fayad, B. M. Kinsey, K. S. R. Sastry, R. A. Taube, and S. J. Adelstein, Radiotoxicity of ^{125}I in mammalian cells. *Radiat. Res. 111*:305–318 (1987).
190. Y. Sugiyama, F.-A. Chen, H. Takita, and R. B. Bankert, Selective growth inhibition of human lung cancer cell lines bearing a surface glycoprotein gp160 by ^{125}I-labeled anti-gp160 monoclonal antibody. *Cancer Res. 48*: 2768–2773 (1988).
191. A. I. Kassis, S. J. Adelstein, C. Haydock, K. S. R. Sastry, K. D. McElvany, and M. J. Welch, Lethality of Auger electrons from the decay of bromine-77 in the DNA of mammalian cells. *Radiat. Res. 90*:362–373 (1982).
192. E. R. DeSombre, P. V. Harper, A. Hughes, R. C. Mease, S. J. Gatley, O. T. DeJesus, and J. L. Schwartz, Bromine-80m radiotoxicity and the potential for estrogen receptor-directed therapy with Auger electrons. *Cancer Res. 48*:5805–5809 (1988).
193. E. Mizusawa, H. L. Dahlman, S. J. Bennett, D. M. Goldenberg, and M. F. Hawthorne, Neutron-capture therapy of human cancer: *In vitro* results on the preparation of boron-labeled antibodies to carcinoembryonic antigen. *Proc. Natl. Acad. Sci. USA 79*:3011–3014 (1982).
194. F. Alam, A. H. Soloway, and R. F. Barth, Boronation of antibodies with mercaptoundecahydro-closo-dodecaborate (2-) anion for potential use in boron neutron capture therapy. *Appl. Radiat. Isot. 38*:503–506 (1987).
195. J. N. Lowder, The current status of monoclonal antibodies in the diagnosis and therapy of cancer. *Curr. Prob. Cancer 10*:485–551 (1986).
196. D. A. Vallera and F. M. Uckun, Immunoconjugates. In *Biological Response Modifiers and Cancer Research* (J. W. Chiao, ed.), Marcel Dekker, New York, 1988, pp. 17–50.
197. C.-W. Vogel (ed.), *Immunoconjugates. Antibody Conjugates in Radioimaging and Therapy of Cancer.* Oxford University Press, New York, 1987, pp. 3–308.
198. A. E. Frankel (ed.), *Immunotoxins.* Kluwer Academic Publishers, Boston, 1988, pp. 1–565.
199. J. M. Manske, D. J. Buchsbaum, D. E. Hanna, and D. A. Vallera, Cytotoxic effects of anti-CD5 radioimmunotoxins on human tumors *in vitro* and in a nude mouse model. *Cancer Res. 48*:7107–7114 (1988).
200. D. Buchsbaum, J. Sinkule, P. Brubaker, D. Hanna, V. Terry, F. Burns, J. Spicker, and S. Howell, Radionuclide and drug immunoconjugates:

Localization and therapy studies in human colon cancer nude mouse models. *Tumor Biol. 8*:356 (1987).

201. J. A. Sinkule, D. J. Buchsbaum, K. Foon, M. Kaminski, and R. A. Miller, Therapeutic potential of doxorubicin/antibody/radionuclide conjugates in a B-cell lymphoma model. *Proc. AACR 28*:397 (1987).
202. V. K. Langmuir and R. M. Sutherland, Radiobiology of radioimmunotherapy: Current status. *Antibody, Immunoconjugates, and Radiopharm. 1*:195–211 (1988).
203. K. P. Borlinghaus, D. A. Fitzpatrick, N. D. Heindel, J. A. Mattis, B. A. Mease, K. J. Schray, D. J. Shealy, H. L. Walton, Jr., and D. V. Woo, Radiosensitizer conjugation to the carcinoma 19-9 monoclonal antibody. *Cancer Res. 47*:4071–4075 (1987).
204. G. Morstyn, R. Miller, A. Russo, and J. Mitchell, 131-iodine conjugated antibody cell kill enhanced by bromodeoxyuridine. *Int. J. Radiat. Oncol. Biol. Phys. 10*:1437–1440 (1984).
205. R. E. Bird, K. D. Hardman, J. W. Jacobson, S. Johnson, B. M. Kaufman, S.-M. Lee, T. Lee, S. H. Pope, G. S. Riordan, and M. Whitlow, Single-chain antigen-binding proteins. *Science 242*:423–426 (1988).
206. J. W. Greiner, F. Guadagni, P. Noguchi, S. Pestka, D. Colcher, P. B. Fisher, and J. Schlom, Recombinant interferon enhances monoclonal antibody-targeting of carcinoma lesions *in vivo*. *Science 235*:895–898 (1987).

part four

SPECIALIZED THERAPEUTICS APPLICATIONS

10

From Monoclonal Antibodies to Immunonanospheres in the Pharmaceutical Sciences

ALAIN P. ROLLAND*
Ciba-Geigy Pharmaceuticals, Horsham, England

DOMINIQUE BOUREL
Centre Régional de Transfusion Sanguine, Rennes, France

I. REVIEW—CLINICAL APPLICATIONS OF MONOCLONAL ANTIBODIES: CANCER DIAGNOSIS AND THERAPY

Since the development of hybridoma technology in 1975 to produce monoclonal antibodies (MoAbs) [1], many improvements in the technique have been made in order to provide sufficient quantities of pure, homogeneous MoAbs that bind specifically to single epitopes on antigenic molecules [2].

A. Diagnosis

With the advent of monoclonal antibodies that selectively bind to tumor cells in vitro, and in vivo in immunodeprived mice with animal or human tumors as xenografts, radioimmunolocalization studies have developed considerably [3-6]. To date, there has also been a wide range of effective tumor radioimmunodetection using MoAbs, in patients with breast cancer [7-9], colorectal tumors [10-12], melanoma [13,14], ovarian cancer [15,16], and other tumors [17-22].

**Current affiliation*: Centre International de Recherches Dermatologiques, Valbonne, France

B. Therapy

Monoclonal Antibodies

Over the past few years, many attempts have been made to use MoAbs in cancer therapy [23-25]. Although the actual mechanism of tumor cell killing by MoAbs in vivo is not completely known, some of the ways in which MoAbs could be applied to the treatment of neoplastic diseases are: (1) passive cytotoxic or (2) regulatory therapy.

1. There are several means by which MoAbs may interact with the host immune system to produce a cytotoxic antitumor effect:

Complement-mediated cytotoxicity, where the binding of MoAbs to cells expressing the target antigen is followed by the fixation of complement and cell lysis

Antibody-dependent monocyte toxicity [26-29], produced either by the binding of effector cells, via Fc receptors, to the target cells to which the MoAbs are attached or by the attachment to circulating MoAbs by the same process and transport to the target cell.

In both cases, cell lysis results from the antibody-effector cell conjugate.

Another mechanism involves the phagocytosis of MoAbs-opsonized circulating tumor cells by cells of the mononuclear phagocytic system. MoAbs may also increase the antitumor effect of activated macrophages, natural killer cells, killer T cells, or other cells that mediate immunity [30].

Unfortunately, most of the clinical trials using MoAbs for passive anticancer therapy have been disappointing, generally because of problems due to the immunogenicity of the murine MoAbs injected into patients [30,31], and to antigenic modulation, which is a loss of the target antigen from the surface of the tumor cells [31-34]. Nevertheless, some successful clinical trials have been reported, which demonstrate the ability of MoAbs to destroy circulating tumor cells [35-41] and solid, nonhematological malignancies [42-44]. There are, however, many pitfalls or limitations in passive cytotoxic antibody therapy, including:

Immunogenicity of MoAbs

Cross-reactivity of MoAbs with normal cells

Lack of expression or transient loss or change (immune modulation) of target antigen on some tumor cells

Possible inaccessibility of MoAbs to the specific sites

Interactions of MoAbs with circulating target antigens before they can reach their target

2. Apart from their cytotoxic effect on the target cells, MoAbs may have an antitumor activity via a regulatory mechanism. For instance, anti-idiotype MoAbs may be used for the diagnosis and regulatory therapy of B-cell lymphoproliferative disorders [45]. Specific therapy of patients with advanced B-cell lymphoma using anti-idiotype monoclonal antibodies has shown a remarkable efficacy that may be due to the regulation of the immune network [36,46,47], since the idiotypic region is crucial to the regulation of a proliferating B-cell clone.

MoAbs that specifically bind to growth-factor receptors, such as transferrin receptor [48,49], may also provide a regulatory therapy, in the sense that transferrin is necessary for the growth of cells and its receptor is predominantly present on proliferating cells.

Immunoconjugates

Since the first theory of drug targeting by means of "bodies which have a particular affinity for a certain organ," described by P. Ehrlich (1906) as "magic bullets" [50], site-specific drug delivery has become an important area of research and development in pharmaceutics. The advent of monoclonal antibodies has stimulated a growing interest in the use of immunoconjugates in the form of radioisotopes, anticancer drugs, or toxins coupled to MoAbs.

Radioisotopes. Radiation therapy in the treatment of cancers can be improved using MoAbs, which bind to tumor-associated antigens, in order to target radionuclides to tumors. An important consideration in radioimmunotherapy is the optimal choice of radionuclide. Isotopes emitting alpha particles, such as astatine-211, have been poorly exploited owing to their instability, although radiolabeled MoAbs with ^{211}At have been successfully used in vitro and in vivo [51,52].

Targeting high-energy beta emitters, such as ^{131}I and ^{32}P [53,54] is more attractive because the path lengths of the emissions are long, thus destruction of inaccessible cells, e.g., in poorly vascularized tumors, or cells that lack the target antigen, may be possible. Furthermore, as intracellular localization of the radionuclide is not required to achieve cell killing, this represents a great advantage compared to drugs or toxins.

Clinical responses to radioimmunotherapy have been reported in patients with advanced melanoma [55,56] and primary hepatic cancer [57] using ^{131}I-labeled antibodies.

Unfortunately, the major drawback of this approach is that radioisotopes provide a nonspecific irradiation of normal tissues during their distribution throughout the body.

Anticancer Drugs. Due to the lack of site-specific drug delivery, the efficacy of conventional chemotherapy in the treatment of cancer patients is generally hindered by the toxic effects of cytotoxic drugs on normal tissues.

An approach for overcoming these limitations is to couple anticancer drugs to antibodies directed to surface-associated, tumor-specific antigens. Such antibody-targeted drugs would be expected to provide improved therapeutic effects and reduced toxic side effects.

The most commonly investigated drugs are the alkylating agents chlorambucil and melphalan, the antimetabolite methotrexate, the DNA-intercalating agents doxorubicin and daunorubicin, and the plant alkaloid vindesine.

In some early studies, noncovalent conjugates prepared by simple adsorption of drugs, e.g., chlorambucil, to antitumor antibodies were shown to be more effective than free drug or antibody alone, both in vitro and in vivo [58,59]. However, part of the reported antitumor activity could have been due to dissociation of the conjugate and to a synergistic toxic action of free drug and antibody [60,61].

Thus, covalent binding of anticancer drugs to monoclonal antibodies has been used to produce stable conjugates with an improved specific activity.

1. *Chlorambucil.* Chlorambucil is rarely used as a free drug in cancer chemotherapy, because of undesirable side effects (renal toxicity, marrow aplasia, pulmonary fibrosis, gastrointestinal disorders) due to its lack of specificity for cancer cells.

Although noncovalent complexes and covalent conjugates have been shown to retain antibody specificity and drug activity both in vitro and in vivo [62-65], few attempts have been made to treat human malignant diseases with such constructs [66-69]. None of the clinical trials published so far has used monoclonal antibodies for preparing conjugates with chlorambucil.

In addition, in the case of noncovalent conjugates, the fact that release of free drug could occur and that antibody and chlorambucil could react separately but synergistically with tumor cells leads to the question of what the real mechanism of action of such conjugates may be in vivo.

2. *Methotrexate.* Since the early study in 1958 in which Mathé [70] coupled methotrexate to an antibody against L1210 leukemia cells and observed an improved activity of the conjugate in tumor-bearing mice, the folic acid antagonist methotrexate has been extensively used to prepare immunoconjugates [71-78].

From studies on the mechanism of action of methotrexate conjugated to monoclonal antibodies, it seems that the conjugate first binds to the tumor cell surface via tumor-associated antigens and is then internalized by endocytosis and degraded by lysosomal enzymes, providing active forms of the drug [77-79].

3. *Daunomycin and adriamycin.* The anthracycline antibiotics, daunomycin and adriamycin, have been coupled to antibodies by direct cross-linking using either glutaraldehyde or carbodiimide [80-82], resulting in problems such as

aggregation of the conjugate and loss of drug activity. Therefore, direct coupling of anthracyclines to MoAbs has been performed by periodate oxidation, giving carbonyl groups able to react with amine groups on the protein. This technique leads to conjugates exhibiting both drug and antibody activity [82].

Specific cytotoxic effects of daunomycin-antibody conjugates have been demonstrated both in vitro and in vivo with animal models [82-84].

4. *Vindesine*. The vinca alkaloid, vindesine, has been coupled directly to antibody by reacting desacetyl vinblastine azide with amine groups on the antibody surface. After early studies using polyclonal antibodies [85,86], conjugates were prepared by coupling vindesine to monovalent monoclonal antibodies [87-89] or by using bispecific MoAbs [90].

From one antibody to another, variations of the amount of vindesine that could be coupled to MoAbs have been observed, and specific in vitro vindesine activity and localization to tumor cells using MoAbs have been demonstrated [88,91].

Vindesine-MoAb conjugates have also been proved to be effective in vivo against solid human tumors growing in immune-deprived animals [92,93]. Nevertheless, such conjugates are often less potent than the parenteral free drug.

In another way, specific targeting of vinblastine to tumor xenografts implanted on nude mice has been also achieved by prior administration of a hybrid-hybrid monoclonal antibody (bispecific MoAb) with specificity for both carcinoembryonic antigen (CEA) and the cytostatic vinca alkaloid drugs, e.g., vindesine, vinblastine, vincristine. This technique enhances the efficacy of vinblastine, which may be by specific localization and increased concentration of circulating drug at the tumor mass and subsequent release of free drug in a pharmacologically active form [94].

5. *Melphalan*. Although melphalan, an aromatic alkylating agent derived from phenylalanine, has been coupled to antibody for use in preliminary clinical trials [95], its chemical structure has subsequently been modified in order to reduce its toxicity [96]. The amino group of melphalan was modified to form an N-acetyl derivative, thus blocking uptake of the drug, normally involving an active carrier-mediated process via an amino acid transport system [97]. N-acetyl melphalan has therefore two major advantages compared to melphalan: It is at least 20 times less toxic, and it can be more easily coupled to MoAbs [98].

It was shown that for N-acetyl melphalan-MoAb conjugates, the drug entered the cells only via the MoAb, by endocytosis, and these conjugates displayed a higher antitumor activity in vivo and a lower nonspecific toxicity than the free drug [96].

Toxins. Recent interest in the use of antibody-toxin conjugates, or immunotoxins, has arisen from their intrinsic potency, maximizing the chance of killing cancer cells that are inaccessible to conventional chemotherapy or that do not express specific antigens at high density.

Toxins are powerful toxic proteins of bacterial and plant origin; for example, a human cell can be killed by a single molecule of the diphtheria toxin, secreted by *Corynebacterium diphtheriae*, or of the protein ricin, if the toxin enters the cytosol [99,100].

Toxins are composed of two polypeptide chains linked by a disulfide bond: The A chain carries the enzymatic activity and the B chain confers specificity by binding to surface cellular receptors [101,102]. After cell internalization, the A chain inactivates a protein called elongation factor 2 (EF-2), resulting in cell death [103-105]. In one day, a single diphtheria toxin molecule can inactivate most of the two million EF-2 molecules in a typical animal cell [102].

The use of intact toxins for coupling to antibodies, however, is hampered by nonspecific toxicity, due to the interactions between B chains and almost any cells, resulting in similar killing efficiencies for normal cells as for cancer cells expressing the target antigen.

To overcome the ability of toxins to bind to their own receptors, an approach has been to couple only toxin A chains to MoAbs. This results in a high degree of cell type selectivity in vitro [106,107], but in poor toxicity to solid tumors in vivo [108,109]. This low toxicity may be due in part to slow transport of A chain to the cytosol [110,111], as A chains coupled to antibodies seem to reach the cytoplasm with lower efficiency than intact toxin molecules do, B chain being essential in receptor binding and membrane insertion [102].

Thus target cell toxicity of immunotoxins could be ideally increased by including the B-chain region required for membrane insertion without the one for receptor binding, or by adding B chain separately [112,113].

A first simple and effective way of blocking the nonspecific binding of intact ricin-MoAb conjugates in vitro is to hinder the ricin receptor-binding site by adding high concentrations of lactose [114,115]. However, this technique does not seem clinically viable and is therefore limited to in vitro bone marrow purging [116].

Another approach is to chemically modify the B chain so that receptor-binding activity is removed but membrane-insertion ability is retained [117-119]. Another promising alternative is to manipulate the diphtheria toxin gene to eliminate the nucleotide sequences that correspond to the receptor-binding site on the B chain [120]. Gelonin, a toxin consisting of a single protein chain whose property is ribosomal inactivation, with no ability to bind to intact cells, has also shown a selective activity when coupled to antibodies [121-124].

Although efficacy of immunotoxins has been demonstrated in vitro [107, 113,120,125-132] and in vivo against a wide variety of animal tumor models [133-138], clinical trials of the in vivo administration of immunotoxins to patients have been to date very limited [139-141].

Immunoconjugates with Intermediate Carriers

Although direct conjugation of drugs to antibodies has been reported to be efficient in many in vitro/in vivo experiments, this technique is limited by several drawbacks; for example, linking of drugs to antibodies may result in inactivation of antibody and/or drug [142] and there are also only a small number of functional groups available per antibody molecule that can be used for direct coupling with drugs without significant loss of antibody specificity.

One means of overcoming these drawbacks of direct conjugation is to use an intermediate carrier to increase the drug:antibody molar ratio without harming the antigen-binding activity. The intermediate can carry many more drug molecules than the antibody alone and is subsequently conjugated with monoclonal antibodies.

Human Serum Albumin (HSA) *and Dextran as Intermediates.* Over the past few years, this approach has been extensively developed, mainly to prepare drug-carrier-MoAb conjugates.

Immunoconjugates have been prepared, for instance, by linking MoAbs to methotrexate, using human serum albumin (HSA) as a carrier. These conjugates have presented an increased cytotoxicity compared to the free drug, while retaining the MoAb specificity [143-150]. In addition, methotrexate-HSA-MoAb conjugates have been proved to be more effective and less toxic than free drug when injected in immunodeprived mice bearing human tumors as xenografts [85,145,146].

Mitomycin C has also been conjugated with monoclonal antibodies through two different carriers, HSA and dextran [150], and both immunoconjugates were more cytotoxic than free drug against murine bladder tumor cells.

Daunomycin-dextran-antibody conjugates have similarly been prepared to enhance the drug:antibody molar ratio thus increasing cytotoxicity, and to allow a higher drug exposure on the conjugate surface with less steric hindrance thus resulting in higher efficacy [82].

Liposomes as Intermediates. Liposomes have also been investigated as intermediate carriers, monoclonal antibodies being coupled to their surface in order to obtain immunoliposomes. Although liposomes can be used as drug carriers for passive targeting corresponding to the natural distribution of the carrier [151-154], in order to overcome the problem of nonspecific targeting, the binding of MoAbs to these drug-delivery systems has been extensively studied.

The use of liposomes as intermediates presents several advantages over those previously described:

A relatively large amount of drug can be encapsulated into the liposomal carrier, so the drug is protected from premature inactivation during transport to the site of action.

Since liposomes act as an envelope, the body is protected from the drug, resulting in reduction of toxic side effects.

The number of antibodies per liposome can be varied up to a few hundred, so immunoliposomes can be highly multivalent. It is therefore possible to enhance the affinity of drug targeting by modifying the valency of the immunoliposomes [155].

Immunoliposomes have been prepared with different methodologies and have shown high specific affinity to the target cells in vitro [155–159]. Since it is essential to understand the mechanisms of interaction of immunoliposomes with target cells for their potential use as a site-specific drug delivery system, their binding and consequent uptake by cells have been investigated. It has been demonstrated, for instance, that immunoliposomes can bind to the target cell surface and then be internalized by the cells, via an endocytic pathway [160,161]. It is clear that if endocytosis is the mechanism involved in immunoliposome uptake by the target cells, the efficacy of the liposome-encapsulated drug will be related to its stability in the lysosomal environment and to its ability to escape to the intracellular target sites, if either the drug is active in its native form or the site of action is not the lysosomes.

Although encapsulated drugs can reach intracellular sites of action after the binding event, either by leakage from the immunoliposomes or by fusion of the liposomal bilayer with the cell membrane and consequent release of drugs into the cytoplasm, the endocytic pathway for immunoliposome-encapsulated drug uptake by the target cells has been proved to be important.

Methotrexate encapsulated in liposomes bearing a monoclonal antibody against the murine major histocompatibility antigen, H-2K^k, exhibits, for instance, a target cell-specific cytotoxicity in vitro, owing to the endocytosis of immunoliposomes and release of free drug from the lysosomes into the cytoplasm [160]. Chloroquine, a lysosomotropic amine known to raise the intracellular pH of mammalian cells, partially reversed this cytotoxicity. This phenomenon could be due to the increase of the intralysosomal pH above the pKa of the acidic groups on methotrexate, thus blocking the partition of the weakly acidic drug molecule into the cytoplasm from the lysosomes; the lysosomal hydrolases may be also inhibited by the chloroquine treatment. Other studies have demonstrated the specificity of MoAb-bearing liposomes containing methotrexate and the ability of the drug molecules to penetrate the cell by endocytosis of the liposomes [162–164].

Conversely, cytosine-β-D-arabinofuranoside (ara-C) encapsulated into immunoliposomes did not induce target cell-specific cytotoxicity, probably because the drug was rapidly degraded by the lysosomal enzymes after endocytosis of the carrier system [160].

Other approaches have attempted to avoid this lysosomal inactivation of cytotoxic nucleoside analogs by using different types of liposomes. Liposomes that become unstable and fuse with endosomes upon acidification around an endosomal pH of 5.0–6.5 have been prepared [165–167], so that most of the endocytosed liposomes released the drug molecules into the target-cell cytoplasm before reaching the lysosomes [167–170].

The inhibition of virus replication and reduced cell toxicity were also observed relatively to the free drug [171] when using target-sensitive immunoliposomes for the delivery of cytotoxic drugs of nucleoside analogs, ara-C and acycloguanosine (acyclovir), into herpes simplex virus-infected cells. This significant enhancement of antiviral potency by these immunoliposomes was shown to result from a target cell-induced release of high drug concentrations at the cell surface, followed by a rapid drug transport into the cytosol by an active transport mechanism. Therefore, to be efficient, site-specific delivery systems that target and release drug molecules at the cell surface would probably require an active transport mechanism of the released drug across the cell membrane.

Immunoliposomes containing ara-C lipophilic prodrug have also exhibited cell-specific targeting in vitro, and since ara-C prodrug-liposome showed in vivo an increased antitumor activity against L1210 lymphoid leukemia compared to free drug, these effects might be improved using immunoliposomes [172].

Liposomes containing the anticancer drug, actinomycin D, coated with anti-mouse mammary tumor MoAbs or anti-human bladder cancer MoAbs and so-called chemoimmunoliposomes [173], bound specifically to the target cells in vitro and were selectively active, independently of an endocytic pathway. A single intraperitoneal (I.P.) injection of actinomycin D encapsulated into liposomes bearing anti-mouse mammary tumor MoAbs were highly effective against I.P. transplanted tumor cells bearing antigens and resulted in many cures of mice. A single intravenous injection of these immunoliposomes into mice was also effective against subcutaneous transplanted tumors bearing the antigen, but only when the reticuloendothelial system (RES) clearance was first blocked by injecting a preloading dose of unmodified liposomes. Ricin-containing small unilamellar (SUV) and large unilamellar (LUV) liposomes bearing anti-CEA antibody have also been prepared, in order to achieve a selective targeting and to protect the toxin from enzymatic degradation in vivo, as observed with anti-CEA MoAb ricin-A-chain conjugate [174].

It was shown that LUV immunoliposomes containing ricin exerted their cytotoxicity in vitro by increasing the toxin concentration around the tumor cells but not through fusion or internalization, since the same LUV carriers containing ricin A chain were ineffective. Conversely, ricin A chain encapsulated in SUV immunoliposomes was specifically highly cytotoxic, suggesting that, unlike the LUV immunoliposomes, these smaller SUV site-specific drug delivery systems were internalized by tumor cells.

Polymeric Particles as Intermediates. Antibodies have been associated with polymeric particles either by direct adsorption [175-178] or via protein A adsorption [179-181], leading generally to cell-specific targeting in vitro but to competitive displacement of the adsorbed antibodies by blood components in vivo, resulting in a nonspecific recognition of the carrier system [182].

An alternative method for overcoming the reversibility of MoAb adsorption is to covalently attach MoAbs to the particle surface. To date, immunomicrospheres have been designed mainly as reagents for cell receptor mapping and cell separation [183-188], but a few papers have discussed the potential use of covalently bound MoAb-drug-loaded nanoparticles as a new site-specific drug delivery system, and this subject will be covered in the second part of this chapter.

II. CD3 MONOCLONAL ANTIBODIES COVALENTLY COUPLED TO POLYMETHACRYLIC NANOPARTICLES: IMMUNONANOSPHERES AS A POTENTIAL SITE-SPECIFIC DRUG DELIVERY SYSTEM

A. Polymethacrylic Nanoparticles for Passive Drug Targeting

Site-specific drug delivery aims to ensure that drugs will arrive at their sites of action in a way that is optimal for interaction with the pharmacological receptors and in a manner that avoids toxic reactions at nontargeted sites and protects the drug from any premature inactivation. In recent years, there has been a growing interest in the use of particulate carriers for drug targeting [189-192], and we have focussed our investigations on the development and potential applications of polymethacrylic nanoparticles as a site-specific drug delivery system [193,194].

When administered intravenously, colloidal particles of a size that are not able to extravasate are generally cleared within a few minutes from the bloodstream, by the cells of the mononuclear phagocyte system (MPS) [193]. Among the MPS cells, the macrophages of the liver, called Kupffer cells, are normally the most efficient cells for the phagocytosis of injected particles [193]. Nevertheless, particle uptake by the MPS has been shown to be related to many parameters, such as size, surface characteristics (charge, hydrophilicity), and particle stability [192,193,195]. MPS clearance can be regarded as either a critical barrier to drug targeting via the vascular compartment, or as an opportunity for a passive delivery of drugs for the treatment of some diseases that involve the MPS or for macrophage activation [196-198].

Polymethacrylic nanoparticles have thus been designed for passively targeting the anticancer drug, doxorubicin (DXR), to the liver, with the aim of reducing the toxic effects, e.g., acute hematologic disorders and dose-dependent cardiac toxicity, while increasing the therapeutic activity.

After intravenous administration of DXR-loaded nanoparticles into rabbits, a prolonged elevation in DXR plasma levels and a reduction of the total clearance were observed [199]. In addition, 1 h after injection of DXR-bound nanospheres, the concentration of DXR in the liver was found to be 75% higher and the levels of both metabolites, doxorubicinol and doxorubicinone, were twofold lower than those corresponding to the free injected drug. The observed modifications of pharmacokinetics and tissue distribution of DXR coupled to the nanospheres were thought to be due to the passive hepatic targeting of the drug-carrier system into the Kupffer cells, followed by a slow intracytoplasmic release of DXR from the lysosomes and thence a diffusion into the hepatocytes [200].

The lysosomotropism and enhanced antitumor activity of DXR-loaded polymethacrylic nanoparticles on a human monocyte-like cancer cell line U-937 have been demonstrated [201], and this new site-specific DXR delivery system is now being investigated, in preliminary clinical trials, for the treatment of patients with hepatocarcinoma [202].

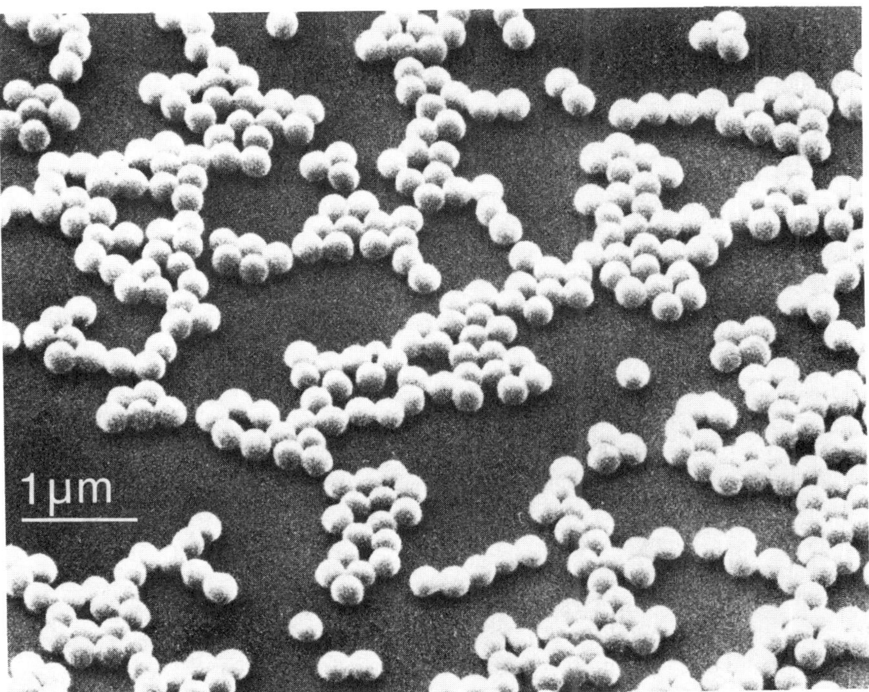

Figure 1 Scanning electron micrograph of methacrylic copolymer nanoparticles (mean diameter = 250 nm). (From Ref. 205.)

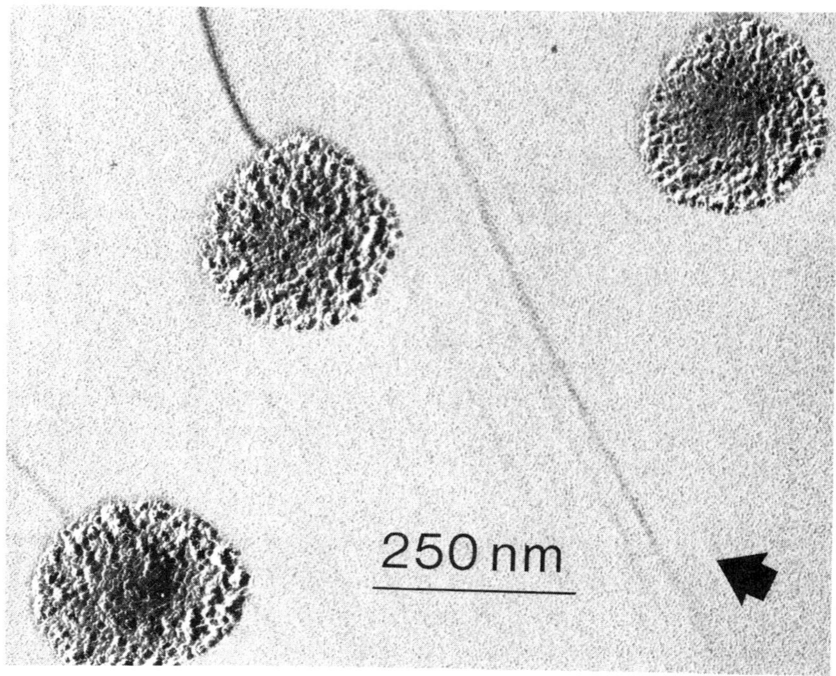

Figure 2 Transmission electron micrograph of freeze-fractured nanospheres. (From Ref. 205.)

Apart from therapeutic applications, these nanoparticles could probably be used diagnostically for the passive targeting of contrast agents, gadolinium (Gd) for instance, for contrast enhancement in magnetic resonance imaging (MRI). As suggested elsewhere [203], Gd-DTPA-nanoparticle systems might give a durable and increased MRI enhancement of the liver, owing to the passive targeting of the contrast agent to this specific organ.

Although passive targeting can be successful for delivering drugs to certain specific organs and cellular sites, the cells of the MPS may not be the desired target. Strategies must be employed for avoiding particle uptake by the MPS [195, 204]. Among the possibilities, active targeting (referring to a modification in the natural distribution pattern of the drug carrier) has been investigated by coupling monoclonal antibodies to polymethacrylic nanoparticles [205].

B. CD3 Immunonanospheres: In Vitro Active Targeting to Human T Lymphocytes

Preparation of the CD3 Immunonanospheres

Manufacture of Polyalkylmethacrylate Nanoparticles. Polymethacrylic nanoparticles were prepared by aqueous emulsion copolymerization of the methacrylic monomers: methyl methacrylate, 2-hydroxypropyl methacrylate, methacrylic acid, and ethylene glycol dimethacrylate (as previously described in [194, 206,207]). After polymerization the nanoparticle suspension was purified by filtration and dialysis [208].

Using a Coulter Nano-Sizer and scanning electron microscopy, the mean particle diameter was determined to be around 250 nm. As seen in Figure 1, the nanospheres are spherical, homogeneous, and discrete, with a very narrow size distribution. The internal structure, observed by transmission electron microscopy after freeze-fracturing, appears to be porous, with a dense core surrounded by a granulous shell (Figure 2).

Methacrylic copolymer nanoparticles, with an average size of 250 nm, have an external surface of 1.96×10^5 nm^2 and a volume of 8.2×10^6 nm^3 per particle, corresponding to about 2×10^{12} particles/ml suspension.

Hydroxyl and carboxyl groups were detected in the nanospheres by nuclear magnetic resonance and infrared spectrometry [207,208], and hydrogen ion titration measurements indicated approximately 10^5 carboxyl groups on the surface of each particle. The molecular weight of the methacrylic copolymers constituting the nanoparticles was estimated by liquid exclusion chromatography to be 270,000. At neutral pH the particles were negatively charged, with a zeta potential of about -40 mV.

The nanoparticle suspension could be sterilized by autoclaving, and freeze-dried without any effect on the physicochemical stability for at least 1 year after treatment. Due to their physicochemical characteristics, the nanospheres were able to efficiently adsorb a wide variety of molecules, such as drugs (doxorubicin, daunorubicin, bleomycin, morphin) or fluorescent agents (ethidium bromide, acridine orange).

Production of the X.35 Monoclonal Antibody. A CD3 MoAb (X.35) was selected for its high affinity to human T lymphocytes [209,210].

The X.35 hybrid secreted murine IgG2a molecules at high concentrations (10 μg/ml in culture supernatant and 8-10 mg/ml in ascites) and reacted with all the human peripheral T lymphocytes, mature thymocytes, some acute T-lymphoblastic leukemias, and most chronic T-lymphoid leukemias. No reactivity was observed with B lymphocytes and monocytes.

With X.35 MoAb, 70 ± 9% of the human peripheral blood mononuclear cells, isolated from whole blood by Ficoll gradient centrifugation and composed on

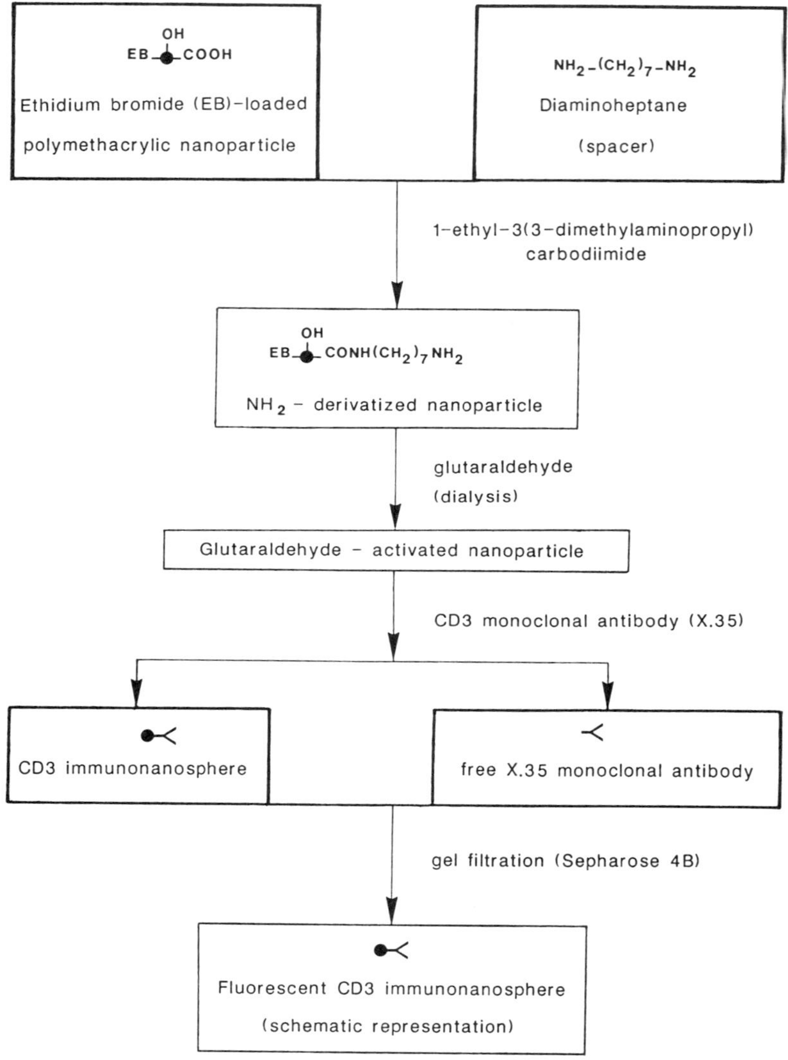

Figure 3 Preparation scheme of CD3 immunonanospheres.

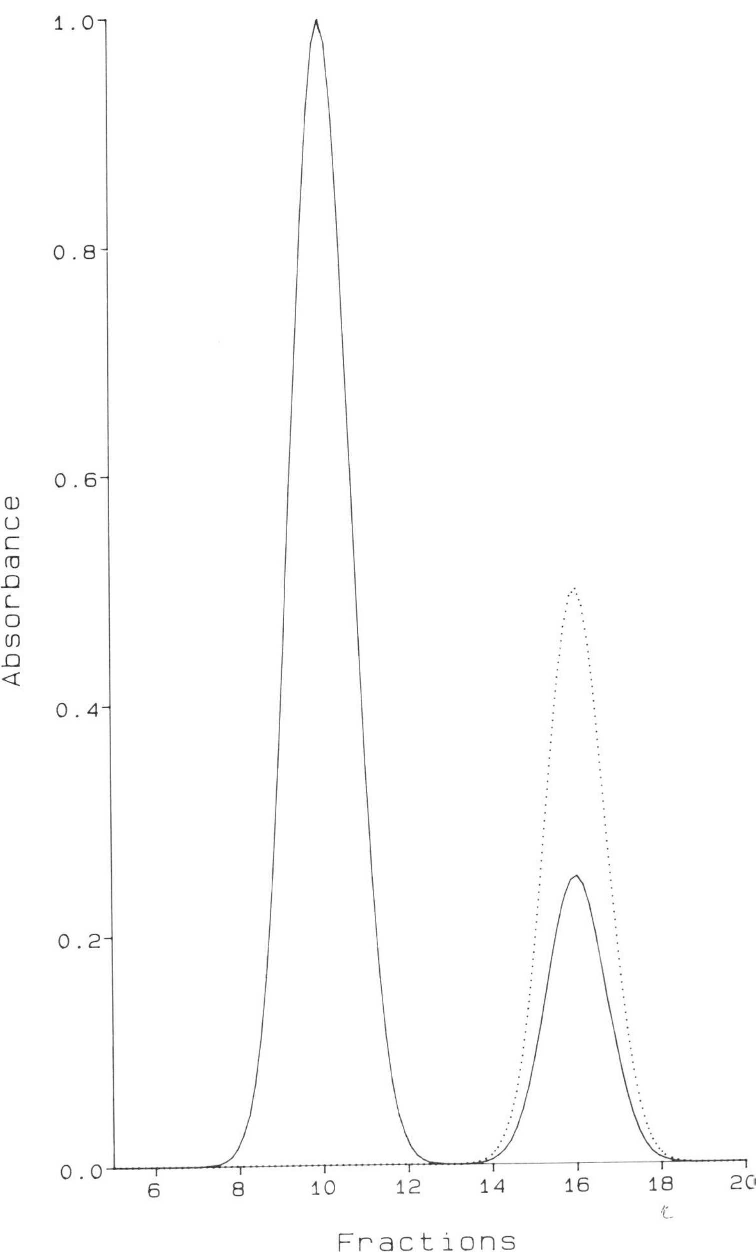

Figure 4 Purification of CD3 immunonanospheres by gel filtration. Fractions 8–12, purified immunonanospheres; fractions 14–18, unbound X.35 MoAb; dotted line, elution profile of pure X.35 MoAb. (From Ref. 205.)

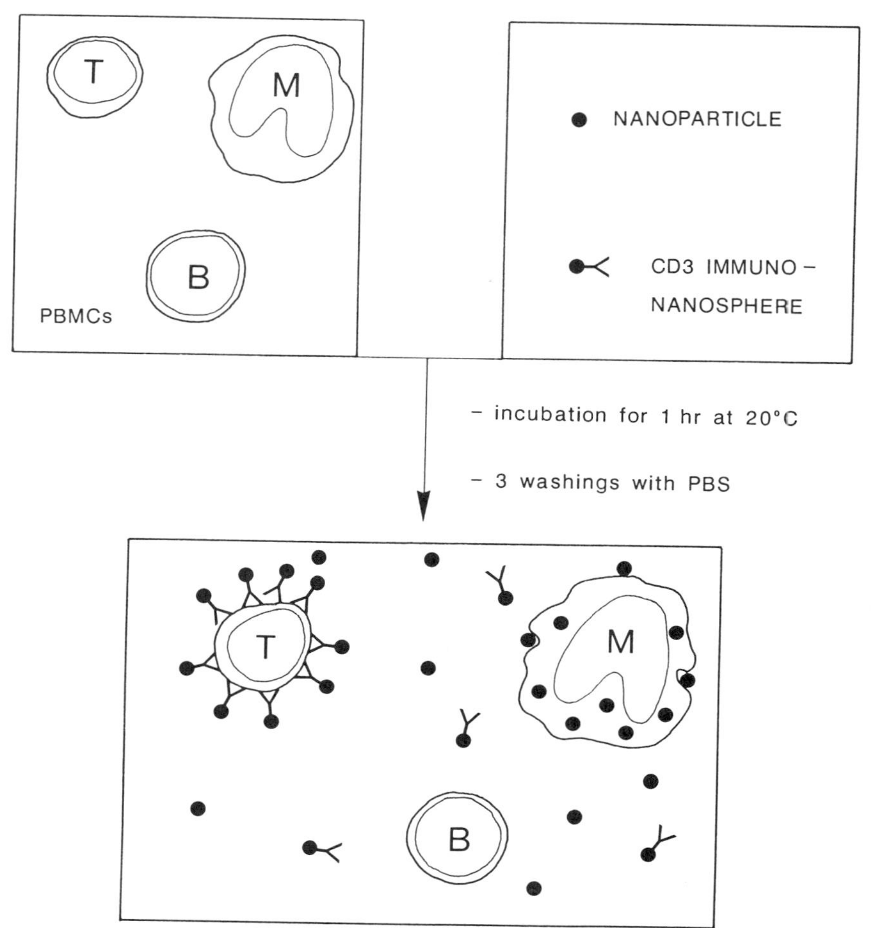

Figure 5 Representation of in vitro specific targeting of CD3 immunonanospheres to human T lymphocytes. (PBMCs = peripheral blood mononuclear cells; T and B = T and B lymphocytes; M = monocyte).

average of 70% T lymphocytes, 15% B lymphocytes, 5% monocytes, and 10% null cells, were labeled (indirect immunofluorescence test using goat anti-mouse Ig).

CD3 Monoclonal Antibody-Nanoparticle Binding. CD3 immunonanospheres were prepared according to Figure 3 [205]. Fluorescent particles were first obtained by adsorbing ethidium bromide at a concentration of 2.5 mM with a 100% binding yield.

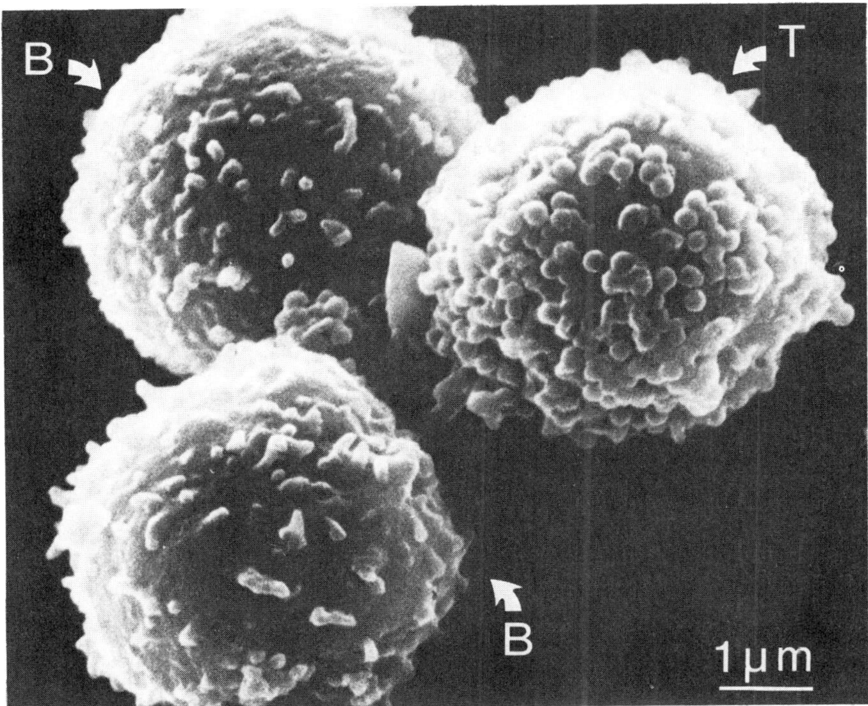

Figure 6 Active targeting of CD3 immunonanospheres to human T lymphocytes observed by scanning electron microscopy. (T and B = T and B lymphocytes). (From Ref. 205.)

Ethidium bromide-loaded nanoparticles were subsequently modified by coupling a spacer, 1,7-diaminoheptane, using the carbodiimide reaction [211]. Amino-derivatized nanospheres were thus obtained and stored at 4°C in the dark until tested for MoAb binding.

X.35 MoAbs were bound to NH_2-derivatized particles by covalent coupling using the glutaraldehyde method in two steps [185,205,210].

Fluorescent immunonanospheres were separated from unbound antibodies by gel filtration on Sepharose 4B (Figure 4), and the quality of the purification was confirmed by an indirect immunofluorescence test using fluoresceinated goat anti-mouse Ig.

In Vitro Specific Targeting of CD3 Immunonanospheres to Human T Lymphocytes

Prior to the preparation of CD3 immunonanospheres, two different assays were performed, either with passive adsorption of MoAbs onto fluorescent particles or with direct covalent binding of MoAbs onto the particles without using any spacer.

With MoAb-coated nanoparticles, no evidence of active targeting could be shown, perhaps because of displacement of the adsorbed antibodies and/or binding of the antibodies onto the particle surface via the antigen binding sites, Fab, resulting in a loss of MoAb-coated nanosphere specificity.

By directly and covalently binding MoAbs onto particles, even if sometimes some specific targeting was observed [193], aggregates of both immunonanoparticles and labeled cells were observed. In addition, since MoAbs have both amino and carboxyl groups, inter- and intramolecular cross-linking of these proteins can occur.

In the test procedure, peripheral blood mononuclear cells (PBMCs) were isolated from whole blood by Ficoll gradient centrifugation and the cells were incubated with the fluorescent CD3 immunonanospheres at 20°C. Previous elimination of free X.35 MoAb from the immunonanosphere suspension, by gel filtration, was crucial, since competitive binding between immunonanospheres and CD3 MoAb was demonstrated.

After 1 h incubation, the cells were washed three times by centrifugation with PBS to remove the unbound nanoparticles. The cells, labeled in one step, were observed by fluorescence and scanning electron microscopy. Unbound, NH_2-derivatized and glutaraldehyde-activated fluorescent nanoparticles were used as controls.

Using purified ethidium bromide-loaded CD3 immunonanospheres, 70 ± 9% of PBMCs, corresponding to the T lymphocytes, were labeled, while B lymphocytes remained unlabeled as determined by fluorescence microscopy [205, 210]. All of the controls were negative, since they did not exhibit any specific recognition of T lymphocytes.

Surprisingly, some monocytes exhibited an intracellular fluorescence. As shown in Figure 5, if the binding of MoAbs to the fluorescent particles is not complete, some free nanospheres can be available for phagocytosis by the human monocytes, as described previously [212,213]. In addition, it is possible that some CD3 MoAbs may bind onto a particle via the Fab site, thus exposing the Fc part on the surface of the particle, which can be then recognized by membrane receptors of the monocytes. Therefore, some of the CD3 immunonanospheres could be taken up by macrophages by phagocytosis mediated by Fc receptors.

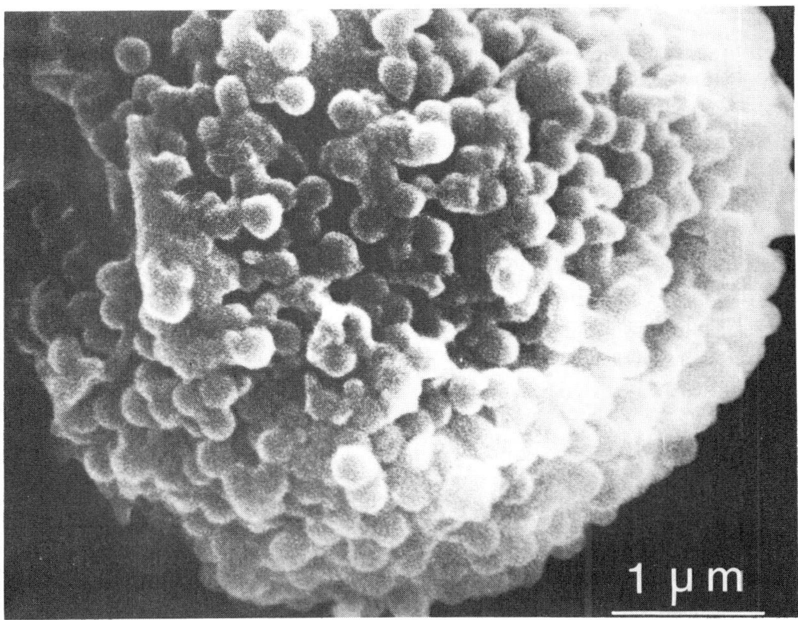

Figure 7 Scanning electron micrograph of a T lymphocyte covered with surface-targeted CD3 immunonanoparticles.

Although active targeting of fluorescent nanoparticles to human T lymphocytes, via CD3 MoAbs, was demonstrated, a heterogeneous distribution of the immunonanospheres from one positive cell to another one was observed [205].

Using scanning electron microscopy (Figure 6), active targeting to T lymphocytes was clear while B lymphocytes remained unlabeled. As shown in Figure 7, a large number of particles could be targeted to one specific cell using covalently coupled MoAbs, and since these particles are porous and present a large surface area, a large number of drug molecules could be associated with the carriers and targeted to cells expressing the corresponding antigen (e.g., 4.5×10^5 doxorubicin molecules can be carried by a single nanosphere).

Active targeting, proved in vitro by using CD3 immunonanospheres, illustrates the stability of the MoAb-nanoparticle linkage, the retention of MoAb specificity, and the possibility of targeting a large number of molecules (ethidium bromide in the above case) to a specific cell population.

III. CONCLUSION–PERSPECTIVES

The development of monoclonal antibody technology, which was initiated by Kohler and Milstein in 1975, has confirmed the early speculations of Paul Ehrlich about the existence of the body's natural "magic bullets."

Monoclonal antibodies have been proposed mainly for cancer diagnosis and therapy, in native and conjugated form (with or without intermediates).

Although successful clinical trials in cancer patients have been obtained with native MoAbs, there are some problems due to the immunogenicity of the murine MoAbs injected in patients, to antigenic modulation and to the limited natural capabilities of antibodies.

The use of immunoconjugates that are radioisotopes, anticancer drugs or toxins bound to MoAbs without intermediates, leads to the same limitations as native antibodies and to some additional problems. Although radioimmunotherapy has given clinical responses in patients with cancer, the major drawback of such an approach is that radioisotopes can damage normal tissues during their distribution to the target site.

Direct conjugation of drugs or toxins to MoAbs has been reported to be efficient in many in vitro/in vivo experiments, but to date very few published clinical trials have used such immunoconjugates for the treatment of cancer patients.

To overcome some problems related to direct binding of drugs (or toxins) to MoAbs, such as inactivation of antibody and/or drug, and also limited drug: antibody molar ratio, immunoconjugates with intermediate carriers (e.g., HSA, dextran, liposomes) have been extensively developed over the past few years.

More recently, some polymeric particles have been designed as intermediate carriers, and a few attempts have been made to bind MoAbs to particles in order to achieve cell-specific targeting.

The experimental part of this chapter describes a new carrier system consisting of submicroscopic polymethacrylic particles, which could be used for either passive or active drug targeting.

Although passive targeting has been used successfully for delivering drugs such as the anticancer agent doxorubicin to desired sites (mainly to cells of the MPS), in order to modify the natural distribution pattern of the drug carrier active targeting has been investigated using immunonanospheres, namely MoAbs covalently coupled to polymethacrylic nanoparticles.

A CD3 monoclonal antibody (an anti-human T mature lymphocyte antibody) was covalently bound to ethidium bromide-loaded nanoparticles. Peripheral blood mononuclear cells were used as targets, and the specific targeting of CD3 immunonanospheres to T lymphocytes was demonstrated by fluorescence and scanning electron microscopy.

The results obtained in vitro look very promising for active targeting of drugs to specific cells. Drug molecules can bind to polymethacrylic nanoparticles either at the surface or inside their porous structure, thus protecting the drugs from premature inactivation and also preventing toxic reactions at nontargeted sites. Using such carriers, the number of transported molecules can be varied up to a few thousand, by which immunonanospheres can target a large amount of drug molecules to a specific cell.

Hopefully, intravenously administered immunonanospheres will exhibit a different distribution pattern than nanoparticles, which are quantitatively and rapidly taken up by the MPS cells. However, nonspecific immunonanosphere uptake by macrophages of the MPS remains a possibility, unless the MPS clearance is first blocked by injecting a preloading dose of placebo colloid.

To minimize the uptake of immunonanospheres by macrophages, perhaps seen in vitro for some CD3 immunonanospheres having possibly MoAbs bound on the surface via the Fab site, thus having the Fc part available for recognition by membrane receptors on macrophages, another alternative could be to bind Fab fragments only to the nanoparticles.

Another approach to modifying the natural distribution of nanoparticles, already successful in animal models, is to change their surface characteristics in order to avoid interactions with macrophages; coating the particle surface with a nonionic surfactant, for example [177,195], can provide a steric stabilization effect, resulting in an avoidance of particle uptake by the liver.

Immunonanospheres that avoid capture by the MPS could represent a new generation of site-specific drug delivery systems, especially for an active targeting of drug molecules to intravascular targeted cells. For extravascular target sites, the potential use of immunonanospheres will be limited by their ability to extravasate, under anatomical and (patho)physiological control, and also under carrier- and/or disease-related effects [204].

IV. ACKNOWLEDGMENTS

The authors would like to thank R. Le Verge (Laboratoire de Pharmacie Galénique, Université de Rennes) and B. Genetet (Centre Régional de Transfusion Sanguine, Rennes).

We are also indebted to E. Tomlinson and to colleagues of the Advanced Drug Delivery Research Centre (Ciba-Geigy Pharmaceuticals) for suggestions and valuable discussions.

V. REFERENCES

1. G. Kohler and C. Milstein, *Nature 256*:495 (1975).
2. F. M. Brodsky, *Pharm. Res. 5*(1):1 (1988).

3. V. L. Alvarez, M. L. Wen, C. Lee, A. D. Lopes, J. D. Rodwell, and T. J. McKearn, *Nucl. Med. Biol. 13*(4):347 (1986).
4. C. L. Viswanathan and A. V. Damle, *Indian Drugs 24*(2):82 (1986).
5. C. Ford and A. Casson, *Cancer Chemother. Pharmacol. 17*:197 (1986).
6. R. L. Vessella, V. Alvarez, R. K. Chiou, J. Rodwell, M. Elson, D. Palme, R. Shafer, and P. Lange, *NCI Monographs 3*:159 (1987).
7. R. M. Rainsbury, R. J. Ott, J. H. Westwood, T. S. Kalirai, R. C. Coombes, V. R. McCready, A. M. Neville, and J. C. Gazet, *Lancet 2*:934 (1983).
8. C. H. Thompson, M. Lichtenstein, S. A. Stacker, M. J. Leyden, N. Salehi, J. T. Andrews, and I. F. C. McKenzie, *Lancet 2*:1245 (1984).
9. M. R. Williams, A. C. Perkins, F. C. Campbell, M. V. Pimm, J. G. Hardy, M. L. Wastie, R. W. Blamey, and R. W. Baldwin, *Clin. Oncol. 10*:375 (1984).
10. P. A. Farrands, A. C. Perkins, M. V. Pimm, J. D. Hardy, M. J. Embleton, R. W. Baldwin, and J. D. Hardcastle, *Lancet 2*:397 (1982).
11. D. M. Goldenberg, E. E. Kim, S. J. Bennett, M. O. Nelson, and F. H. Deland, *Gastroenterology 84*:524 (1983).
12. M. R. Price, M. V. Pimm, C. M. Page, N. C. Armitage, J. D. Hardcastle, and R. W. Baldwin, *Br. J. Cancer 49*:809 (1984).
13. S. M. Larson, J. P. Brown, P. W. Wright, J. A. Carrasquillo, I. Hellstrom, and K. E. Hellstrom, *J. Nucl. Med. 24*:123 (1983).
14. J. L. Murray, M. G. Rosenblum, R. E. Sobol, R. M. Bartholomew, C. E. Plager, T. P. Haynie, M. F. Jahns, H. J. Glenn, L. Lamki, R. S. Benjamin, N. Papadopoulos, A. W. Boddie, J. M. Frincke, G. S. David, D. J. Carlo, and E. M. Hersh, *Cancer Res. 45*:2376 (1985).
15. A. A. Epenetos, S. Mather, M. Granowska, C. C. Nimmon, L. R. Hawkins, K. E. Britton, J. Shepherd, J. Taylor-Papadimitriou, H. Durbin, and J. S. Malpas, *Lancet 2*:999 (1982).
16. A. A. Epenetos, J. Shepherd, K. E. Britton, S. Mather, J. Taylor-Papadimitriou, M. Granowska, H. Durbin, C. C. Nimmon, L. R. Hawkins, J. S. Malpas, and W. F. Bodmer, *Cancer 55*:984 (1985).
17. C. Berche, J. P. Mach, J. D. Lumbroso, C. Langlais, F. Aubry, F. Buchegger, S. Carrel, P. Rougier, C. Parmentier, and M. Tubiana, *Br. Med. J. 285*:1447 (1982).
18. H. M. Smedley, P. Finan, E. S. Lennox, A. Ritson, F. Takei, P. Wraight, and K. Sikora, *Br. J. Cancer 47*:253 (1983).
19. N. Ishii, K. Nakata, T. Muro, R. Furukawa, K. Kono, Y. Kusumoto, T. Munehisa, T. Koji, S. Nagataki, S. Nishi, Y. Tsukada, and H. Hirai, *Ann. N.Y. Acad. Sci. 417*:270 (1983).
20. K. Nakata, R. Furukawa, K. Kono, T. Muro, A. Sato, Y. Kusumoto, N. Ishii, T. Munehisa, T. Koji, and S. Nagataki, *Tumour Biol. 5*:161 (1984).
21. N. Markham, A. Ritson, O. James, N. Curtin, M. Bassendine, and K. Sikora, *J. Hepatol. 2*:25 (1986).
22. J. F. Bergmann, J. D. Lumbroso, L. Manil, J. C. Saccavini, P. Rougier, M. Assicot, A. Mathieu, D. Bellett, and C. Bohuon, *Eur. J. Nucl. Med. 13*:385 (1987).

23. G. F. Rowland, in *Immunological Intervention in Medicine* (J. Mobray, ed.), Saunders, Philadelphia, Clinics in Immunology and Allergy *3*(2), 1983, pp. 235–257.
24. M. R. Clark and H. Waldmann, *J. Natl. Cancer Inst. 79*:1393 (1987).
25. B. L. Ferraiolo and L. Z. Benet, in *Topics in Pharmaceutical Sciences* (D. D. Breimer and P. Speiser, eds.), Elsevier Science Publishers B.V. (Biomedical Division), 1985, pp. 3–18.
26. W. Uracz, A. Pituch-Noworolska, M. Zembara, T. Popiela, and A. Czupryna, *J. Cancer Res. Clin. Oncol. 104*:181 (1982).
27. G. Schultz, T. F. Bumol, and R. A. Reisfeld, *Proc. Natl. Acad. Sci. USA 80*:5407 (1983).
28. Z. Steplewski, M. D. Lubeck, and H. Koprowski, *Science 211*:865 (1983).
29. K. Takamuku, T. Akiyoshi, and H. Tsuji, *Cancer Immunol. Immunother. 25*:137 (1987).
30. R. O. Dillman and I. Royston, *Br. Med. Bull. 40*:240 (1984).
31. J. Ritz and S. F. Schlossman, *Blood 59*(1):1 (1982).
32. J. Ritz, J. M. Pesando, J. Notis-McConarty, and S. F. Schlossman, *J. Immunol. 125*(4):1506 (1980).
33. K. S. Webb, J. L. Ware, S. F. Parks, W. H. Briner, and D. F. Paulson, *Cancer Immunol. Immunother. 14*:155 (1983).
34. R. Levy and R. A. Miller, *Fed. Proc. 42*(9):2650 (1983).
35. R. L. Edelson, J. Raafat, C. L. Berger, M. Grossman, C. Troyer, and M. Hardy, *Cancer Treat. Rep. 63*:675 (1979).
36. R. A. Miller, D. G. Maloney, R. Warnke, and R. Levy, *N. Engl. J. Med. 306*(9):517 (1982).
37. C. S. Conner, *Drug Intell. Clin. Pharm. 18*:65 (1984).
38. R. A. Miller, D. G. Maloney, J. McKillop, and R. Levy, *Blood 58*:78 (1981).
39. R. O. Dillman, J. C. Beauregard, D. L. Shawler, R. E. Sobol, and I. Royston, in *Protides of the Biological Fluids* (H. Peeters, ed.), Pergamon Press, 1983, pp. 353–358.
40. R. O. Dillman, D. L. Shawler, R. E. Sobol, H. A. Collins, J. C. Beauregard, S. B. Wormsley, and I. Royston, *Blood 59*(5):1036 (1982).
41. R. O. Dillman, D. L. Shawler, J. B. Dillman, I. Royston, and M. Clutter, *Blood 62*:200a (1983).
42. H. F. Sears, J. Mattis, D. Herlyn, P. Hayry, B. Atkinson, C. Ernst, Z. Steplewski, and H. Koprowski, *Lancet 1*:762 (1982).
43. H. F. Sears, D. Herlyn, Z. Steplewski, and H. Koprowski, *J. Biol. Resp. Med. 3*:138 (1984).
44. J. Y. Douillard, *Accomplishments in Oncology 1*:193 (1986).
45. R. S. Geha, *N. Engl. J. Med. 305*:25 (1981).
46. A. Hatzubai, D. G. Maloney, and R. Levy, *J. Immunol. 126*:2397 (1981).
47. T. C. Meeker, D. G. Maloney, K. Thielemans, R. A. Miller, J. Lowder, and R. Levy, *Blood 62*(5, suppl. 1):759 (1983).
48. R. Taetle, J. M. Honeysett, and I. Trowbridge, *Int. J. Cancer 32*:343 (1983).
49. J. Mendelsohn, I. Trowbridge, and J. Castagnola, *Blood 62*:821 (1983).

50. E. Baumler, *Paul Ehrlich. Scientist for Life*, Holmes & Meier, New York and London, 1984.
51. A. T. Vaugham, W. J. Bateman, and D. R. Fisher, *Int. J. Rad. Oncol. Bio. Phys. 8*:1943 (1982).
52. A. T. Vaugham, W. J. Bateman, G. Brown, and J. Cowan, *Int. J. Nucl. Med. Biol. 9*:167 (1982).
53. K. Sikora, H. Smedley, and P. Thorpe, *Br. Med. Bull. 40*:233 (1984).
54. W. D. Bloomer, R. Lipsztein, and J. F. Dalton, *Cancer 55*:2229 (1985).
55. J. A. Carrasquillo, K. A. Krohn, and P. L. Beaumier, *Cancer Treatment Rep. 68*:317 (1984).
56. S. M. Larson, J. A. Carrasquillo, J. C. Reynolds, I. Hellstrom, K. E. Hellstrom, J. C. Mulshine, and L. E. Mattis, *Nucl. Med. Biol. 13*(2):207 (1986).
57. S. E. Order, J. L. Klein, D. Ettinger, P. Alderson, S. Siegelman, and P. Leichner, *Cancer Res. 40*:3001 (1980).
58. D. Blakeslee and J. C. Kennedy, *J. Cancer Res. 34*:882 (1974).
59. A. Guclu, T. Ghose, J. Tai, and M. Mammen, *Eur. J. Cancer 12*:95 (1976).
60. M. Segerling, S. H. Ohanian, and T. Borsos, *Cancer Res. 35*:3195 (1975).
61. G. F. Rowland, D. A. L. Davies, G. J. O'Neill, C. E. Newman, and C. H. J. Ford, in *Immunotherapy of Malignant Diseases* (H. Rainer, ed.), Schattauer-Verlag, Stuttgart, 1977, pp. 316–322.
62. T. Ghose, A. Guclu, and J. Tai, *J. Natl. Cancer Inst. 55*:1353 (1975).
63. G. J. O'Neill, B. A. Pearson, and D. A. L. Davies, *Immunology 28*:323 (1975).
64. C. Vennegoor, D. Van Smeerdijk, and P. Rumke, *Eur. J. Cancer 11*: 725 (1975).
65. L. G. Bernier, M. Page, R. C. Gaudreault, and L. P. Joly, *Br. J. Cancer 49*: 245 (1984).
66. C. J. Oon, H. Apsey, H. Buckleton, K. B. Cooke, I. Hanham, P. Hazarika, J. R. Hobbs, and B. McLeod, *Behring Inst. Mitt. 56*:228 (1974).
67. T. Ghose, S. T. Norvell, A. Guclu, A. Bodurtha, J. Tai, and A. S. MacDonald, *J. Natl. Cancer Inst. 58*:845 (1977).
68. J. D. Everall, P. Dowd, D. A. L. Davies, G. J. O'Neill, and G. F. Rowland, *Lancet 1*:1105 (1977).
69. P. Hazarika, P. G. Elliot, and N. Byron, in *Protides of the Biological Fluids* (H. Peeters, ed.), Proceedings of the Twenty-fifth Colloquium, Brugge, 1977, Pergamon Press, Oxford, 1978, Abstract 156.
70. G. Mathé, T. B. Loc, and J. Bernard, *C. R. Hebd, Séances Acad. Sci. 246*: 1626 (1958).
71. D. A. Robinson, J. M. Whiteley, and N. G. Harding, *Biochem. Soc. Trans. 1*:722 (1973).
72. S. Burstein and R. Knapp, *J. Med. Chem. 20*:950 (1977).
73. Z. A. Latif, B.B. Lozzio, C. J. Wust, S. Krauss, M. C. Aggio, and C. B. Lozzio, *Cancer 45*:1326 (1980).
74. P. N. Kulkarni, A. H. Blair, and T. I. Ghose, *Cancer Res. 41*:2700 (1981).

75. G. F. Rowland, in *Targeted Drugs: Polymers in Biology and Medicine* (E. P. Goldberg, L. G. Donaruma, and O. Vogl, eds.), Vol. 2 in the Wiley-Interscience Series on Polymers in Biology and Medicine, John Wiley, New York, 1983, pp. 57–72.
76. J. Kanellos, G. A. Pietersz, and I. F. C. McKenzie, *J. Natl. Cancer Inst. 75*: 319 (1985).
77. N. Endo, Y. Takeda, K. Kishida, Y. Kato, M. Saito, N. Umemoto, and T. Hara, *Cancer Immunol. Immunother. 25*:1 (1987).
78. P. Uadia, A. H. Blair, and T. Ghose, *Cancer Immunol. Immunother. 16*:127 (1983).
79. P. Uadia, A. H. Blair, T. Ghose, and S. Ferrone, *J. Natl. Cancer Inst. 74*:29 (1985).
80. E. Hurwitz, R. Levy, R. Maron, M. Wilchek, R. Arnon, and M. Sela, *Cancer Res. 35*:1175 (1975).
81. M. Belles-Isles and M. Page, *Br. J. Cancer 41*:841 (1980).
82. R. Arnon and M. Sela, *Immunological Rev. 62*:5 (1982).
83. M. Belles-Isles and M. Page, *Int. J. Immunopharmacol. 3*:97 (1981).
84. M. J. Embleton, *Biochem. Soc. Trans.* (615th Meeting, Belfast) *14*:393 (1986).
85. J. R. Johnson, C. H. J. Ford, C. E. Newman, C. S. Woodhouse, G. F. Rowland, and R. G. Simmonds, *Br. J. Cancer 44*:472 (1981).
86. C. H. J. Ford, C. E. Newman, J. R. Johnson, C. S. Woodhouse, T. A. Reeder, G. F. Rowland, and R. G. Simmonds, *Br. J. Cancer 47*:35 (1983).
87. G. F. Rowland, R. G. Simmonds, J. R. F. Corvalan, R. W. Baldwin, J. P. Brown, M. J. Embleton, C. H. J. Ford, K. E. Hellstrom, I. Hellstrom, J. T. Kemshead, C. E. Newman, and C. S. Woodhouse, in *Protides of the Biological Fluids*, Vol. 30 (H. Peeters, ed.), Pergamon Press, Oxford, 1983, pp. 375–379.
88. M. J. Embleton, G. F. Rowland, R. G. Simmonds, E. Jacobs, C. H. Marsden, and R. W. Baldwin, *Br. J. Cancer 47*:43 (1983).
89. G. F. Rowland, C. A. Axton, R. W. Baldwin, J. R. Brown, J. R. F. Corvalan, M. J. Embleton, V. A. Gore, K. E. Hellstrom, E. Jacobs, C. H. Marsden, M. V. Pimm, R. G. Simmonds, and W. Smith, *Cancer Immunol. Immunother. 19*:1 (1985).
90. J. R. F. Corvalan and W. Smith, *Cancer Immunol. Immunother. 24*:127 (1987).
91. J. R. F. Corvalan, W. Smith, V. A. Gore, and D. R. Brandon, *Cancer Immunol. Immunother. 24*:133 (1987).
92. G. F. Rowland, in *Drug Delivery Systems. Fundamentals and Techniques* (P. Johnson and J. G. Lloyd-Jones, eds.), Ellis Horwood Series in Biomedicine, VCH Publishers, Chichester, 1987, pp. 81–94.
93. G. F. Rowland, R. G. Simmonds, V. A. Gore, C. H. Marsden, and W. Smith, *Cancer Immunol. Immunother. 21*:183 (1986).
94. J. R. F. Corvalan, W. Smith, V. A. Gore, D. R. Brandon, and P. J. Ryde, *Cancer Immunol. Immunother. 24*:138 (1987).

95. D. A. L. Davies, G. J. O'Neill, G. F. Rowland, C. E. Newman, and C. H. J. Ford, in *Tumor-Associated Antigens and Their Specific Immune Response* (F. Spreafico and R. Arnon, eds.), Academic Press, New York, 1979, pp. 305–323.
96. M. J. Smyth, G. A. Pietersz, and I. F. C. McKenzie, *Cancer Res. 47*:62 (1987).
97. V. J. Stella and K. J. Himmelstein, *J. Med. Chem. 23*:1275 (1980).
98. L. Brent, T. Horsburgh, G. Rowland, and P. Wood, *Transplantation* (Baltimore) *29*:280 (1980).
99. M. Yamaizumi and E. Mekada, *Cell 15*:245 (1978).
100. K. Eiklid, S. Olsnes, and A. Pihl, *Exp. Cell Res. 126*:321 (1980).
101. S. Olsnes and A. Pihl, *Biochemistry 12*:3121 (1973).
102. R. J. Collier and D. A. Kaplan, *Sci. Am. 251*(1):44 (1984).
103. S. Olsnes, K. Refsnes, and A. Pihl, *Nature 249*:627 (1974).
104. L. Barbieri and F. Stirpe, *Cancer Surv. 1*:489 (1982).
105. S. Olsnes and A. Pihl, in *Molecular Action of Toxins and Viruses* (P. Cohen and S. Van Heyningen, eds.), Elsevier Biomedical Press, New York, 1982, pp. 51–105.
106. T. W. Griffin, L. R. Haynes, and J. A. DeMartino, *J. Natl. Cancer Inst. 69*(4):799 (1982).
107. R. D. May, E. S. Vitetta, G. Moldenhauer, and B. Dorken, *Cancer Drug Delivery, 3*(4):261 (1986).
108. F. K. Jansen, H. E. Blythman, D. Carriere, P. Casellas, O. Gros, P. Gros, J. C. Laurent, F. Paolucci, B. Pau, P. Poncelet, G. Richer, H. Vidal, and G. A. Voisin, *Immunol. Rev. 62*:185 (1982).
109. R. J. Youle and M. Colombatti, in *Monoclonal Antibodies in Cancer: Advances in Diagnosis and Treatment* (J. A. Roth, ed.), M. D. Futura Publishing Company, Mount Kisco, N.Y., 1986, pp. 173–213.
110. R. J. Youle and D. M. Neville, *J. Biol. Chem. 257*:1598 (1982).
111. P. Casellas, B. J. P. Bourrie, P. Gros, and F. K. Jansen, *J. Biol. Chem. 259*: 9359 (1984).
112. D. P. McIntosh, D. C. Edwards, A. J. Cumber, G. D. Parnell, C. J. Dean, W. C. J. Ross, and J. A. Forrester, *FEBS Lett. 164*:17 (1983).
113. H. J. Thiesen, H. Juhl, and R. Arndt, *Cancer Res. 47*:419 (1987).
114. R. J. Youle and D. M. Neville, *Proc. Natl. Acad. Sci. USA 77*:5483 (1980).
115. P. E. Thorpe, A. J. Cumber, N. Williams, D. C. Edwards, W. C. J. Ross, and A. J. S. Davies, *Clin. Exp. Immunol. 43*:195 (1981).
116. A. H. Filipovich, R. J. Youle, D. M. Neville, D. A. Vallera, R. R. Quinones, and J. H. Kersey, *Lancet* 469 (1984).
117. J. E. Leonard, Q-c Wang, N. O. Kaplan, and I. Royston, *Cancer Res. 45*: 5263 (1985).
118. E. S. Vitetta, *J. Immunol. 136*(5):1880 (1986).
119. R. J. Youle and M. Colombatti, *J. Biol. Chem. 262*(10):4676 (1987).
120. L. Greenfield, V. G. Johnson, and R. J. Youle, *Science 238*:536 (1987).
121. F. Stirpe, S. Olsnes, and A. Pihl, *J. Biol. Chem. 255*:6947 (1980).

122. P. E. Thorpe, A. N. F. Brown, W. C. J. Ross, A. J. Cumber, S. I. Detre, D. C. Edwards, A. J. S. Davies, and F. Stirpe, *Eur. J. Biochem. 116*:447 (1981).
123. J. M. Lambert, P. D. Senter, A. Yau-Young, W. A. Blattler, and V. S. Goldmacher, *J. Biol. Chem. 260*(22):12035 (1985).
124. C. F. Scott, J. M. Lambert, V. S. Goldmacher, W. A. Blattler, R. Sobel, S. F. Schlossman, and B. Benacerraf, *Int. J. Immunopharmacol. 9*(2): 211 (1987).
125. D. G. Gilliland, Z. Steplewski, R. J. Collier, K. F. Mitchell, T. H. Chang, and H. Koprowski, *Proc. Natl. Acad. Sci. USA 77*:4539 (1980).
126. I. S. Trowbridge and D. L. Domingo, *Nature 294*:171 (1981).
127. V. Raso, J. Ritz, M. Basala, and S. F. Schlossman, *Cancer Res. 42*:457 (1982).
128. J. C. Merriam, H. S. Lyon, and D. H. Char, *Cancer Res. 44*:3178 (1984).
129. M. J. Bjorn, D. Ring, and A. Frankel, *Cancer Res. 45*:1214 (1985).
130. K. Tsukazaki, E. G. Hayman, and E. Rusolahti, *Cancer Res. 45*:1834 (1985).
131. R. K. Oldham, in *Immunity to Cancer* (A. E. Reif, and M. S. Mitchell, eds.), Academic Press, 1985, pp. 575-585.
132. M. J. Bjorn, G. Groetsema, and L. Scalapino, *Cancer Res. 46*:3262 (1986).
133. T. F. Bumol, Q. C. Wang, R. A. Reisfeld, and N. O. Kaplan, *Proc. Natl. Acad. Sci. USA 80*:529 (1983).
134. M. I. Bernhard, K. A. Foon, T. N. Oeltmann, M. E. Key, K. M. Hwang, G. C. Clarke, W. L. Christensen, L. C. Hoyer, M. G. Hanna, and R. K. Oldham, *Cancer Res. 43*:4420 (1983).
135. K. M. Hwang, K. A. Foon, P. H. Cheung, J. W. Pearson, and R. K. Oldham, *Cancer Res. 44*:4578 (1984).
136. G. Weil-Hillman, W. Runge, F. K. Jansen, and D. A. Vallera, *Cancer Res. 45*:1328 (1985).
137. P. E. Thorpe, A. N. F. Brown, J. A. G. Bremmer, B. M. J. Foxwell, and F. Stirpe, *J. Natl. Cancer Inst. 75*:151 (1985).
138. C. F. Scott, V. S. Goldmacher, J. M. Lambert, R. V. J. Chari, S. Bolender, M. N. Gauthier, and W. A. Blattler, *Cancer Immunol. Immunother. 25*:31 (1987).
139. G. Laurent, J. Pris, J. P. Farcet, P. Carayon, H. Blythman, P. Casellas, P. Poncelet, and F. K. Jansen, *Blood 67*:1680 (1986).
140. L. E. Spitler, M. Del Rio, A. Khentigan, N. I. Wedel, N. A. Brophy, L. L. Miller, W. S. Harkonen, L. L. Rosendorf, H. M. Lee, R. P. Mischak, R. T. Kawahata, J. B. Stoudemire, L. B. Fradkin, E. E. Bautista, and P. J. Scannon, *Cancer Res. 47*:1717 (1987).
141. A. A. Hertler, D. M. Schlossman, M. J. Borowitz, G. Laurent, F. K. Jansen, C. Schmidt, and A. E. Frankel, *J. Biol. Response Mod. 7*(1):97 (1988).
142. M. V. Pimm, J. A. Jones, M. R. Price, J. G. Middle, M. J. Embleton, and R. W. Baldwin, *Cancer Immunol. Immunother. 12*:125 (1982).
143. M. C. Garnett, M. J. Embleton, E. Jacobs, and R. W. Baldwin, *Int. J. Cancer 31*:661 (1983).

144. M. C. Garnett, M. J. Embleton, E. Jacobs, and R. W. Baldwin, *Anti-Cancer Drug Design 1*:3 (1985).
145. R. W. Baldwin, *Method of Surv. Biochem. Anal. 15*:299 (1985).
146. T. Hara and Y. Takahashi, *Cancer Drug Delivery 2*:231 (1985).
147. M. C. Garnett and R. W. Baldwin, *Cancer Res. 46*:2407 (1986).
148. M. J. Embleton and T. H. Ho, *IRCS Med. Sci. 14*:1163 (1986).
149. M. J. Embleton, *Med. Sci. Res. 15*:1107 (1987).
150. S. Iwasa, E. Konishi, K. Kondo, T. Suzuki, H. Akaza, and T. Niijima, *Chem. Pharm. Bull. 35*(3):1128 (1987).
151. R. L. Juliano, G. Lopez-Berestein, R. Hopfer, R. Mehta, K. Mehta, and K. Mills, *Ann. N.Y. Acad. Sci. 446*:390 (1985).
152. G. Gregoriadis, *Pharm. Int.* 33 (1983).
153. A. Supersaxo, W. Rubas, H. R. Hartmann, H. Schott, H. Hengartner, and R. A. Schwendener, *J. Microencaps. 5*(1):1 (1988).
154. G. Poste, R. Kirsh and P. Bugelski, in *Novel Approaches to Cancer Chemotherapy* (P. S. Sunkara, ed.), Academic Press, New York, 1984, pp. 165–230.
155. K. S. Houck and L. Huang, *Biochem. Biophys. Res. Commun. 145*(3): 1205 (1987).
156. A. Huang, L. Huang, and S. J. Kennel, *J. Biol. Chem. 255*:8015 (1980).
157. J. Barbet, P. Machy, and L. D. Leserman, *J. Supramol. Struct. Cell. Biochem. 16*:243 (1981).
158. L. D. Leserman, J. Barbet, and F. Kourilsky, *Nature 288*:602 (1980).
159. J. Sunamoto, T. Sato, M. Hirota, K. Fukushima, K. Hiratani, and K. Hara, *Biochim. Biophys. Acta 898*:323 (1987).
160. A. Huang, S. J. Kennel, and L. Huang, *J. Biol. Chem. 258*(22):14034 (1983).
161. O. V. Trubetskaya, V. S. Trubetskoy, S. P. Domogatsky, A. V. Rudin, N. V. Popov, S. M. Danilov, M. N. Nikolayeva, A. L. Klibanov, and V. P. Torchilin, *FEBS Lett. 228*(1):131 (1988).
162. L. D. Leserman, P. Machy, and J. Barbet, *Nature 293*:226 (1981).
163. P. Machy, J. Barbet, and L. D. Leserman, *Proc. Natl. Acad. Sci. USA 79*: 4148 (1982).
164. T. D. Heath, J. A. Montgomery, J. R. Piper, and D. Papahadjopoulos, *Proc. Natl. Acad. Sci. USA 80*:1377 (1983).
165. L. Huang and S. S. Liu, *Biophys. J. 45*:72a (1984).
166. R. Nayar and A. J. Schroit, *Biochemistry 24*:5967 (1985).
167. R. M. Straubinger, N. Duzgunes, and D. Papahadjopoulos, *FEBS Lett. 179*:148 (1985).
168. J. Connor and L. Huang, *Cancer Res. 46*:3431 (1986).
169. R. J. Y. Ho, B. T. Rouse, and L. Huang, *Biochem. Biophys. Res. Commun. 138*(2):931 (1986).
170. R. J. Y. Ho, B. T. Rouse, and L. Huang, *Biochemistry 25*:5500 (1986).
171. R. J. Y. Ho, B. T. Rouse, and L. Huang, *J. Biol, Chem. 262*(29):13973 (1987).

172. R. Schwendener, H. Schott, B. Leitner, and H. Hengartner, *Experientia 43*:709 (1987).
173. Y. Hashimoto, *Cancer Drug Delivery 2*:231 (1985).
174. Y. Watanabe and T. Osawa, *Chem. Pharm. Bull. 35*(2):740 (1987).
175. P. Bagchi and S. M. Birnbaum, *J. Colloid Interface Sci. 83*(2):460 (1981).
176. L. Illum, P. D. E. Jones, J. Kreuter, R. W. Baldwin, and S. S. Davis, *Int. J. Pharm. 17*:65 (1983).
177. L. Illum, P. D. E. Jones, and S. S. Davis, in *Microspheres and Drug Therapy. Pharmaceutical, Immunological and Medical Aspects* (S. S. Davis, L. Illum, J. G. McVie, and E. Tomlinson, eds.), Elsevier, Amsterdam, 1984, pp. 353–364.
178. C. Kubiak, L. Manil, and P. Couvreur, *Int. J. Pharm. 41*:181 (1988).
179. P. Couvreur and J. Aubry, in *Topics in Pharmaceutical Sciences* (D. D. Bremer, ed.), Elsevier, Amsterdam, 1983, pp. 305–316.
180. P. Couvreur, *J. Pharm. Belg. 39*:249 (1984).
181. T. Laakso, J. Andersson, P. Artursson, P. Edman, and I. Sjoholm, *Life Sci. 38*:183 (1986).
182. L. Illum, P. D. E. Jones, R. W. Baldwin, and S. S. Davis, *J. Pharmacol. Exp. Therap. 230*:733 (1984).
183. A. Rembaum, *Pure & Appl. Chem. 52*:1275 (1980).
184. A. Rembaum, S. P. S. Yen, and R. S. Molday, *J. Macromol. Sci.–Chem. A13*(5):603 (1979).
185. A. Rembaum and W. J. Dreyer, *Science 208*:364 (1980).
186. A. Rembaum, R. C. K. Yen, D. H. Kempner, and J. Ugelstad, *J. Immunol. Methods 52*:341 (1982).
187. S. Margel, U. Beitler, and M. Ofarim, *J. Cell Sci. 56*:157 (1982).
188. T. Suzuta, in *Controlled Drug Delivery, Volume II, Clinical Applications* (S. D. Bruck, ed.), CRC Press, Boca Raton, Fla., 1983, pp. 149–188.
189. S. S. Davis, L. Illum, J. G. McVie, and E. Tomlinson (eds.), in *Microspheres and Drug Therapy, Pharmaceutical, Immunological and Medical Aspects,* Elsevier, Amsterdam, 1984.
190. L. Illum and S. S. Davis (eds.), *Polymers in Controlled Drug Delivery,* John Wright, Bristol, 1987.
191. E. Tomlinson and S. S. Davis (eds.), *Site-Specific Drug Delivery*, John Wiley, 1986.
192. E. Tomlinson, in *Drug Delivery Systems. Fundamentals and Techniques* (P. Johnson and J. G. Lloyd-Jones, eds.), VCH Publishers, Chichester, 1987, pp. 32–65.
193. A. Rolland, Ph.D. thesis, Université de Rennes I, Faculté de Pharmacie, Rennes, France, No. 1, 1987.
194. R. Le Verge and A. Rolland, European Patent No. 0240424 (1987).
195. S. S. Davis and L. Illum, *Biomaterials 9*:111 (1988).
196. C. R. Alving, *Pharmacol. Ther. 22*:407 (1983).
197. S. L. Croft, *Pharm. Int. 9*:229 (1986).
198. P. Stjarnkvist, P. Artursson, A. Brunmark, T. Laakso, and I. Sjoholm, *Int. J. Pharm. 40*:215 (1987).

199. A. Rolland, D. Gibassier, and R. Le Verge, in *4th International Conference on Pharmaceutical Technology* (APGI, ed.), Paris, 1986, Vol. 4, pp. 206–214.
200. A. Rolland, *Int. J. Pharm. 42*:145 (1988).
201. A. Astier, B. Doat, M. J. Ferrer, G. Benoit, J. Fleury, A. Rolland, and R. Le Verge, *Cancer Res. 48*:1835 (1988).
202. A. Rolland, P. L. Etienne, P. Kerbrat, and R. Le Verge, *Cancer Commun. 2*(2):7 (1988).
203. A. Rolland and R. Le Verge, in *Contrast Agents for MRI Tissue Characterization: Basic Principles and Research Methodology* (J. De Certaines and F. Podo, eds.), EEC, Luxembourg, 1988, pp. 77–94.
204. E. Tomlinson, *Adv. Drug Delivery Rev. 1*:87 (1987).
205. A. Rolland, D. Bourel, B. Genetet, and R. Le Verge, *Int. J. Pharm. 39*:173 (1987).
206. A. Rolland, D. Gibassier P. Sado, and R. Le Verge, *J. Pharm. Belg. 41*(2): 83 (1986).
207. A. Rolland, D. Gibassier, P. Sado, and R. Le Verge, in *4th International Conference on Pharmaceutical Technology* (APGI, ed.), Paris, 1986, Vol. 4, pp. 183–192.
208. A. Rolland, D. Gibassier, P. Sado, and R. Le Verge, *J. Pharm. Belg. 41*(2): 94 (1986).
209. D. Bourel, N. Genetet, E. Carosella, G. Merdrignac, J. Armand, and B. Genetet, in *Advances in Immunomodulation* (B. Binzzini and E. Bonmassar, eds.), Pythagora Press, Roma, 1988, pp. 279–288.
210. D. Bourel, A. Rolland, R. Le Verge, and B. Genetet, *J. Immunol. Methods 106*:161 (1988).
211. S. P. S. Yen, A. Rembaum, R. W. Molday, and W. Dreyer, in *Emulsion Polymerization* (I. Piirma and J. Gardon, eds.), ACS Symposium Series, ACS, Washington, D.C., 1979, Vol. 24, pp. 236–257.
212. A. Rolland, A. Guillouzo, J. M. Begue, G. Merdrignac, B. Genetet, A. Astier, and R. Le Verge, in *4th International Conference on Pharmaceutical Technology* (APGI, ed.), Paris, 1986, Vol. 4, pp. 193–205.
213. A. Rolland, G. Merdrignac, J. Gouranton, D. Bourel, R. Le Verge, and B. Genetet, *J. Immunol. Methods. 96*:185 (1987).

11

Antibodies as Drug Carriers for Solid Tumors: Evaluation of Drug–Anti-SSEA-1 Conjugates in the treatment of Teratocarcinoma

WEI-CHIANG SHEN and STEFANO PERSIANI
University of Southern California School of Pharmacy, Los Angeles, California

BYRON BALLOU and THOMAS R. HAKALA
University of Pittsburgh School of Medicine, Pittsburgh, Pennsylvania

I. INTRODUCTION

Since their development more than 13 years ago [1], monoclonal antibodies have been touted as Ehrlich's "magic bullets." Targeted drug delivery mediated by monoclonal antibodies is especially attractive for the treatment of cancer, where the majority of anticancer agents are highly toxic, the diseased tissues are often disseminated, and the tumors are not significantly different from their normal counterparts. It is hoped that by targeted delivery, the adverse side effects of anticancer drugs can be decreased, and therapeutic indices improved.

During the last several years, there have been many reports on the use of monoclonal antibodies either as anticancer drugs [2] or as targeted drug carriers [3]. Despite the initial high expectations, the overall results, especially in clinical trials, are generally disappointing [4]. The lack of convincing results has raised doubt about the feasibility of this approach. Some of the limitations suggested by many investigators are summarized in Table 1. These limitations are mostly derived from our current knowledge of cell biology and immunology. The suggested impermeability of endothelial cells to antibodies would be especially troublesome in the treatment of solid tumors because the targets would

Table 1 Some of the Suggested Limitations in Antibody-Mediated Drug Targeting in Cancer Therapy

1. Lack of tumor-specific antigens
2. Impermeability of endothelium to immunoglobulins
3. Insufficient drug carried by an antibody molecule
4. Low *in vivo* cytotoxicity of antibodies
5. Antigenic modulation and heterogeneity
6. Systemic toxicity of drug-antibody conjugates

not be accessible to the I.V.-injected drug-antibody conjugates. Consequently, most of the reports showing favorable responses in monoclonal antibody therapy were from treatment of either leukemia or lymphoma. Here we will try to reconsider the limitations listed in Table 1 by reviewing our recent studies on the usage of monoclonal anti-stage-specific embryonic antigen-1 (anti-SSEA-1) antibody as a drug carrier in the treatment of teratocarcinomas as solid tumor models in experimental animals.

II. SSEA-1 AS A TUMOR-ASSOCIATED ANTIGEN

A. Anti-SSEA-1 as a Monoclonal Antitumor Antibody

Anti-SSEA-1 was first developed as a monoclonal antibody against mouse teratocarcinoma cells [5]. The antigen, SSEA-1, to which the antibody binds includes the terminal determinant lacto-N-fucopentaose III [6]. The antigen has been found on many human and mouse tumor cells [7]. Subsequent studies, however, revealed that the antibody reacted histochemically with many normal tissues, presumably because of the presence of SSEA-1 in nontumor cells [8]; no significant differences in antibody avidity for tumors and normal tissues were found (unpublished data). Moreover, quantitative studies showed that normal kidney contained two to eight times as much antigen per gram as the tumor. Therefore, SSEA-1 is an antigen that is tumor-associated but not specific to tumors. Besides the lack of tumor specificity, another apparent disadvantage of anti-SSEA-1 is that this monoclonal antibody is an IgM; the molecular weight of IgM, about 960,000, is more than fivefold larger than that of IgG. It has been long assumed that IgM possesses very poor extravascular transport because of this molecular size. Furthermore, IgM-class antibodies have been shown to be ineffective in the inhibition of tumor cell growth in vivo, presumably because they do not induce antibody-dependent cell-mediated cytotoxicity

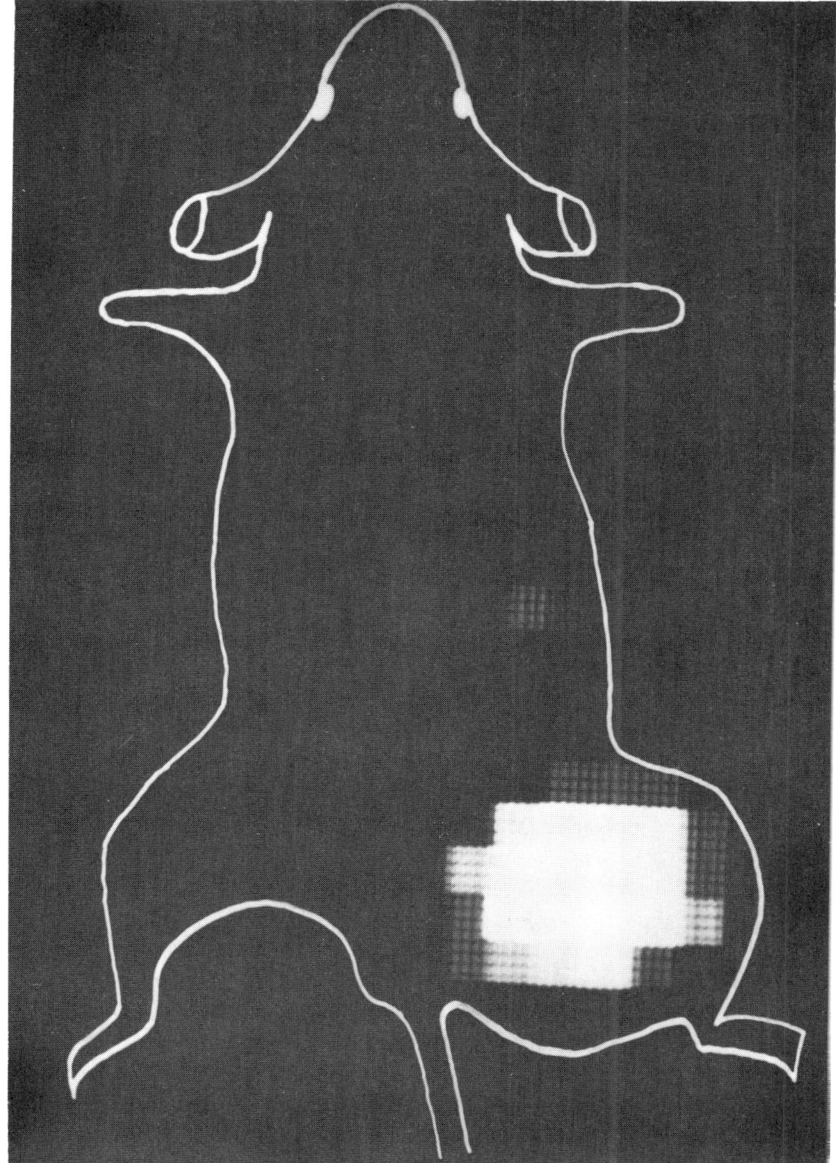

Figure 1 MH-15 tumor image of BALB/c mouse with ^{131}I-anti-SSEA-1. Scintigraphy was done with a gamma camera 5 days after administration of 150 μCi of ^{131}I-anti-SSEA-1. Right thigh contains MH-15 teratocarcinoma (25 mm); left thigh contains P3/X63–Ag 8 myeloma (10 mm) as control. (From Ref. 23.)

[9]. In our hands, anti-SSEA-1 alone had no significant effect on tumor growth. Thus, anti-SSEA-1 is not tumor-specific, should extravasate poorly, and has no antitumor effect when used alone.

B. Tumor Localization by Anti-SSEA-1

Early studies by Ballou et al. [10] indicated that intravenously administered ^{131}I-anti-SSEA-1 selectively localized in solid tumors inoculated in the thighs of mice. These studies included the MH-15 mouse teratocarcinoma in BALB/c mice and the human BeWo choriocarcinoma in nude mice. Neither antigen-negative tumors, nor mice injected with control (nonbinding) IgM showed significant tumor localization. Pharmacokinetic studies indicated that anti-SSEA-1 was not only highly concentrated in the antigenic tumors, but also cleared at a slower rate from tumor than from other tissues [11]. The high retention in the tumor produced a clear tumor image in scintigraphy (Figure 1). Biodistribution and imaging studies indicated that anti-SSEA-1, despite being an IgM molecule, reaches its target teratocarcinoma cells, and that SSEA-1-positive normal tissues were not targeted. Thus good tumor localization did not require either low-molecular-weight antibody or absolute tumor specificity.

III. DRUG TARGETING IN TERATOCARCINOMA CELL CULTURES

A. Conjugation of Methotrexate and Daunomycin to Anti-SSEA-1

Drugs can be conjugated to monoclonal antibodies by either direct linkages or linkages using spacer molecules. Methotrexate (MTX) was conjugated to anti-SSEA-1 by a simple carbodiimide-catalyzed coupling reaction between a carboxyl group in the drug molecule and an amino group on the antibody. A conjugate with 45 MTX per anti-SSEA-1 was obtained [12] in a reaction with water-soluble carbodiimide. For daunomycin (DM), an acid-sensitive cis-aconityl linkage was used to prepare active conjugates [13]. An acid-sensitive linkage will release active DM only after the conjugate is internalized by the antigen-bearing cells and transported to intracellular acidic compartments, i.e., endosomes and lysosomes [14]. In both MTX and DM conjugates, the antibody's antigen-binding activity was fully preserved [12,13]. The stability of antibody binding in anti-SSEA-1 conjugates is very different from that of IgG-class conjugates. When 10 molecules of MTX were conjugated to an IgG antibody by carbodiimide reaction, the antigen-binding activity decreased to less than 50% of its original value [15]. Since an IgM molecule is about five times larger than an IgG, it is not surprising that each IgM molecule can carry more drug molecules than

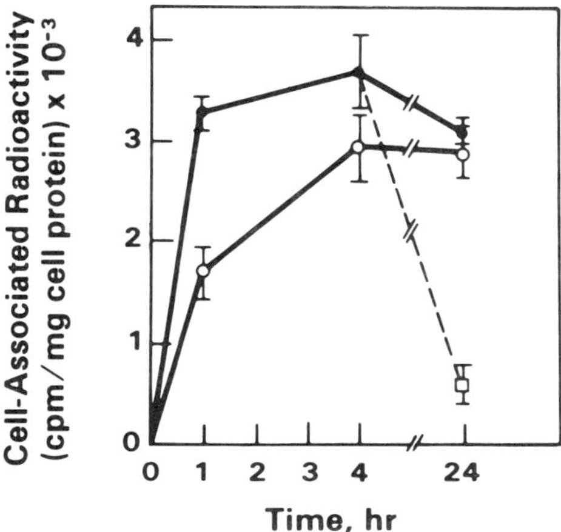

Figure 2 Cell-associated ^{3}H-MTX–anti-SSEA-1 in F-9 cells with or without preincubation in unlabeled atni-SSEA-1. F-9 cells were seeded at 1×10^6 cells/flask. After 24 h, cell cultures were changed to fresh growth medium, either with (○) or without (●) 6 μg/ml of unlabeled anti-SSEA-1. After 1 h incubation at 37°C, medium from each flask was removed and cells were washed once with 5 ml of PBS. Cells were then incubated in 5 ml of growth medium containing 6 μg/ml of ^{3}H-MTX-anti-SSEA-1. After 1, 4, and 24 h, monolayers were washed with PBS and detached with trypsin-EDTA. Cell pellets after centrifugation were washed three times each with 5 ml PBS before dissolving in 1 ml of 0.1% Triton X-100 in H_2O). Two flasks after 3 h incubation with ^{3}H-MTX-anti-SSEA-1 were washed and reincubated with fresh growth medium (□) for an additional 20 h. Cell-associated radioactivity was measured by counting the cell extracts in a liquid scintillation counter and is given as cpm/mg cell protein in each flask. Bars indicate the range of duplicate measurements for each point. (From Ref. 12.)

an IgG. The stability of the anti-SSEA-1 conjugates is probably due to the fact that IgMs have 10 binding sites, and that not all binding sites are occupied when interacting with cell surface antigens. An IgM conjugate with half of the binding sites inactivated may still bind to cell surface antigens nearly as effectively as the native antibody. For IgG, destruction of one of the two antigen-binding sites would markedly decrease the avidity of the antibody. We have found that anti-SSEA-1 can be conjugated with more than 65 MTX molecules and still maintain full antigen-binding activity.

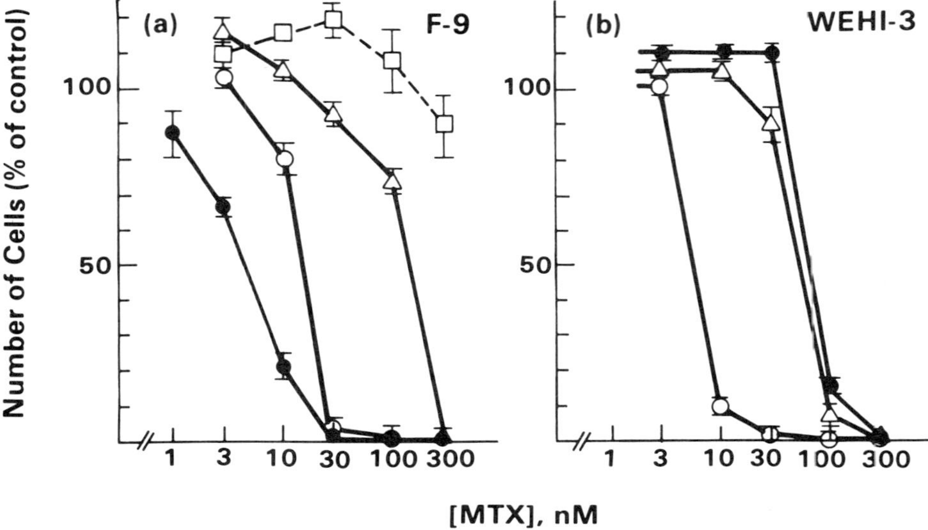

Figure 3 Effects of MTX and its conjugates on the growth of (a) F-9 and (b) WEHI-3 cells. Cells (5×10^4 cells per T-25 flask) were treated with various concentrations of MTX given as free drug (○), anti-SSEA-1 conjugate (●), or MOPC 104E-IgM conjugate (△). In SSEA-1 positive F-9 cells, unconjugated anti-SSEA-1 was also tested for cytotoxicity at antibody concentrations corresponding to those of conjugates containing 3 to 300 nM MTX (□). Cells were exposed to the drugs for 3 days and grown in drug-free medium for 2 additional days before they were detached from the flasks and counted in a Coulter counter. The number of cells in each flask is given as percent of untreated controls. (From Ref. 12.)

B. Antigenic Modulation

The binding of anti-SSEA-1 on SSEA-1-positive F-9 teratocarcinoma cells demonstrated antigenic modulation. When cells were preincubated with anti-SSEA-1 for 1 h prior to treatment with ^{3}H-MTX-anti-SSEA-1, the cell-associated radioactive antibody was much less than that in cells without preincubation. However, tumor cells seem to be able to regenerate the surface antigens efficiently. Four hours after removal from "cold" antibody, cells with and without preincubation bound equal amounts of cell-associated radioactive antibodies (Figure 2). The mechanism of the antigenic modulation and regeneration is not known. It is possible that the internalized SSEA-1 antigens are recycled back to the cell surface while the antigen-bound antibodies are slowly dissociated to either the inside or the outside of target cells. Another possibility is that new antigenic

Table 2 Effects of Leucovorin and Thiamine Pyrophosphate on the Growth Inhibition in F-9 Cells by MTX and its Conjugates

Drug (nM)	Number of cells (% of controls ± S.D., n = 2)		
	No addition	Leucovorin	Thiamine pyrophosphate
Control	100	99.5 ± 5.0	117.4 ± 4.4
MTX			
30	0.5 ± 0.2	84.7 ± 2.4	96.0 ± 8.0
300	0.3 ± 0.1	72.0 ± 3.9	0.4 ± 0.1
MXT–anti-SSEA-1			
30	1.1 ± 0.1	99.0 ± 0.2	1.3 ± 0.1
300	0.4 ± 0.1	35.7 ± 0.7	1.0 ± 0.4
MTX–IgM			
30	110.5 ± 2.1	108.2 ± 1.0	119.5 ± 2.9
300	0.7 ± 0.1	100.0 ± 2.8	107.2 ± 2.2

F-9 cells were treated with 30 or 300 nM of MTX as the free, anti-SSEA-1 conjugated, or MOPC 104E IgM-conjugated drug for 5 days as described in Figure 3. Leucovorin (3 μM) or thiamine pyrophosphate (300 μM) was added to some flasks at the same time as MTX. The cell number in each treatment is given as percent of untreated controls. (From Ref. 12.)

determinants are shifted from an intracellular pool to the cell surface, or synthesized de novo.

C. Effects of MTX–Anti-SSEA-1 on the Growth of Teratocarcinoma Cells in Culture

MTX–anti-SSEA-1 conjugates have been tested for the inhibition of tumor cell growth in tissue culture. The effects of MTX–anti-SSEA-1 were compared with those of free MTX and of MTX–MOPC 104E, a conjugate prepared from a nonspecific IgM secreted by MOPC-104E myeloma cells. As shown in Figure 3, MTX–anti-SSEA-1 effectively inhibited the growth of SSEA-1-positive F9 teratocarcinoma cells. At 50% inhibition, the conjugate was 3-fold and 30-fold more effective than free MTX and MTX–MOPC 104E, respectively, per mole of MTX. Anti-SSEA-1 alone showed no significant effect on the growth of F-9 cells. When tested on SSEA-1-negative WEHI-3 leukemia cells, MTX–anti-SSEA-1 and the MTX–MOPC 104E were equally ineffective. The coincidence of the two dose-response curves for MTX–anti-SSEA-1 and MTX–MOPC 104E in SSEA-1-negative cells indicates that anti-SSEA-1 is recognized by nonantigenic cells as a nonspecific IgM molecule.

Table 3 Effect of Leupeptin (50 μg/ml) on the Growth Inhibition and Intracellular Degradation of ^{3}H-MTX–Anti-SSEA-1 in F-9 Cells

Treatment	No. of cells (%)[a]	Percent of degradation[b]
Without leupeptin	1.1	18.7
With leupeptin	20.8	11.2

[a]Cell number in each flask (percent of the control flask) after a 4-day treatment with 3×10^8 M of ^{3}H-MTX–anti-SSEA-1.

[b]Percentage of degraded ^{3}H-MTX-containing molecules inside F-9 cells that have been treated with 3×10^7 M of ^{3}H-MTX–anti-SSEA-1 for 24 h.

To further study the selective effect of MTX–anti-SSEA-1 on F-9 cells, we compared the growth inhibition by MTX, MTX–anti-SSEA-1, and MTX–MOPC 104E in the presence of various agents. Table 2 shows that F-9 cells responded differently toward the MTX transport inhibitor, thiamine pyrophosphate (TPP), when treated with three forms of MTX. TPP inhibited the effect of free MTX as well as MTX–MOPC 104E, but did not change the efficacy of MTX–anti-SSEA-1. On the other hand, leucovorin (LV), an antagonist of MTX, abolished the inhibitory effects of all three forms of MTX. These results indicated that the specific antibody conjugate differs in its transport mechanism from the nonspecific conjugate and free MTX, but all three are similar in their ultimate mode of action.

When F-9 cells were treated with MTX–anti-SSEA-1 in the presence of leupeptin (LP), a lysosomal protease inhibitor, a partial reduction of cytotoxicity was observed (Table 3). The observation that the LP-treated cells contained less degradation products of ^{3}H-MTX–anti-SSEA-1 than cells without LP treatment (Table 3) implies that this protective effect was due to the inhibition of intracellular degradation of the conjugate. Because LP cannot inhibit all lysosomal proteases, the partial effects shown in both the growth inhibition and the degradation of MTX–anti-SSEA-1 were expected. When the cells were treated with either MTX or MTX–MOPC 104E, this protection by LP was not observed. Therefore, lysosomal proteolysis is required only for the activity of MTX–anti-SSEA-1, but not of MTX and MTX–MOPC 104E.

From results obtained in the above studies, the cellular mechanism of MTX–anti-SSEA-1 targeting is summarized in Figure 4. Free MTX is transported by a carrier-mediated active transport system, which can be blocked by TPP [16]. MTX conjugates of an irrelevant IgM do not have any affinity for the plasma membrane of F-9 teratocarcinoma cells, and therefore are only poorly transported and are ineffective. The small activity of MTX–MOPC 104E is also sensitive to TPP, indicating that IgM conjugates are degraded in the culture medium

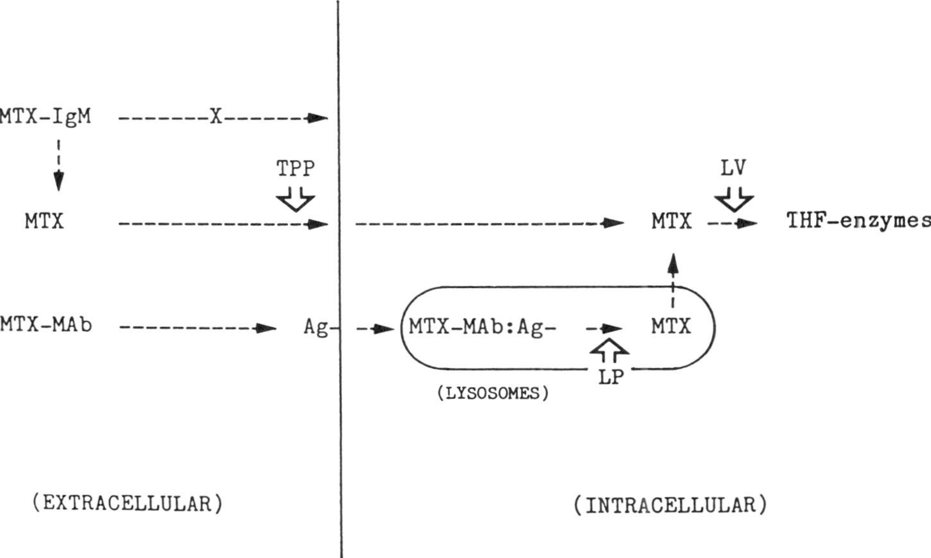

Figure 4 Cellular transport and processing of MTX–anti-SSEA-1. MTX–MAb: MTX–anti-SSEA-1 conjugate; MTX–IgM: MTX conjugate of irrelevant MOPC-104E IgM; Ag: SSEA-1 antigen; TPP: thiamine pyrophosphate; LV: leucovorin; LP: leupeptin.

and the degraded products are transported into F-9 cells through the MTX-transport system on the plasma membrane. The anti-SSEA-1 conjugates (MTX-MAb), on the other hand, are bound to the SSEA-1 antigen determinants on the cell surface, and subsequently internalized by F-9 cells via endocytosis. The bound antibodies (MTX–MAb:Ag) are transported to lysosomes, where they are digested by lysosomal proteases. This proteolytic process can be partially inhibited by the protease inhibitor (LP). The MTX-containing degradation products from MTX–MOPC 104E and intracellular degradation products from MTX–MAb all have the same mode of action, presumably on tetrahydrofolate-related enzymes (THF enzymes), e.g., dihydrofolate reductase, and therefore the effects of all three forms of MTX can be abrogated by the MTX-antagonist LV.

Table 4 Effects of MTX and Its Antibody Conjugates on the size of MH-15 Teratocarcinoma in Mice

Treatment[a]	Mean tumor size (cm^2 ± SEM)[b]
PBS	7.3 ± 1.3
MTX	8.52 ± 0.12
MTX + anti-SSEA-1	9.09 ± 0.09
MTX–anti-SSEA-1	2.3 ± 0.3
Anti-SSEA-1	8.82 ± 0.15

[a]All mice were treated with 15 mg MTX/kg body weight except for anti-SSEA-1 alone, which was 1 g antibody/kg.
[b]Tumors were measured on the 22nd day after the treatment.

IV. DRUG TARGETING IN TERATOCARCINOMA-BEARING MICE

A. Treatment of Teratocarcinoma-Bearing Mice with MTX–Anti-SSEA-1

Solid tumors of MH-15 teratocarcinoma were grown in the right thighs of 7-week-old BALB/c mice. Treatment was started 2 weeks after the inoculation of tumor cells, when a solid tumor approximately 0.8 cm^2 was measured. Treatment with MTX–anti-SSEA-1 (15 mg MTX/kg body weight) using five I.V. injections at 3-day intervals resulted in significant tumor growth inhibition with no lethal toxicity. Free MTX, anti-SSEA-1 alone, an irrelevant conjugate (MTX–MOPC 104E), and a mixture of anti-SSEA-1 with free MTX all showed no significant antitumor activity at the same drug dose (Table 4). The drug–anti-SSEA-1 conjugate also had a greater antitumor activity than free MTX, even at higher doses of the free drug (Table 5). At doses that caused 80% death of the treated animals, 22 mg/kg for MTX–anti-SSEA-1 and 420–600 mg/kg for free MTX, the conjugate-treated mice showed no trace of tumor at the time of their death, while mice treated by free MTX retained large tumor masses (average 2 cm^2). In other studies, MTX–anti-SSEA-1 at 10- and 15-mg/kg doses showed significant inhibition of tumor growth, and increased the life span by 40% and 100%, respectively (Table 5).

Even though antigenic modulation was observed in cultured teratocarcinoma cells (Figure 2), the drug conjugates inhibited tumor growth in mice. We do not know whether antigenic modulation occurs in solid tumors in vivo, but if it does, our dose and dose schedule must circumvent any such problem. Another possible difficulty is antigenic heterogeneity. MH-15 teratocarcinomas are

Table 5 Comparison of the Effects of MTX and MTX–Anti-SSEA-1 on the Survival of MH-15 Teratocarcinoma-Bearing Mice

Treatment	Dose, mg/kg MTX	Mean survival time (days)
PBS	–	11
MTX	300	10
	420	9.5[a]
	600	7[a]
MTX–anti-SSEA-1	10	15
	15	22[b]
	22	9[a]

[a]Death caused by the toxicity.
[b]$p < 0.05$ that the average life span $\leq$ PBS control group.

Table 6 Biodistribution of ^{3}H-MTX-Anti-SSEA-1 in Tumor-Bearing Mice

	Percent dose/g tissue (± SEM)			
Tissue	10 h	(tumor/tissue)	48 h	(tumor/tissue)
Heart	1.8 ± 0.2	(3.7 ± 0.81)	0.9 ± 0.05	(4.1 ± 0.72)
Spleen	5.8 ± 0.1	(1.1 ± 0.11)	1.1 ± 0.09	(3.3 ± 0.66)
Lung	2.4 ± 0.07	(2.7 ± 0.36)	0.8 ± 0.05	(4.5 ± 0.45)
Kidney	2.4 ± 0.1	(2.7 ± 0.33)	0.9 ± 0.08	(3.9 ± 1.02)
Jejunum	2.2 ± 0.5	(3.1 ± 0.38)	0.9 ± 0.61	(4.1 ± 0.62)
Ileum	1.9 ± 0.5	(3.9 ± 1.03)	1.0 ± 0.1	(3.9 ± 0.99)
Brain	0.4 ± 0.03	(14.7 ± 0.6)	0.5 ± 0.07	(8.1 ± 2.69)
Tumor[a]	6.5 ± 0.7	(1)	3.7 ± 0.6	(1)
Liver	22.7 ± 4.0	(0.29 ± 0.03)	3.7 ± 0.6	(1.0 ± 0.25)
Blood	15.3 ± 2.6	(0.46 ± 0.13)	1.1 ± 0.2	(3.5 ± 0.99)

[a]MH-15 teratocarcinoma in right thighs of Balb/c mice.

heterogeneous, and a significant fraction of the cells (15–50%) are SSEA-1-negative. However, the lack of the antigen on some of the cells did not allow growth of SSEA-1-negative tumor in MTX–anti-SSEA-1-treated mice. In the group that received 22 mg/kg of the antibody conjugate (Table 5), no trace of tumor could be detected at the end of the treatment.

B. Biodistribution of MTX–Anti-SSEA-1 in MH-15 Teratocarcinoma-Bearing Mice

MH-15 tumor-bearing mice were injected intravenously with ^{3}H-MTX–anti-SSEA-1. Mice were sacrificed at 10 and 48 h after the injections; tumors and normal organs were dissected from each mouse. The radioactivity was measured by liquid scintillation counting and expressed as percent of injected dose/gr of normal tissue or tumor (Table 6). The half-life of MTX–anti-SSEA-1 in the bloodstream was about 12 h, which was significantly longer than that of free MTX, that is, 3–4 h [17]. All organs, except tumor and liver, showed a marked decrease of radioactivity between 10 and 48 h. A high uptake with a long retention of radioactivity was found in the tumors, which may explain the selective inhibition of tumor growth in the MTX–anti-SSEA-1-treated mice (Table 5). The liver also retained a large amount of radioactivity. Because the liver is SSEA-1-negative [8], the uptake of MTX–anti-SSEA-1 was probably due either to nonspecific trapping of antigen–antibody complexes in the reticuloendothelial system, or to catabolism of drug–antibody conjugates by liver cells.

V. DISCUSSION

A. Antigenic Specificity

The original concept of antibody targeting of drugs in cancer therapy assumed that there are antigens specific to tumor cells. However, the existence of tumor-specific antigens in human cancers remain to be demonstrated [18]. The results from the study of anti-SSEA-1 antibody suggest that a strict tumor specificity is not a prerequisite to the selection of in vivo target. A nonspecific tumor-associated antigen may be a good target for antibody-mediated drug delivery if the tumor possesses any of the following properties: (1) The antigen is accessible to antibody in the tumor but not in normal tissues; (2) the antigen-bearing normal cells are not vital; (3) The tumor has a longer retention time for the drug or the conjugate than do normal tissues; or (4) tumor cells are more sensitive to the antibody-carried drug than are normal cells. In the case of anti-SSEA-1, the selectivity in drug delivery is probably due to the difference in the accessibility of the antigen in tumors and normal tissues [11]. The increased retention time of drug-antibody conjugates in tumor cells, shown in Table 6, may also play an important role in the effectiveness of MTX-anti-SSEA-1.

B. Antibody Cytotoxicity

Antitumor antibodies have been considered in cancer therapy both as cytotoxic agents in their own right and as targeted carriers for anticancer drugs. In our studies, anti-SSEA-1 alone or mixed with free MTX did not show any inhibitory effect on the growth of teratocarcinoma cells (Figure 3 and Table 4). These results indicate that antibodies can be used as drug carriers to obtain a good antitumor activity without potentiating any immunological reaction. Since the in vivo cytotoxicity of an antibody is a very complicated and often unpredictable process [2], tumor-localizing antibodies might be more practical as drug carriers rather than used alone.

C. Drug-Carrying Capacity of Antibody

One of the limitations in using an antibody as a drug carrier is that the number of drug molecules carried per antibody molecule is usually too low to be highly effective. Because chemical modification often leads to the denaturation of proteins, increasing the number of drug molecules conjugated to an antibody will eventually destroy its antigen-binding activity. This limitation has been an obstacle to the development of drug conjugates that would be more active than free drugs in the inhibition of tumor growth. However, results from anti-SSEA-1 studies demonstrated that more than 65 molecules of MTX can be conjugated to an IgM antibody without any effect in antigen-binding activity. Furthermore, at least some of our drug-IgM conjugates do not differ significantly from the native antibody in their biodistribution in experimental animals [13], unlike those conjugates prepared using carriers (e.g., dextran or serum albumin) to increase the drug-carrying capacity of IgG [19]. The only apparent differences in biodistribution between ^{3}H-MTX-conjugated and ^{131}I-labeled anti-SSEA-1 is the low uptake of radioiodinated antibody in liver [13]. This difference is probably due to the dehalogenation of iodinated proteins in liver cells, which has been reported in the comparison of ^{125}I- and ^{111}In-labeled antibodies [20].

D. Blood-Tumor Barriers

Regardless of the mechanism of its antitumor activity, a drug-antibody conjugate must be transported to the tumor site to be effective. It is generally believed that a reduction in molecular weight will increase transport across the endothelial barrier of blood capillaries. However, digestion of an IgG to its antigen-binding fragments will undoubtedly decrease the drug-carrying capacity, may decrease the antigen affinity, and will decrease the circulating lifetime [21]. In this chapter we demonstrated that an IgM antibody, which is five times larger than an IgG, can effectively reach an antigen-positive solid tumor. The conclusion is based on the results from both tumor imaging and drug targeting studies of anti-SSEA-1 on teratocarcinoma-bearing mice ([14] and this paper). The exact mechanism by which the antibody crosses endothelial cells is not well understood. However, it has been suggested that the capillaries in some tumors are more "leaky" than those in normal tissues [22]. This may increase the circulation of plasma proteins through the tumor and thus allow rapid accumulation of antibodies at antigenic tumor sites. Unfortunately, we cannot compare the permeability to IgG and IgM antibodies in the teratocarcinoma studies, because no anti-SSEA-1 IgG is currently available. Furthermore, it is possible that the permeability of capillary endothelium may be different in each type of tumor.

E. Systemic Toxicity

Despite the encouraging results obtained from tumor localization [14] and from the inhibition of tissue-cultured tumor cell growth [13], the treatment of MH-15 teratocarcinoma in mice with MTX–anti-SSEA-1 conjugates was only partially successful. On one hand, the drug conjugate was effective at much lower doses than free MTX in reducing the tumor size, and significantly increased the lifespan in those tumor-bearing mice that responded to treatment. On the other hand, complete tumor remission was observed only in mice that received a toxic dose of the conjugate. The cause of death in the apparently tumor-free mice in this group was probably gastrointestinal (GI) toxicity, as evidenced by the marked deterioration of the intestine found on autopsy. No damage was observed in tissues that are known to be SSEA-1-positive, e.g., kidney and central nervous system [8]. Free methotrexate equivalent to that in the IgM conjugates had no observable toxicity; therefore, the gut toxicity is unlikely to be due to free drug. Neither radioiodinated anti-SSEA-1 nor drug conjugates showed significant targeting to the GI system, so there is no reason to suspect specific conjugate localization. Preliminary biodistribution data using anti-SSEA-1 conjugated to ^{3}H-MTX suggest a possible reason for the GI toxicity. The liver was the only organ other than the tumor that retained large amounts of MTX–anti-SSEA-1, and hepatotoxicity was not observed. However, the gall bladder showed the highest concentration of radioactivity of all organs measured. We hypothesize that the GI toxicity was caused by liver uptake, followed by the transport of degradation products from liver to GI epithelial cells through bile secretion. If this is correct, then unless methods can be developed to block liver uptake of the antibody conjugates, or to selectively rescue the GI epithelial cells, systemic toxicity will be a serious problem in treatment using MTX–anti-SSEA-1 conjugates.

VI. CONCLUSION

We have demonstrated that a monoclonal antibody against a tumor-associated antigen that is present in large quantities in normal tissues can effectively deliver anticancer drugs to a tumor. We have also shown that an IgM antibody is capable of reaching solid tumors outside the bloodstream. MTX-anti-SSEA-1 conjugates are more effective than the free drug in the inhibition of SSEA-1-positive solid tumor growth in syngeneic mice. Although the therapeutic efficacy is hampered by a nonantigen-specific toxicity of the drug-antibody conjugates, results from these studies offer hope for the use of monoclonal antibodies as targeted carriers in the treatment of solid tumors. Finally, our data suggest that IgMs may be better than IgGs as drug carriers, given the number of drug molecules that can be conjugated while preserving antigen-binding activity.

VII. REFERENCES

1. G. Kohler and C. Milstein, *Nature 256*:495 (1975).
2. I. D. Bernstein, C. C. Badger, E. Denkers, and J. Ledbetter, *Immunity to Cancer* (A. E. Reif and M. S. Mitchell, eds.), Academic Press, New York, p. 561 (1985).
3. T. Ghose and A. H. Blair, *Critical Rev. in Therap. Drug Carriers System* CRC Press), 3:263 (1987).
4. R. D. Dillman and I. Royston, *Br. Med. Bull. 40*:240 (1984).
5. D. Solter and B. B. Knowles, *Proc. Natl. Acad. Sci. USA 75*:5565 (1978).
6. S. Hakamori, E. Nudelman, S. Levery, D. Solter, and B. B. Knowles, *Biochem. Biophys. Res. Commun. 100*:1578 (1981).
7. B. Ballou, J. Reiland, G. Levine, D. Solter, R. Taylor, and T. R. Hakala, *Fed. Proc. 40*:1144 (1981).
8. N. Fox, I. Damjanov, B. B. Knowles, and D. Solter, *Cancer Res. 43*:669 (1983).
9. I. D. Bernstein, R. C. Nowinski, M. R. Tam, B. McMaster, L. L. Houston, and E. A. Clark, in *Monoclonal Antibodies* (R. Kennett, T. McKearn, and K. Bechtol, eds.), Plenum Press, New York, 1980, p. 275.
10. B. Ballou, G. Levine, T. R. Hakala, and D. Solter, *Science 206*:844 (1979).
11. B. Ballou, R. Jaffe, R. J. Taylor, D. Solter, and T. R. Hakala, *J. Immunol. 132*:2111 (1984).
12. W. C. Shen, B. Ballou, H. J.-P. Ryser, and T. R. Hakala, *Cancer Res. 46*: 3912 (1986).
13. B. Ballou, J. M. Reiland, G. Levine, W. C. Shen, H. J.-P. Ryser, D. Solter, and T. R. Hakala, *J. Surg. Oncol. 31*:1 (1986).
14. W. C. Shen and H. J.-P. Ryser, *Biochem. Biophys. Res. Commun. 102*: 1048 (1981).
15. P. N. Kulkarni, A. H. Blair, and T. I. Ghose, *Cancer Res. 41*:2700 (1981).
16. G. B. Henderson, B. Grzelakowska-Sztabert, E. M. Zevely, and F. M. Huennekens, *Arch. Biochem. Biophys. 202*:144 (1980).

17. E. S. Henderson, R. H. Adamson, C. Denham, and V. T. Oliverio, *Cancer Res. 25*:1008 (1965).
18. H. Schreiber, P. L. Ward, D. A. Rowley, and J. J. Strauss, *Ann. Rev. Immunol. 6*:465 (1988).
19. M. C. Garnett and R. W. Baldwin, *Cancer Res. 46*:2407 (1986).
20. P. L. Hagan, S. E. Halpern, A. Chen, L. Krishnan, J. Frinck, R. M. Bratholomew, G. S. David, and D. Carlo, *J. Nucl. Med. 26*:1418 (1985).
21. D. G. Covell, J. Barbet, O. D. Holton, C. D. V. Black, R. J. Parker, and J. N. Weinstein, *Cancer Res. 46*:3969 (1986).
22. S. E. Halpern and R. O. Dillman, *J. Biol. Resp. Modifiers 6*:235 (1987).
23. G. Levine, B. Ballou, J. Reiland, D. Solter, L. Gumerman, and T. Hakala, *J. Nucl. Med. 21*:570 (1980).

12

Chemistry and Biology of Immunotargeted Liposomes

ALAN L. WEINER*
The Liposome Company, Inc., Princeton, New Jersey

"Nearly all men die of their medicines, not of their diseases."

Molière
Le Malade Imaginaire, 1673

I. INTRODUCTION

Too often the practicing clinician is faced with an unresolvable dilema. To effectively achieve therapy against disease it becomes necessary to balance the devastation wrought by the illness against the noxious effects conferred by the drugs used to treat it. Although therapy should represent a complete tolerance to the dose regimen, in absolute terms it frequently delineates a compromise position, thereby effecting only an "acceptable" level of treatment. In the absence of preventative measures to combat the onset of disease, the next ideal alternative is to design a drug that specifically recognizes the origin of the disorder and then corrects it. This is not a new concept, with credit typically attributed to the teachings of Paul Ehrlich [1]. The "magic bullet" proposition of Ehrlich sought to link the action of drugs (and eventually antibodies) to specific receptor-mediated events. These concepts still serve as the primary foundation for the continuing development of drugs and antibodies.

Admittedly, this figurative description by Ehrlich was not intended to forecast the development of "bullet"-shaped drug microcarriers such as liposomes, invented some 60 years later. Given the complex level of technology needed to solve even the most rudimentary form of biological targeting by a small mole-

* *Current affiliation*: Escalon Ophthalmics, Inc., Rocky Hill, New Jersey

cule as Ehrlich had envisioned, it would seem a simplistic viewpoint to consider macrostructures such as liposomes as having an equivalent targeting potential. Yet, despite this observation, liposomes continue to be the subject of intensive research dedicated to this sole concept.

In practice, the physiological obstacles confronting the homing of particulate or colloidal materials to specific sites from the circulation are not trivial [2]. Among the first barriers encountered by a particle destined to reach the extravascular space is the vasculature itself. The breaching of a capillary wall by a microparticle will be governed by the type of endothelial lining and basement membrane found along the specific capillary bed of interest. Depending on whether the capillary is continuous, fenestrated, or of sinusoidal origin, the limiting feature to extravasation of particulate matter will be the operative size of the particle. The permeability of vascular endothelia is controlled by effective pore sizes ranging between 300 and 1000 Å. Therefore, in considering the design of therapeutic targeted liposomal drug carriers, it is of importance to recognize the size restriction of artificially constructed vesicles. Specifically, physical constraints on bilayer curvature limit the field to formulation of carriers possessing diameters between approximately 300 Å and 10 μm [3]. A further restriction to extravasation of liposomes occurs by the utilization of multilamellar or large unilamellar vesicles that have minimum diameters of approximately 0.1 μm [4].

Of equal importance in evaluating the potential of liposomes to reach a desired target is the recognition of these structures as foreign particles by reticuloendothelial cells [5-8]. It has been acknowledged that accumulation of liposomal particles in liver, spleen, and lung is a major barrier preventing delivery of effective concentrations of the encapsulated therapeutic agent to other desired sites. However, in many instances this "passive targeting" feature of unmodified liposomes has been successfully exploited to shuttle immunomodulators or anti-infective agents to the reticuloendothelial system (RES) [9-13].

Despite these obvious constraints toward achievement of site-specific targeting with liposomes, advances in this field continue to be made. In part these strides can be attributed to a better understanding of the physiological changes that occur at diseased tissue sites. In particular, it has been observed that hyperpermeability of endothelial barriers can occur during disease states such as inflammation, hypertensive vascular lesions, and tumor development [14]. These angiogenic changes are mediated via actions of histamine, bradykinin, and lymphokines. As a result, the efficient collection of unmodified liposomal particles in inflamed joints of rheumatoid arthritis patients [15] or around injured vessels in regions of myocardial infarction [16] has been demonstrated. The leakage of liposomal particles into the vasculature of tumors is also known [17], although the heterogeneity and variable morphology of these tissues may limit the operation of such a mechanism to a smaller subset of tumor types [14, 18].

Another factor contributing to the more efficient use of liposome formulations in targeting applications is the accrued collective experience in designing compositions that are engineered to avoid recognition by the reticuloendothelial system. Several approaches have met with varying degrees of success. As one can surmise from the preceding discussion on architecture of the endothelium, size-reduced preparations can dramatically prolong the circulating half-life of entrapped drugs following intravenous injection as well as promote distribution to nonreticuloendothelial tissues [19-21]. Altering the surface charge was also found to extend the circulation time [19,22]. An alternative approach is to modify the surface of the vesicle membrane with biomaterials conferring a property of stealth with respect to the mononuclear phagocytes. For example, this can be achieved using compounds that enable the liposome to mimic the facade of a natural particle such as the erythrocyte. Materials such as sialoglycoproteins [23-25], sialic acid [26], or gangliosides [20] have been useful for this purpose. As the scope broadens in the search for and application of new raw materials for construction of lamellar systems [27], additional species that bestow an invisibility to the vesicles from the RES will undoubtedly surface.

Although liposomes are now under evaluation in human clinical trials for various applications, the testing of an immunomodified formulation in humans is still some years away. Thus far, the primary focus of investigations on immunotargeted liposomes in animals has been for the delivery of anticancer agents to tumor cells. Since the promise for development of a human pharmaceutical with similar properties has such vital impact, the activity in this area has been fairly intensive. The targeting potential of surface-modified liposomes in many of these animal model systems has been examined in a number of excellent review articles [2,12,14,19,20,28-35]. Recent targeting trials to other cell types such as erythrocytes suggests that the field for immunotargeted liposomes may be substancially broadened in the future. This review will present an updated status of the technology and will provide further perspectives on the commercial utility of immunotargeted liposomes as judged by both chemical and biological criteria.

II. CHEMICAL COUPLING OF ANTIBODIES TO LIPIDS AND LIPOSOMES

A. Noncovalent Association

Simple Mixtures of Lipid and Antibody Components

Various trials have been reported in which antibodies have been attached noncovalently to liposomal membranes by virtue of procedures that involve mixing or sonicating the two components [36-40]. Generally, compositions that include charged lipids such as phosphatidic acid or stearylamine are most effective for binding of antibodies. Liposomes prepared by this method have been shown to bind more efficiently to target cells in vitro [37-40] or to tumor sites in vivo [36]. These procedures, however, are not viewed as commercially viable techniques, since (1) prolonged sonication can cause aggregation and protein denaturation; (2) final protein-to-lipid ratio is variable and difficult to control between preparations; (3) dissociation of components can occur in vivo; and (4) the technique is limited to those proteins that bind nonspecifically.

Aggregated Immunoglobulins

An increase in the hydrophobic interaction between antibodies and lipid membranes can be effected by inducing the aggregation of immunoglobulins either chemically or with heat treatment [41–43]. While this method ensures a more efficient, controllable incorporation of antibody into the vesicles, additional variables are introduced that make the process less attractive in a commerical context. These variables include (1) destabilization of the liposome membrane with concurrent increase in solute permeability, (2) insufficient stability of the complex in vivo, and (3) alteration in the proper presentation of the binding regions of the antibodies.

Biotin-Avidin Coupling

Avidin, a tetrameric protein of molecular weight 68,000 daltons, possesses a tight affinity ($K = 10^{15}$) for the coenzyme biotin. This strong but noncovalent interaction has been exploited for the conjugation of antibodies to compounds including lipids and liposomes [44–47]. Although the lipid-antibody interaction is noncovalent, the total process encompasses covalent coupling steps for conjugation of either the biotin or avidin to the appropriate ligand. For example, biotin or avidin can be chemically coupled by succinimide-, carbodiimide-, or glutaraldehyde-type reactions to either the lipid or antibody component. This allows for a number of different schemes by which antibodies and lipids are linked (Figure 1). Torchilin and collaborators [48–50] have recently proposed a novel variation of these procedures to include an "in vivo" coupling method wherein biotinylated antibody is first targeted to the receptor cell, which then recognizes avidin-conjugated liposomes. In this way, the manufacture of liposomes is simplified by eliminating variables attributed to the presence of antibodies on the vesicle surface. However, although tested successfully in cell culture [50], the demonstration of this approach in vivo is still pending. A bacterially derived streptavidin has also been introduced as an improvement for reducing some of the nonspecific binding associated with mammalian-derived avidin [47]. Thus far, the biotin-avidin coupling procedure holds the most promise for targeting of liposomes by noncovalent affiliations. These methods, however, are still subject to the disadvantages of the chemical coupling steps needed

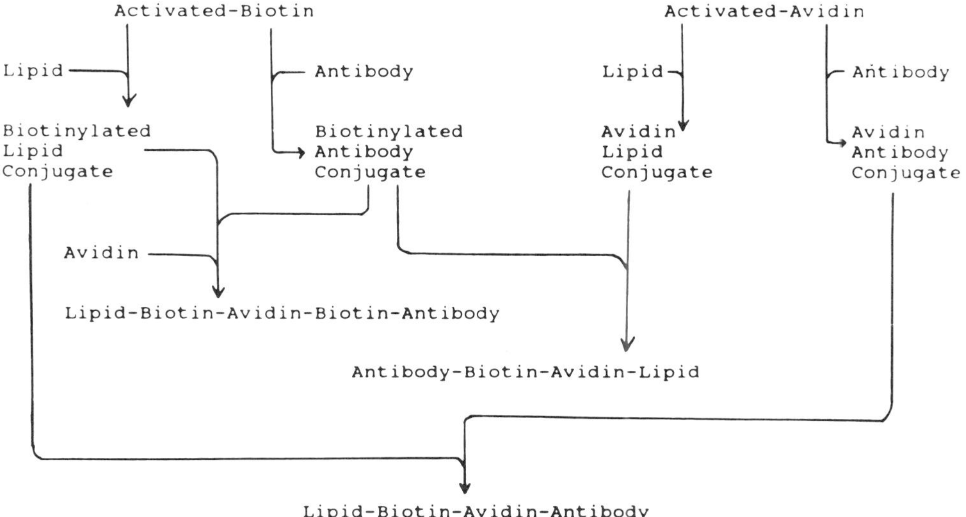

Figure 1 Schemes for coupling antibodies to lipid using biotin or avidin conjugates.

to link the biotin or avidin to their respective ligands. A discussion of these limitations follows.

B. Covalent Coupling Methods

Lipids with Reactive Moieties

It becomes readily apparent that the coupling of antibodies to liposomes is intrinsically flexible by simply examining the diverse numbers of lipids that possess reactive groups. Some of the commonly used lipids are listed in Table 1. It is important to note, however, that not all of the lipids outlined in this table represent choice candidates when considering commerical or manufacturing concerns [51,52]. This issue is particularly germane with respect to certain naturally derived lipids that may be in low abundance and are cost-prohibitive to manufacture on a larger scale.

One of the popular choices for conjugate preparation is lipid containing reactive amine functions. A primary selection in this category is phosphatidylethanolamine (PE), a common and relatively abundant natural phospholipid that is usually produced as a by-product of phosphatidylcholine purification. Egg-derived PE is nontoxic and will align itself naturally within the vesicle bilayer. Certain

Table 1 Lipids with Reactive Groups

Amine
Phosphatidylethanolamine
Stearylamine
Phosphatidylserine
Carboxyl
Fatty acids
Hydroxyl
Phosphatidylglycerol
Fatty alcohols
Phosphatidylserine
Carbohydrate
Phosphatidylinositol
Gangliosides
Cerebrosides
Glycolipids (mono or digalactosyl diglyceride)
Phosphate
Phosphatidic acid
Dicetylphosphate

forms of PE, such as the highly unsaturated types extracted from soybean or synthetically made dioleoyl PE, can assume a hexagonal orientation [53], which may be disruptive to maintenance of the bilayer configuration. Another useful amine-containing lipid is the single-chain saturated lipid stearylamine. It is conceivable that conjugates containing stearylamine will have more limited usage due to some of the known biological toxicities observed upon administration in vivo [27,52]. What is not clear at the present time, however, is whether a reduction in toxic effects will occur upon covalent modification.

The coupling of lipids via carboxyl groups is possible with single-chain fatty acids (FA). Materials such as palmitic, stearic, or oleic acid are obtainable in bulk supply for minimal expense. A possible disadvantage to use of single fatty chains in liposome formulations is that they can destabilize the vesicle bilayer at high mole ratios. A bilayer forming diacyl lipid with an available carboxyl function is phosphatidylserine (PS). Because of its zwitterionic nature, it is not usually a primary choice, particularly since its available amine function can cause cross-linkage.

A variety of lipids exhibit carbohydrate functions that can be modified. The list includes phosphatidylinositol, gangliosides, and cerebrosides from brain tissue, and glycosyl glycerides from plants, such as digalactosyl diglyceride. These materials are generally in low supply in nature when compared to other phospha-

Table 2 Coupling Sites and Reagents for Lipid-Antibody Conjugates

Function required on lipid	Reagent	Function required on protein
Amine	SPDP	Sulfhydryl
Amine	MPB	Sulfhydryl
Amine	Glutaraldehyde	Amine
Amine	Carbodiimide	Carboxyl
Amine	Iodoacetate	Sulfhydryl
Amine	Iminothiolane/DTNB	Sulfhydryl
Benzylamine	$NaNO_2$/HCl	Phenolic
Carboxyl	NHS	Amine
Carboxyl	Acyl halide/anhydride	Amine
Hydroxyl	Periodate/borohydride	Amine

tides. Since the carbohydrate function possesses more than one reactive hydroxyl residue, the likelihood of creating multiple products increases, thereby making the chemistry somewhat more complicated. This same problem would be encountered for attachment to phosphatidylglycerol (PG), which presents three open hydroxyls. A lone hydroxyl residue can be found on the single-chain fatty alcohols. These have not been exploited to any degree, probably due to the toxicity associated with this class of compounds.

It is possible to create phosphate esters using phosphatidic acid (PA) or dicetyl phosphate (DCP). Although not yet used extensively in antibody conjugate reactions, selection of PA is attractive since this compound is readily available, nontoxic, and incorporates efficiently into the vesicle bilayer.

Reactive Sites on Antibodies

As is the case for proteins in general, antibody molecules possess four possible reactive moieties for conjugation. These are amines, carboxyls, sulfhydryls, and hydroxyls (including phenolic functions). The design of an antibody conjugate must correlate both with the functionality of the molecule as well as with the resultant physical behavior with respect to the liposomal carrier. For example, coupling at the antigen combining site may not only prevent subsequent binding to the target site but could possibly hide this region within the bilayer where it would be shielded from antigen. It is for this reason that sulfhydryl conjugation techniques tend to be favored, since lipids can be attached to the free sulfhydryls generated from dissociation of the disulfide bonds that link the large chains

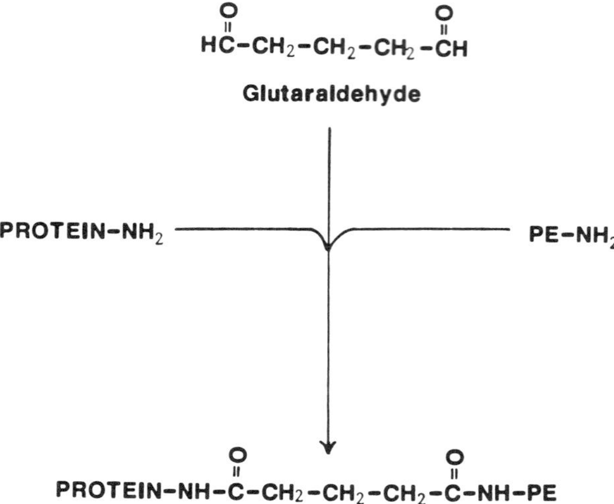

Figure 2 Cross-linkage of protein and lipid with the homobifunctional agent gluteraldehyde.

of the immunoglobulin molecule. Nonetheless, other coupling methods have yielded successful results. Table 2 outlines the different coupling schemes that have been used. These reactions are described in greater detail below.

Conjugation to the Amine Residues of Antibody

One of the first coupling reagents to be tested for liposome applications was the bifunctional cross-linker glutaraldehyde [54-55]. Show in Figure 2, this dialdehyde reacts by hooking together adjacent amine residues. A severe drawback to the scheme shown in Figure 2 is an extrapolation to include cross-bridging between two adjacent protein molecules or, similarly, two adjacent lipid molecules. This limitation, which results in multiple products and loss of material, restricts the eventual commercial utility of the method.

A more useful procedure, developed primarily by Huang and co-workers [56, 57], makes advantage of the N-hydroxysuccinimide (NHS) function (Figure 3). Fatty acid modified by carbodiimide can be reacted with NHS to produce an activated fatty acid capable of attaching to amine residues of the antibody.

Fatty anhydrides or halides can be conjugated successfully with amine portions of the antibody [58,59]. An example of the reaction with dodecanoic anhydride is illustrated in Figure 4.

The amine resudes of the protein can also be involved in a Schiff base reaction with free aldehydes. A common approach has been to oxidize the hydroxyl

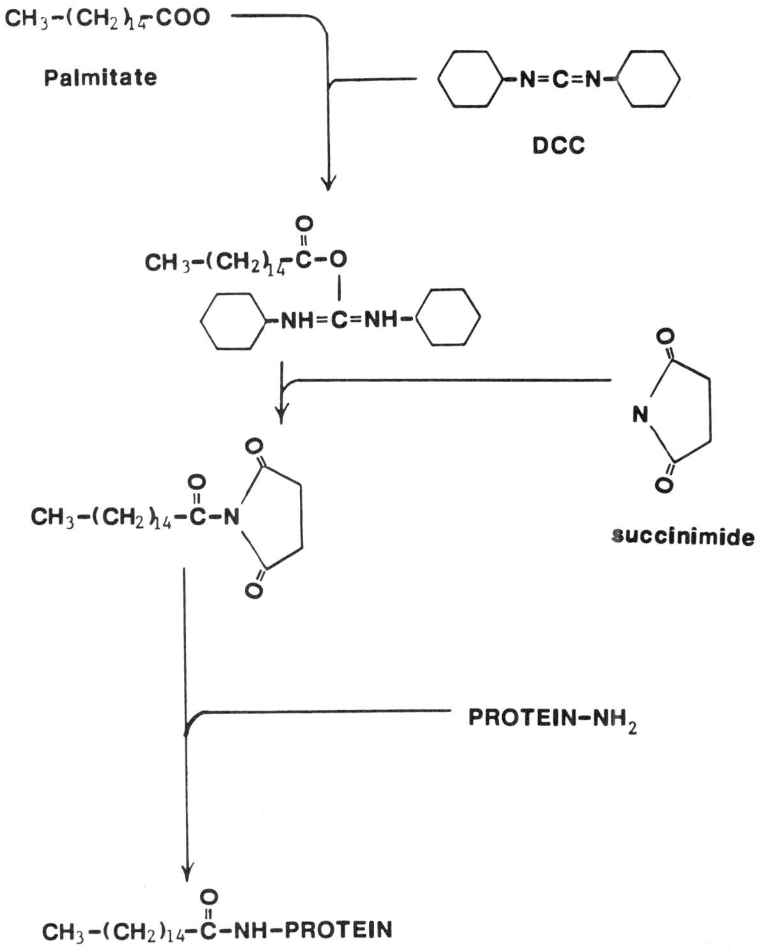

$CH_3-(CH_2)_{14}-COO$

Palmitate

$-N=C=N-$

DCC

$CH_3-(CH_2)_{14}-C(=O)-O$

$-NH=C=NH-$

N

succinimide

$CH_3-(CH_2)_{14}-C(=O)-N$

$PROTEIN-NH_2$

$CH_3-(CH_2)_{14}-C(=O)-NH-PROTEIN$

Figure 3 Coupling of the amine residue of protein to fatty acids via N-succinimide.

groups of an appropriate glycolipid to produce reactive aldehydes [45,60,61]. The amide bond is then stabilized by reduction with borohydride. This scheme is depicted in Figure 5. A variation to this approach is to use the periodate to create aldehyde groups on the protein, which can then react with amine-containing [62] or hydrazinated lipids [63].

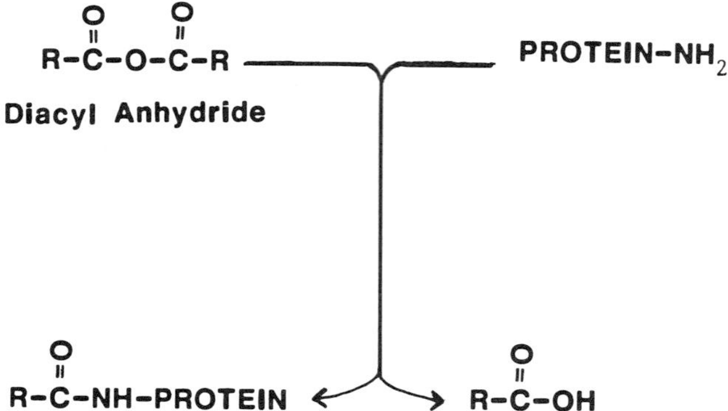

Figure 4 Reaction of diacyl anhydrides such as dodecanoic anhydride with the amino function of protein.

Conjugation to the Carboxyl Residues of Antibody

Carbodiimide reagents have been used extensively in coupling reactions, including those involving proteins [64]. Activation of the carboxyl functions of amino acids with carbodiimide allows for reaction with amine groups of lipid [65]. The water-soluble 1-(3-dimethylaminopropyl)3-ethylcarbodiimide is typically selected for reactions of this type (Figure 6). As with glutaraldehyde, the carbodiimides can cause cross-linkage of the protein between the internal carboxyls and amines, a problem alleviated by prior citraconylation of the antibody [66]. The instability of the carbodiimides in aqueous solution is considered to be a major disadvantage of this method.

Conjugation to the Phenolic Group of Tyrosine

The diazotization reaction has been used as a means to couple lipid to tyrosine residues of protein [67,68]. This procedure (Figure 7) avoids the homopolymerization side reactions encountered with other bifunctional agents and does not necessitate thiolation of the protein, a procedure often required with sulfhydryl reactive compounds (see below).

Conjugation to Sulfhydryl Groups of Antibody

As noted earlier, the interchain disulfide linkage of immunoglobulins or corresponding F(ab′) fragments presents a useful target for sulfide-reactive agents without disruption of immune reactivity. A number of these have been developed for use with lipids. A general prerequisite to employment of thiol modifiers is prior treatment of the antibodies with disulfide-reducing agents such as

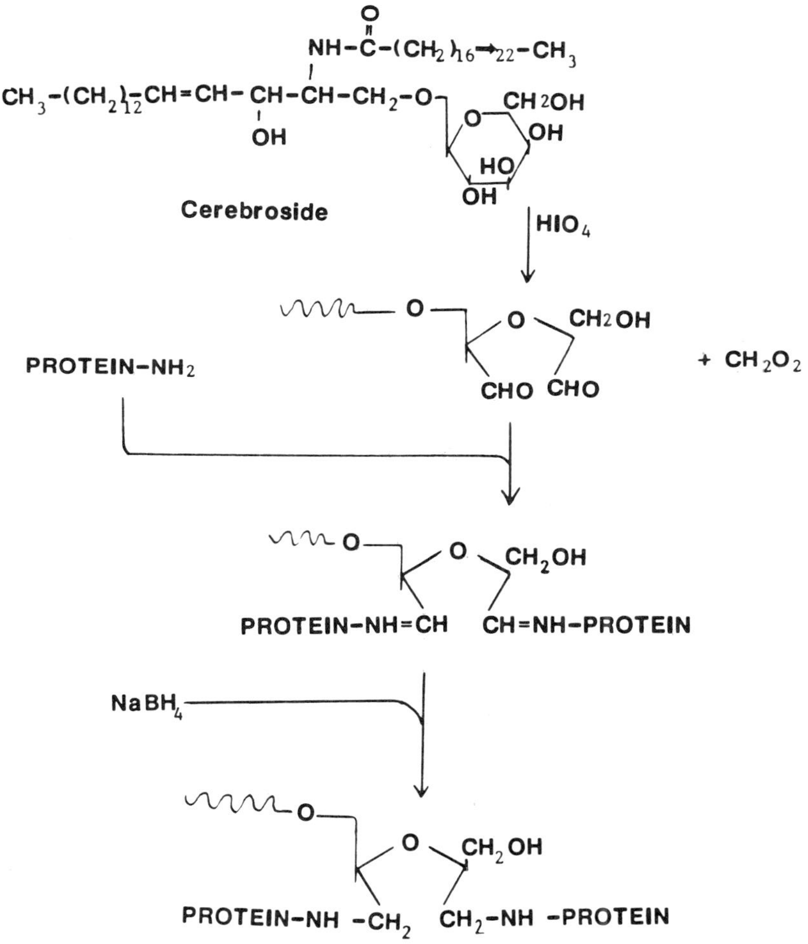

Figure 5 Schiff base reaction between the periodate-activated carbohydrate function of a glycolipid and amine group of protein.

dithiothreitol to expose primary sulfhydryls. One can avoid the use of reducing compounds by introducing new sulfhydryls into the protein with activated thioacetyl compounds such as SATA [69] or SAMSA [70], which attach to available amines. Exposure of free sulfhydryls on the conjugated SATA molecule can be initiated by the addition of hydroxylamine, which need not be removed in further conjugation reactions with lipid (Figure 8).

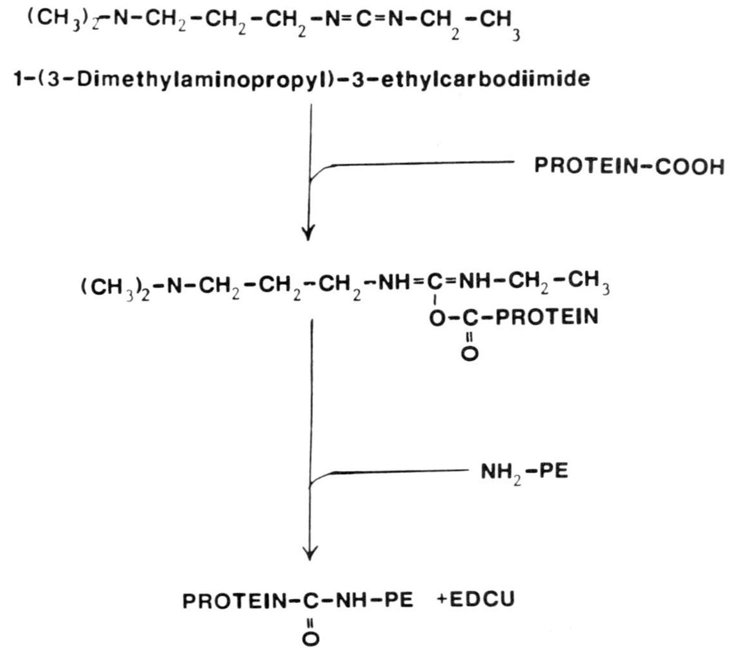

$(CH_3)_2\text{-}N\text{-}CH_2\text{-}CH_2\text{-}CH_2\text{-}N{=}C{=}N\text{-}CH_2\text{-}CH_3$

1-(3-Dimethylaminopropyl)-3-ethylcarbodiimide

↓ PROTEIN-COOH

$(CH_3)_2\text{-}N\text{-}CH_2\text{-}CH_2\text{-}CH_2\text{-}NH{=}C(\text{-}O\text{-}C({=}O)\text{-}PROTEIN){=}NH\text{-}CH_2\text{-}CH_3$

↓ NH_2-PE

PROTEIN-C(=O)-NH-PE +EDCU

Figure 6 Water-soluble carbodiimide coupling between the carboxyl function of the protein and the amine terminus of lipid.

The common sulfhydryl detection agent 5,5′-dithio-bis-2-nitrobenzoic acid (Ellmans reagent) can be coupled to lipid that has been activated at the amine residue with 2-iminothiolane [71] (Figure 9). Disulfide coupling with available sulfhydryl groups on the protein occurs with production of thionitrobenzoate. The water solubility of the 2-iminothiolane reagent is an advantage over some of the other agents.

Sulfhydryl groups of proteins can also be attacked by alkyl halides such as iodoacetate. With careful blocking of the appropriate groups, Sinha and Karush [72] were able to link protein sulfhydryl groups to an iodoacetate-modified lysine-PE conjugate (Figure 10).

To date the most useful sulfhydryl linkers for liposome modification are the heterobifunctional agents SPDP and SMBP (Figures 11 and 12), which present an N-hydroxysuccinimide function on one end of the molecule and either a pyridine or maleimide group on the other [73-76]. The succinimide moiety reacts with reactive amines on the lipid to form a conjugate with an activated end. Whereas PDP-lipid conjugate then forms a dithiol bond with reactive sulf-

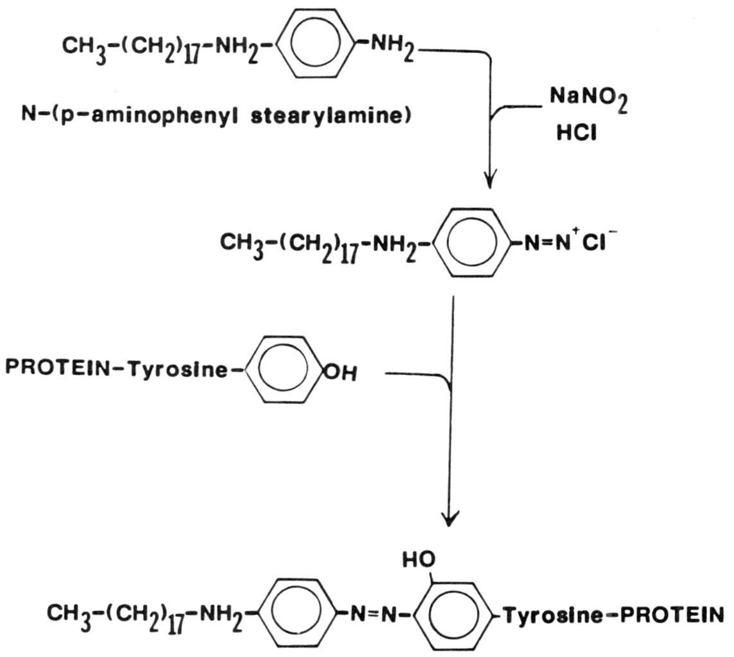

$CH_3-(CH_2)_{17}-NH_2-C_6H_4-NH_2$

N-(p-aminophenyl stearylamine)

$NaNO_2$
HCl

$CH_3-(CH_2)_{17}-NH_2-C_6H_4-N{=}N^+Cl^-$

PROTEIN-Tyrosine-C_6H_4-OH

HO

$CH_3-(CH_2)_{17}-NH_2-C_6H_4-N{=}N-C_6H_3$-Tyrosine-PROTEIN

Figure 7 Diazotization reaction with phenolic residues of tyrosine.

hydryl groups on the protein, the MBP-lipid conjugate forms stable thioether linkages that are not susceptible to cleavage by reducing agents.

Considerations for Choice of Coupling Method

Two basic modes exist for construction of antibody-modified liposomes. The first approach relies on coupling the protein directly to lipid prior to manufacture of liposomes. An endeavor of this kind requires the manipulation of reaction mixtures containing two phases, one to dissolve the lipid and the other for solubilization of the protein. The organic or detergent phase necessary to keep lipid in soluble form may act adversely on proteins by altering proper conformation. However, reaction of the two individual components offers the following advantages: (1) easier and more complete purification of reactants and by products; (2) exposure of the total amount of lipid in the preparation to the coupling reagents; (3) incorporation of the modified lipid effected throughout all the bilayers and not restricted to the outer shell; and (4) chemically harsh reaction conditions precede the formation of the liposome bilayers and thus are not detrimental to the stability of the vesicle structure or lipid integrity. A per-

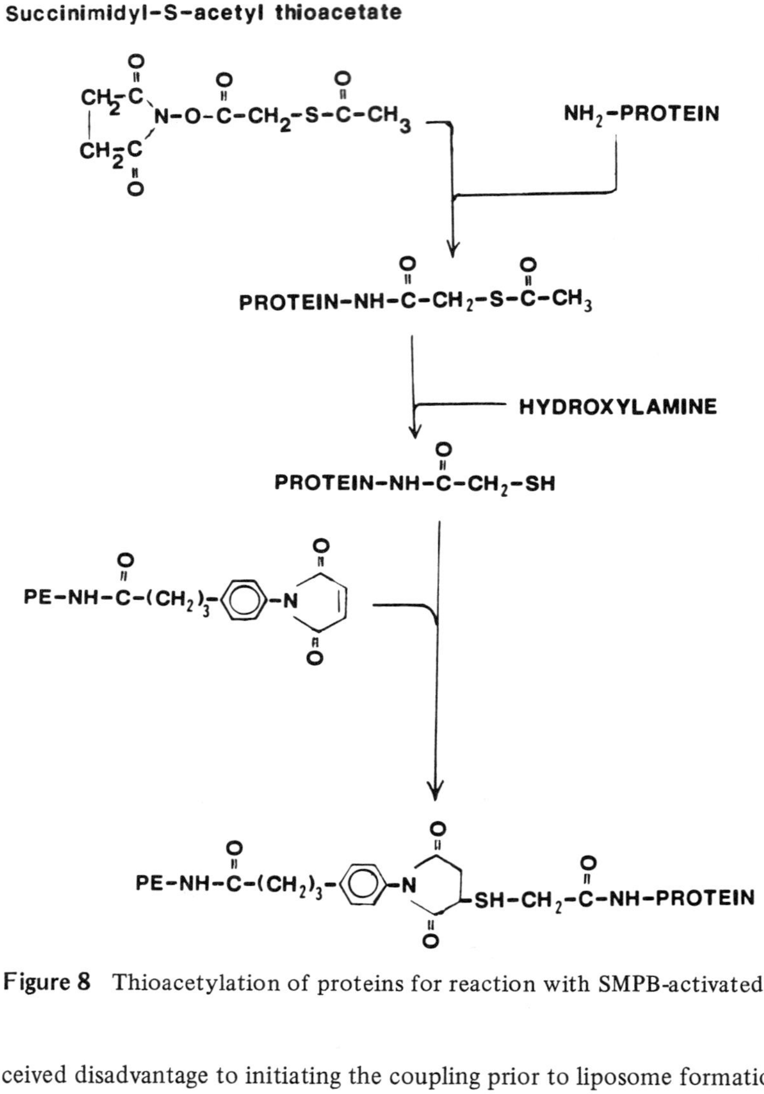

Succinimidyl-S-acetyl thioacetate

$$\text{Succinimidyl-}N\text{-O-}\overset{O}{\overset{\|}{C}}\text{-CH}_2\text{-S-}\overset{O}{\overset{\|}{C}}\text{-CH}_3 \xrightarrow{\text{NH}_2\text{-PROTEIN}} \text{PROTEIN-NH-}\overset{O}{\overset{\|}{C}}\text{-CH}_2\text{-S-}\overset{O}{\overset{\|}{C}}\text{-CH}_3$$

$$\xrightarrow{\text{HYDROXYLAMINE}} \text{PROTEIN-NH-}\overset{O}{\overset{\|}{C}}\text{-CH}_2\text{-SH}$$

$$\text{PE-NH-}\overset{O}{\overset{\|}{C}}\text{-(CH}_2)_3\text{-C}_6\text{H}_4\text{-N(maleimide)} \longrightarrow \text{PE-NH-}\overset{O}{\overset{\|}{C}}\text{-(CH}_2)_3\text{-C}_6\text{H}_4\text{-N(succinimide)-SH-CH}_2\text{-}\overset{O}{\overset{\|}{C}}\text{-NH-PROTEIN}$$

Figure 8 Thioacetylation of proteins for reaction with SMPB-activated lipid.

ceived disadvantage to initiating the coupling prior to liposome formation is the means by which conjugate is then incorporated into the vesicle membrane [77]. Previous approaches have employed either detergent dialysis techniques [57] or reverse evaporation using diethyl ether [56]. Removal of detergent has proven to be difficult and time-consuming, while procedures requiring ether are potentially hazardous for both the protein and the process engineer working with it. However, one must consider some of the more recent solvent- or detergent-free

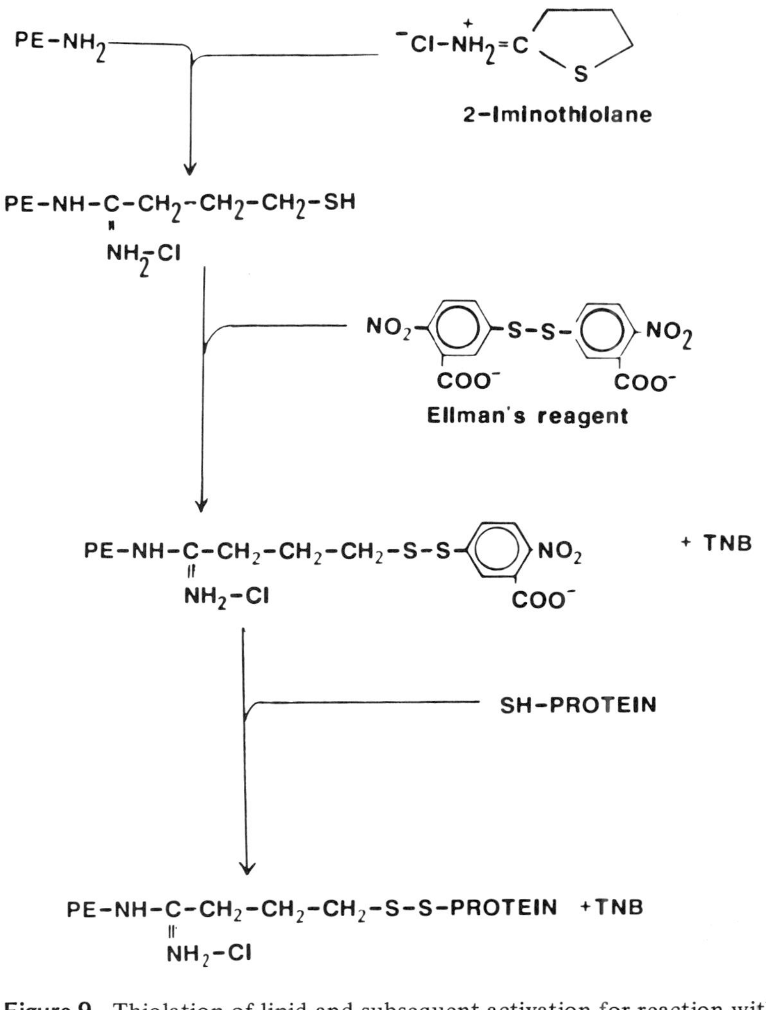

$PE-NH_2$ + $^{-}Cl-\overset{+}{N}H_2=C$ (2-Iminothiolane)

$PE-NH-C(=NH_2{-}Cl)-CH_2-CH_2-CH_2-SH$

+ Ellman's reagent: NO_2–(ring, COO^-)–S–S–(ring, COO^-)–NO_2

$PE-NH-C(=NH_2{-}Cl)-CH_2-CH_2-CH_2-S-S$–(ring, COO^-)–NO_2 + TNB

+ SH-PROTEIN

$PE-NH-C(=NH_2{-}Cl)-CH_2-CH_2-CH_2-S-S-$PROTEIN + TNB

Figure 9 Thiolation of lipid and subsequent activation for reaction with the sulfhydryl functions of proteins.

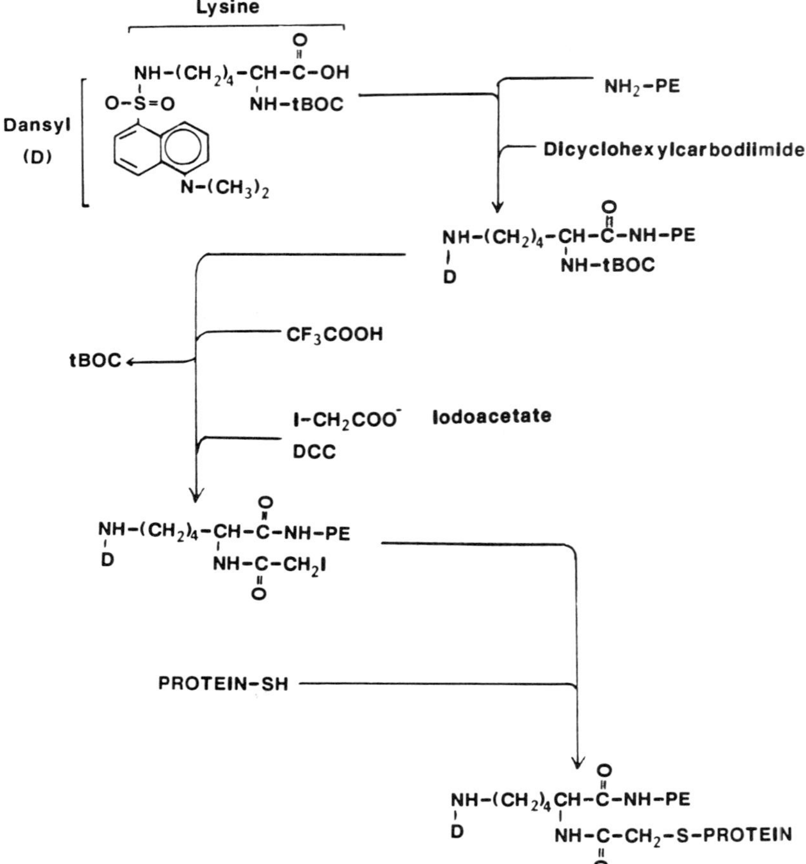

Figure 10 Reaction of protein sulfhydryl groups with iodoacetate-activated lipid-lysine conjugate.

procedures such as freeze-thaw [78] or dehydration [79] as useful techniques for incorporation of protein-lipid conjugates.

The second mode by which antibodies may be attached to vesicles is by reaction with preformed liposomes. In this way protein can be directly linked to the vesicle surface either by homobifunctional reagents or by interaction with preincorporated activated lipids. As mentioned above, this protocol may subject the vesicle to harsh conditions, thereby affecting the stability of the membrane. Furthermore, in this scheme, conjugation takes place only on the outer surface of

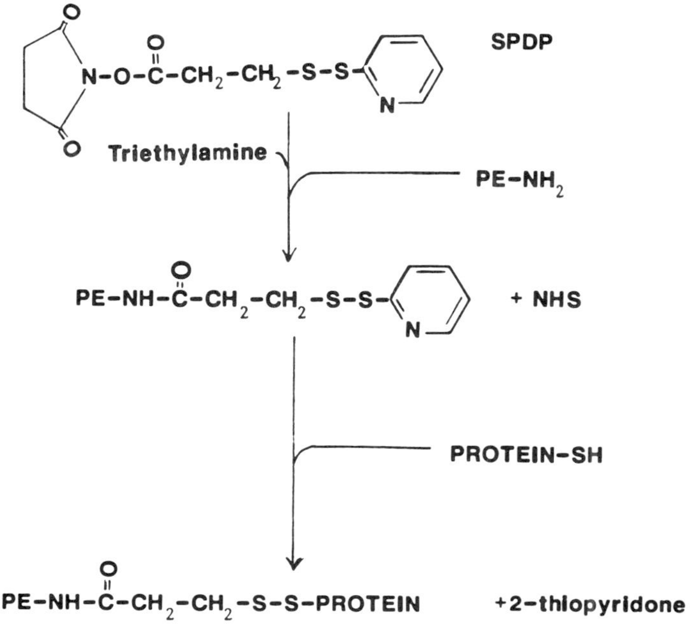

SPDP: N-succinimidyl–O–C(=O)–CH_2–CH_2–S–S–(2-pyridyl)

Triethylamine, $PE{-}NH_2$

↓

$PE{-}NH{-}\overset{O}{\overset{\|}{C}}{-}CH_2{-}CH_2{-}S{-}S{-}$(2-pyridyl) + NHS

PROTEIN-SH

↓

$PE{-}NH{-}\overset{O}{\overset{\|}{C}}{-}CH_2{-}CH_2{-}S{-}S{-}PROTEIN$ +2-thiopyridone

Figure 11 Activation of lipid with the heterobifunctional agent SPDP (N-succinimidyl proprionyl dithiopyridine) for reaction with sulfhydryl groups of protein.

the liposome even if reactive lipid is present in all the bilayers. An important consideration in ths regard is that in vivo degradation or removal of the outer membrane prior to binding of the target will result in loss of targeting capability for the remaining drug-containing vesicle. Nevertheless, attachment of antibody to preformed liposomes does offer the flexibility to design the most appropriate liposome manufacturing process for the drug of interest without giving due consideration to possible effects on the protein.

From a commercial standpoint, it should not be assumed that the issue of liposome stability is a moot point once the antibody has been successfully coupled. Reports have shown that proteins derived from serum can in fact interact with lipid membranes [80,81]. More specifically, bound immunoglobulins can induce flocculation of the vesicles [82]. Recently, Bredehorst et al. [83] demonstrated that the major instability of liposome bilayers containing F(ab) fragments coupled to PE through MPB could be traced to the MPB-lipid rather than the antibody. By decreasing the mole percentage of modified lipid in the total preparation while increasing the number of coupled F(ab) fragments per

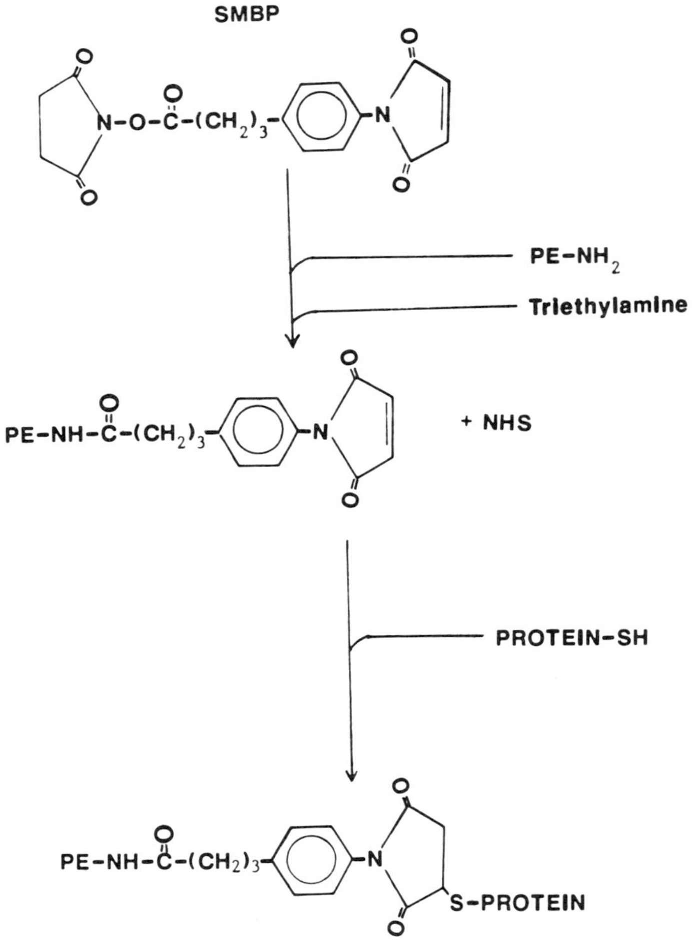

SMBP

N-O-C-$(CH_2)_3$-

PE-NH_2

Triethylamine

PE-NH-C-$(CH_2)_3$-

+ NHS

PROTEIN-SH

PE-NH-C-$(CH_2)_3$-

S-PROTEIN

Figure 12 Activation of lipid with the heterobifunctional agent SMPB [N-succinimidyl 4-(p-maleimidophyeny) butyrate] for reaction with sulfhydryls of protein.

liposome, relative stability could be maintained. However, one must view the term stability in the context of designing pharmaceutically acceptable preparations in which the goal is generally a 2-year shelf life (with specifications that place limits on the degree of lipid and drug degradation as well as the presence of drug in the free state). The advances made in lyophilization technology for liposome formulations over the last several years have essentially eliminated problems of long-term storage [84-86].

III. TARGETING IN VITRO

Cell culture systems are used broadly in liposome targeting studies. While in vitro systems of this type do not provide direct assessments of the clinical or therapeutic utility of the formulations, knowledge can be gained on several levels, which can lead to the engineering of optimally effective preparations for purposes of injection. Three specific types of information can be obtained using cell culture techniques. First, parameters governing the binding of liposome carriers to the cell can be distinguished. Factors including cell receptor density, antibody density of the liposome surface, and specificity of the interaction between the modified liposome and cell type can be investigated. Second, the mechanisms that play a role in controlling uptake and internalization of the liposome once bound to the cell can be probed. An assessment of the contribution of particle size or of events that regulate membrane fusion or active endocytosis is possible from data collected in various cell models. Finally, cell studies furnish perspectives on comparative efficacy between free and encapsulated drug in addition to defining those cell types that may be most responsive to the treatment. These areas will be discussed in greater detail below.

A. Binding of Antibody-Targeted Liposomes to Cells

Thus far it appears that the number of antibody molecules on the liposome surface required to achieve binding to target cells is variable and depends on the specific system. Heath et al. [87] reported that a high antibody density on the liposome surface was optimum for targeting to erythrocytes. Similarly, Shen et al. [56] found that approximately 200 H-2K antibody molecules per liposome proved effective for binding to lymphoma cells. Gregoriadis and collaborators [19,88] concluded that one or two molecules of anti-Thy 1 IgG1 hooked to the surface of a vesicle was sufficient to induce binding to AKR-A cells. While Barbet et al. [89] have indicated that less than 10 antibody molecules per liposome can be effective for cell targeting or protein A precipitation, Weckenmann et al. [90] reported that a minimum of 12 M.2.9.4. antibodies (IgG2a) per liposome were needed for attachment to a human melanoma cell line. In considering the antibody density of vesicles, the concept that "more is better" is not necessarily true. Houck and Huang [91] recently observed that in RDM-4 lymphoma cells induced with proteinase K to express high surface histocompatibility antigen, better binding occurred to liposomes having lower membrane densities of anti H-2K monoclonals. The implications of this study are that cell receptor expression may play as much a role in binding of antibody-modified liposomes as the liposome-to-antibody ratio.

The specificity of binding is particularly explicit in studies that describe liposome targeting to major histocompatibility antigens (MHC) [57,92–95].

Monoclonal antibodies can be constructed to recognize different MHC molecules from cells of the same stain of animal, and therefore differential binding to slightly different MHC antigens is distinguishable. Selectivity between species can also be demonstrated. For example, liposomes modified with anti-rat erythrocyte IgG of $F(ab')_2$ fragments will bind to rat- but not human-derived red blood cells [47]. Similarly, it has been shown that liposomes modified with anti-rat erythrocyte F(ab')2 fragments will not recognize rat leukocytes or mouse or monkey erythrocytes [96]. Most recently, it has been shown that liposomes conjugated to anti-mouse erythrocyte F(ab') will not bind to sheep erythrocytes and that the specificity of binding to mouse erythrocytes can be regulated by the F(ab')-to-phospholipid ratio [97].

B. Uptake and Internalization of Bound Targeted Liposomes

How efficiently liposomes are taken in by cells to which they are bound is a complex equation. The issues that need to be considered are whether (1) binding of a specific antigen with antibody will mediate cellular response mechanisms that signal uptake of that antibody and its conjugated cargo; (2) a bound liposome leaks its content prior to internalization; (3) uptake is regulated by the structure or size of the vesicle; (4) contents of the vesicle are transferred via fusion events or taken up whole and compartmentalized as endocytic vesicles within the cell; and (5) cellular responses are elicited simply by the binding event.

Several reports have eloquently demonstrated that anti-MHC modified liposomes containing methotrexate or methotrexate-gamma aspartate (dihydrofolate reductase inhibitors) are differentially taken up by either B or T lymphocytes that exhibit the same determinant [98–100]. These studies suggest that the physiological basis for internalization is not based on a single global mechanism but rather is a function of individual cellular events not yet fully understood. Berger et al. [101] recently reported that cellular cytotoxicity responses to liposome-conjugated antibodies in the CEM T-cell line were initiated simply by the binding incident and not generated by the uptake of the cytotoxic agent. In this study liposomes conjugated to BE-3 or OKT-4 monoclonals caused discordant losses of antigen on the cell surface yet stimulated similar toxicity responses.

The leakage of vesicle contents external to the cell to which it is bound can be monitored. By indirect means, Machy and Leserman [102] concluded that leakage of methotrexate from liposomes bound to RDM-4 tumor cells was insignificant, since growth of variant cells without expressed surface antigen, which were present in the same cultures, were not affected by the treatment. In another study, however, it was shown that large unilamellar vesicles targeted with anti-carcinoembryonic antigen monoclonals to either human colonic, pancreatic, or lung carcinoma cell lines leaked their contents of ricin A after binding [103]. The activity of liberated ricin A was inhibited by the addition of lactose.

In the previous study [103], the construction of small-sized vesicles with ricin A facilitated both stabilization and uptake of the liposomes. Size reduction of the vesicles has been shown repeatedly to enhance the uptake of vesicle-incorporated materials [19,34,93]. Generally, the low efficiency of drug entrapment in small unilamellar vesicles has been considered a disadvantage in relationship to the effective dose, although properly targeted preparations should concentrate the drug locally and thus improve the feasibility for use of these systems.

The distinction between uptake as denoted by membrane fusion or cellular endocytosis is difficult to define as being mutually exclusive. Some evidence has indicated that the activity of liposome-internalized drugs such as methotrexate are susceptible to alterations in subcellular pH, a finding that implies the drugs were delivered from the acidic environment of the endocytic vesicle [99,104–106]. By incubating liver Kupffer cells in monolayer culture with the lysomotropic agents monensin and chloroquine, Derksen et al. [107] were able to observe changes in the intracellular distribution of labeled cholesterol esters or aqueous markers when delivered from immunoglobulin-conjugated large unilamellar vesicles. The findings show that liposomes are transferred to the lysosomal fraction, where aqueous marker (BSA) is degraded and subsequently released from the cell. These authors also reported a fivefold enhancement in uptake of liposomes that are covalently linked to immunoglobulin as compared to uncoated liposomes [108]. In these experiments, uptake of liposome-encapsulated horseradish peroxidase was traced directly to the intracellular vacuoles.

Although fusion of vesicles may represent a minor component of transfer of encapsulated materials, the mechanism cannot be totally excluded [109].

C. Efficacy of Antibody-Targeted Liposomes in Cell Culture

Notwithstanding the usefulness of cell studies to delineate mechanisms of action of liposomes, the matter at hand is still the topic of therapeutic value. To this end, the relationship of liposome uptake into cells is typically correlated to biochemical changes that ensue as a consequence of the entrapped drug. Thus far the primary focus of immunotargeted liposomes has been for treatment of tumors for which the assay methods usually invoke measurements of cytotoxicity or growth inhibition. Other applications have included the delivery of enzymes to deficient cells or introduction of genetic information for expression of specific protein products.

The incorporation of labeled nucleotides provides a useful means to assess the uptake and activity of anticancer agents when delivered from bound liposomes. For example, Onuma and co-workers [110] reported that small unilamellar vesicles containing adriamycin (doxorubicin) and modified on the suface with monoclonal antibodies to BLS C-KU 1 tumor cells (derived from cattle infected with bovine leukemia virus) effectively inhibited 3H-TdR incorporation in these cells.

A somewhat more direct evaluation of effective drug delivery is facilitated by determining the overall toxicity to the cell line. In this regard, an interesting agent developed for use with targeted lipsomes by Papahadjopoulos and collaborators [34,111] is methotrexate-gamma-aspartate (MGA). This analog of methotrexate possesses equivalent activity to the parent compound but is poorly transported into cells in free form. When presented to L929 fibroblasts or OVCAR-3 human ovarian cancer epithelial cells from liposomes conferring an antibody-targeting determinant for each of these cell types, the MGA exhibited superior growth-inhibiting or cytotoxic activity. This efficacy was not observed against cell types that did not carry the antigenic site. In a similar set of experiments, Boggs et al. [112] showed that the proliferation of T lymphocytes derived from guinea pigs suffering from experimental allergic encephalomyelitis was inhibited by methotrexate-loaded liposomes that were covalently linked to the encephalitogenic nonapeptide determinant of myelin basic protein. In this case the liposome is an "antigen"-modified system while the cell type is antibody-presenting. Inhibition of normal or activated (by phytohemagglutinin, protein, or keyhole limpet hemocyanin) T lymphocytes was not observed, again distinguishing the specificity of the binding and delivery of drug.

Hospenthal et al. [113] attempted to show that targeting of antibody-modified liposomes need not be confined to mammalian cell types. By adsorbing rabbit immunoglobulin onto *Candida albicans*, these authors demonstrated the efficient binding of amphotericin B-containing liposomes modified on the surface with anti-rabbit IgG. The study did not, however, demonstrate a differential effect between free and liposome-delivered amphotericin B on mortality of the organism.

D. Diagnostic Applications of Targeted Liposomes

Antibody- or antigen-modified liposomes have additional diagnostic applications not offered with other targeting strategies. Many of the recently designed test kits have reduced to practice the original observations of Kinsky and colleagues [114–121], who demonstrated the complement-mediated lysis of model membrane systems. These reports outlined the schemes by which leakage from the vesicles could be monitored following binding of the appropriate antibody and complement components to the membrane surface. For the purpose of commercial utility, liposomes can be filled with either colorimetric reagents, fluorescent compounds, enzymes, or enzyme-drug conjugates [122]. In one example, alkaline phosphatase-loaded vesicles can be used to produce an intense yellow colorimetric reaction upon antibody- and complement-mediated lysis of the vesicle [123]. Complement lysis of monoclonal antibody-bound liposomes has also been used for the detection of mycotoxin T-2 in a homogeneous immunoassay [124]. In another example the inhibition of complement-mediated lysis of liposomes with theophylline blocks the release of glucose 6-phosphate dehydrogenase [125]. A similar but simplified procedure allows for detection of serum theophylline levels with a single liquid reagent system [126]. In this method a conjugate of theophylline with glucose 6-phosphate dehydrogenase is encapsulated. Antitheophylline antibody is added externally. This antibody normally inhibits the activity of the enzyme when reacted in free form with the conjugate but has no access to the encapsulated form. Addition of theophylline-containing serum samples with detergent provides the basis for a competitive binding assay. A double-antibody sandwich technique can also be used in conjunction with complement-mediated lysis. Yasuda [127,128] prepared liposomes surface modified with anti-C reactive protein (CRP) IgG. Following the binding of CRP from plasma, the addition of anti-IgG antibodies and complement lyses the vesicles to liberate carboxyfluorescein, which then fluoresces. The use of fluorescent techniques in complement-mediated immunoassays, such as described by Butt et al. [129] for phenytoin determination in serum, is gaining popularity as a convenient procedure for repetitive applications.

Rather than rely on lysis-mediated events, Martin and Kung [82] developed an enhanced agglutination test for detection of various classes of antibodies. Using a sandwich technique in which antibody-modified liposomes and latex particles are bridged between the analyte of interest, a significant increase in flocculation of the solutions could be induced.

An exciting step in utilization of modified liposomes for diagnostics is the development of solid-phase assays such as that described for detection of theophylline, thyroxine, digoxin, or human TSH [130–133]. Two methods have been devised for this solid-phase system. In the first scheme, antibody coated tubes are co-incubated with antigen-expressing, fluorescence-containing liposomes and antigen-containing serum samples. A competitive binding occurs, the incubation mixture is decanted, and fluorescence is quantitated by detergent lysis of the liposomes bound to the tube. Alternatively, antigen in serum can be sandwiched between antibodies that are bound both to the tube and to the liposome surface. This technique appears to be as reliable as radioimmunoassay and has shown sensitivities in the low nanogram-per-milliliter range. Guo et al. [134] also reported the development of a solid-phase assay using cellulose acetate particles bound to substrate incubated with antibody-modified liposomes.

In special cases the binding of serum antibodies to liposomes to elicit a colorimetric change can be accomplished without modification of the liposome surface. We developed one such strategy, which stemmed from the observation that certain classes of anti-DNA antibodies recognized lipids that structurally mimicked the DNA backbone [135]. These lipids, when incorporated into liposome membranes, also functioned as calcium ionophores. Upon binding antibody (present in serum from systemic lupus erythematosis patients), the liposomes lost their ionophoretic capability, thus preventing a color change of an encapsulated calcium-sensitive dye.

IV. TARGETING IN VIVO

The site-specific delivery of therapeutic agents from targeted liposomes has been demonstrated primarily in tumor models. While these models may not precisely reflect the exact indication or utility in humans, an incentive for continuing the development of liposomal formulations for cancer treatment comes from the encouraging results of human clinical trials with liposomes containing doxorubicin [136,137]. Moreover, the success of liposomes containing contrast or radioimaging agents for enhancing the detection of tumors [138–140] in both small animals and humans also emphasizes the practicality of liposomes for this application. These and other ongoing trials have also established the safety of using liposomes in the clinic. The advantages targeting will bring in the future are further reductions in the systemic toxicity of the drug due to lower required dosages, as well as broadening the spectrum of cancer cells against which the liposomes will be efficacious. The latter point will depend on the availability of the appropriate targeting antibody, made possible by monoclonal techniques.

To determine whether targeted liposomes collect more efficiently at the appropriate site in vivo, Papahadjopoulos and co-workers [20,141,142] traced the fate of an encapsulated gallium-67–deferoxamine complex. An enhancement in the collection of monoclonal anti-Thy 1.1 modified liposomes at the tumor site (AKR/SL2 tumor cell induced) in mice (5.1% of injected dose pergram of tissue) was observed over the respective unmodified liposome control (3.7% of injected dose per gram of tissue). When GM1 was not included in the vesicle formulation, the collection of labeled liposomes at the tumor site was reduced even further (1.4% of injected dose per gram of tissue). The gallium-67 label was used also to trace the fate of liposomes bearing antibody raised against the Dalton's lymphoma-associated antigen (DLAA) of mice [143]. Targeted liposomes accumulated more efficiently at solid tumors (15.4% of injected dose per gram of tissue) than did respective unmodified liposome controls (5.3%).

The observed increases in collection of immunoliposomes at the target site appears to be balanced against a greater tendency to be taken up by the liver. In previous studies [40,142], the authors noted that the tissue disposition of antibody-bearing vesicles favored uptake in liver when compared to unmodified controls, in spite of increases in tumor concentrations. Derkesen et al. [108,144]

also reported that coating of liposomes with immunoglobulins enhances the interaction with liver Kupffer cells. This same phenomenon was detected by Gregoriadis et al. [145], who measured the plasma clearance rate of anti-Thy1 IgG1-modified liposomes in both control and AKR-A cell-injected AKR mice. Although the clearance of antibody-modified vesicles was facilitated by a subsequent injection of these antigen-presenting tumor cells, their removal from plasma of mice not receiving tumor cells exceeded that of unmodified liposomes. This suggested the involvement of factors related to the RES system.

Even in view of the findings that RES clearance of antibody-tagged liposomes may be accelerated, the improvements in efficacy against tumor progression are clear. Tadakuma's group [146,147] have been able to show 65–85% reductions in Li-7 hepatoma tumor weights in Balb C nu/nu mice treated with doxorubicin liposomes that displayed surface monoclonal antibodies to the human alpha-fetoprotein secreted by the tumor cells. The regression in tumor size surpassed that achieved by injection of either free doxorubicin or doxorubicin encapsulated in unmodified liposomes. Similar results were achieved with monoclonal C143-modified doxorubicin liposomes when targeted to bovine leukemia virus-induced tumors in Balb C nude mice [110]. More efficient reduction in ascites tumor growth in athymic nude mice has also been shown using monoclonal antibody-modified liposomes containing methotrexate-gamma-aspartate [111].

Actinomycin D is proven to be more effective when delivered from targeted liposomes [32,148]. In these studies, actinomycin D liposomes possessing surface monoclonals to antigens from MM46 mouse mammary cancer cells fully eliminated tumors when injected into C3H/He mice, whereas free drug treatment was ineffective. Within 1 month following the induction of the cancer, a 100% mortality rate was observed in animals treated with either free actinomycin or unmodified liposomal drug. After 2 months, however, 50–80% survival (depending on initial tumor cell inoculum) was recorded in mice treated with the antibody-modified vesicles.

Aside from the treatment of cancer, Gupta and collaborators [96,149,150] have shown the utility of liposomes targeted to red blood cells. Most recently [150], these authors have examined the potential of liposomes conjugated to anti-erythrocyte F(ab′)2 to deliver chloroquine to blood cells of *Plasmodium Berghei*-infected mice. These liposomes were 5 to 10 times more effective than free drug when given 3 days after initiation of the infection. It was also noted in this study that between 15% and 20% of the injected dose actually bound to the target site, with 20–30% delivering the payload within the cells.

V. CONCLUSIONS

In ideal circumstances the development of a targeted delivery should conform to the following guidelines: (1) The conjugation reactions should be quantitative, with 100% yield of product; (2) the resultant conjugate should be stable upon storage for 2 years; (3) the conjugate should not impart instability to the final carrier formulation; (4) development of the targeted carrier should proceed with no loss of drug; (5) 100% of the injected dose should reach and bind to the target site in vivo; and (6) 100% of the drug should be delivered into the target tissue. Needless to say, current liposome formulations fall short of meeting each of these guidelines, as would any of the other targeted systems to date. However, it is important not to lose sight of the ultimate goal of these studies, that is, to further improve the quality of life for those individuals requiring the treatment. In realistic terms, even modest improvements in drug therapy, particularly cancer treatment, may offer benefits that outweigh some of the shortfalls of the system. It will be up to the investigators' discretion as to when the benefit-to-risk ratio is sufficiently convincing. That ratio may not be long off. Certainly the methodical elimination of pharmaceutically relevant problems in liposomology (such as stability, scale-up, sterlization, coupling efficiencies, and entrapment of drug) have offered hope in optimizing the design for targeted applications. Although many biological questions still remain (for example, increasing binding efficiency in vivo, decreasing nonspecific clearance of the carrier, and optimizing transport of bound carrier into the cell), the results to date have been more than encouraging and should foster exciting new developments in the future.

VI. REFERENCES

1. P. Ehrlich, *Collected Studies on Immunology*, John Wiley, New York, 1906, p. 442.
2. G. Poste, *Biol. Cell 47*:19 (1983).
3. K. J. Hwang, in *Liposomes. From Biophysics to Therapeutics* (M. J. Ostro, ed.), Marcel Dekker, New York, 1987, p. 109.
4. S. M. Gruner, in *Liposomes. From Biophysics to Therapeutics* (M. J. Ostro, ed.), Marcel Dekker, New York, 1987, p. 1.
5. G. L. Scherphof, J. Dijkstra, H. H. Spanjer, J. T. P. Derksen, and F. H. Roerdink, *Ann. N.Y. Acad. Sci. 446*:368 (1985).
6. G. Poste, R. Kirsh, and T. Koestler, in *Liposome Technology* (G. Gregoriadis, ed.), CRC Press, Boca Raton, Fla., 1983, p. 1.
7. G. Gregoriadis and A. C. Allison (eds.), *Liposomes in Biological Systems*, John Wiley, New York, 1980.
8. A. H. Stirk and J. D. Baldeschwieler, in *Medical Applications of Liposomes* (K. Yagi, ed.), Japan Scientific Societies Press, Tokyo, 1986, p. 31.
9. I. J. Fidler, in *Cell Biology, Medical and Pharmaceutical Aspects* (E. Tomlinson andS. S. Davis, eds.), John Wiley, London, 1986, p. 111.
10. M. C. Popescu, C. E. Swenson, and R. S. Ginsberg, in *Liposomes. From Biophysics to Therapeutics* (M. J. Ostro, ed.), Marcel Dekker, New York, 1987, p. 219.

11. G. Lopez-Berestein and R. L. Juliano, in *Liposomes. From Biophysics to Therapeutics* (M. J. Ostro, ed.), Marcel Dekker, New York, 1987, p. 253.
12. A. J. Schroit, J. Madsen, and R. Nayar, *Chem. Phys. Lipids 40*:373 (1986).
13. R. Kirsh, P. J. Bugelski, and G. Poste, *Ann. N.Y. Acad. Sci. 507*:141 (1987).
14. E. Tomlinson, *Adv. Drug Del. Rev. 1*:87 (1987).
15. B. D. Williams, M. M. O'Sullivan, G. S. Saggu, K. E. Williams, L. A. Williams, and J. R. Morgan, *Br. Med. J. 293*:1143 (1986).
16. V. J. Caride, W. Taylor, J. A. Cramer, and A. Gottschalk, *J. Nucl. Med. 17*: 1067 (1976).
17. I. Ogihara, S. Kojima, and M. Jay, *J. Nucl. Med. 27*:1300 (1986).
18. J. N. Weinstein, in *Liposomes. From Biophysics to Therapeutics* (M. J. Ostro, ed.), Marcel Dekker, New York, 1987, p. 277.
19. G. Gregoriadis, J. Senior, B. Wolff, and C. Kirby, *Ann. N.Y. Acad. Sci. 446*: 319 (1985).
20. D. Papahadjopoulos and A. Gabizon, *Ann. N.Y. Acad. Sci. 507*:64 (1987).
21. K. Huang, M. M. Padki, D. D. Chow, H. E. Essien, J. Y. Lai, and P. L. Beaumier, *Biochim. Biophys. Acta 901*:88 (1987).
22. J. Senior, J. C. W. Crawley, and G. Gregoriadis, *Biochim. Biophys. Acta 839*:1 (1985).
23. S. Utsumi, H. Shinomiya, J. Minami, and S. Sonoda, *Immunology 49*:113 (1983).
24. V. P. Torchilin, V. R. Berdichevsky, A. A. Barsukov, and N. V. Smirnov, *FEBS Lett. 11*:184 (1980).
25. T. M. Allen and A. Chonn, *FEBS Lett. 223*:42–46 (1987).
26. W. B. Geho and J. R. Lau, U.S. Patent No. 4,501,728 (1985).
27. A. L. Weiner, in *Liposomes. From Biophysics to Therapeutics* (M. J. Ostro, ed.), Marcel Dekker, New York, 1987, p. 339.
28. L. Leserman and P. Macny, in *Liposomes. From Biophysics to Therapeutics* (M. J. Ostro, ed.), Marcel Dekker, New York, 1987, p. 157.
29. G. Gregoriadis, *Drugs 24*:261 (1982).
30. C. A. Alving and R. L. Richards, in *Liposomes* (M. J. Ostro, ed.), Marcel Dekker, New York, 1983, p. 209.
31. V. P. Torchilin, *Adv. Drug Del. Rev. 1*:41 (1987).
32. Y. Hashimoto, M. Sugawara, T. Kamiya, and S. Suzuki, *Methods Enzymol. 121*:817 (1986).
33. G. Gregoriadis (ed.), *Liposome Technology*, Vol. III, CRC Press, Boca Raton, Fla., 1984.
34. D. Papahadjopoulos, T. Heath, K. Bragman, and K. Matthay, *Ann. N.Y. Acad. Sci. 446*:341 (1985).
35. T. D. Heath, *Methods Enzymol. 149*:135 (1987).
36. G. Gregoriadis and E. D. Neerunjun, *Biochem. Biophys. Res. Commun. 65*:537 (1975).
37. G. Gregoriadis, E. Neerunjun, and R. Hunt, *Life Sci. 21*:357 (1977).
38. L. B. Margolis and N. A. Dorfman, *Bull. Exp. Biol. Med. 85*:53 (1977).
39. W. E. Magee, J. H. Cronenberger, and D. E. Thor, *Cancer Res. 38*:1173 (1978).

40. L. Huang and S. J. Kennel, *Biochemistry 18*:1702 (1979).
41. G. Weissmann, A. Brand, and E. C. Franklin, *J. Clin. Invest. 53*:536 (1974).
42. G. Weissmann, D. Bloomgarden, R. Kaplan, C. Cohen, S. Hoffstein, T. Collins, A. Gottlieb, and D. Nagel, *Proc. Natl. Acad. Sci. USA 73*:88 (1975).
43. H. Schieren, G. Weissmann, M. Seligman, and P. Coleman, *Biochem. Biophys. Res. Commun. 82*:1160 (1978).
44. B. Rivnay, E. A. Bayer, and M. Wilchek, *Methods Enzymol. 149*:119 (1987).
45. D. L. Urdal and S. Hakomori, *J. Biol. Chem. 255*:10509 (1980).
46. E. A. Bayer, B. Rivnay, and E. Skutelsky, *Biochim. Biophys. Acta 550*: 464 (1979).
47. H. Loughrey, M. B. Bally, and P. R. Cullis, *Biochim. Biophys. Acta 901*: 157 (1987).
48. V. S. Trubetskoi, V. R. Berdichevski, E. E. Efremov, V. P. Torchilin, and V. N. Smirnov, *Bull. Exp. Biol. Med. 102*:1219 (1986).
49. V. S. Trubetskoi, V. R. Berdichevski, E. E. Efremov, and V. P. Torchilin, *Biochem. Pharmacol. 36*:839 (1987).
50. O. V. Trubetskaya, V. S. Trubetskoi, S. P. Domogatski, A. V. Rudin, N. V. Popov, S. M. Danilov, M. N. Nikolayeva, A. L. Klibanov, and V. P. Torchilin, *FEBS Lett. 228*:131 (1988).
51. A. L. Weiner, J. B. Cannon, and P. Tyle, in *Controlled Release of Drugs: Polymers and Aggregate Systems* (M. Rosoff, ed.), VCH Publishing, New York, 1989, p. 217.
52. A. L. Weiner, *Adv. Drug Del. Rev.* (1989), 3:307.
53. C. P. S. Tilcock, *Chem. Phys. Lipids 40*:109 (1986).
54. V. P. Torchilin, V. S. Goldmacher, and V. N. Smirnov, *Biochem. Biophys. Res. Commun. 85*:983 (1978).
55. V. P. Torchilin, B. A. Khaw, V. N. Smirnov, and E. Haber, *Biochem. Biophys. Res. Commun. 89*:1114 (1979).
56. D.-F. Shen, A. Huang, and L. Huang, *Biochim. Biophys. Acta 689*:31 (1982).
57. A. Huang, L. Huang, and S. J. Kennel, *J. Biol. Chem. 255*:8015 (1980).
58. V. P. Torchilin, A. N. Klibanov, and V. N. Smirnov, *FEBS Lett. 138*:117 (1982).
59. E. Claassen and N. van Rooijen, *Prep. Biochem. 13*:167 (1983).
60. T. D. Heath, R. T. Fraley, and D. Papahadjopoulos, *Science 210*:539 (1981).
61. T. D. Heath, B. A. Macher, and D. Papahadjopoulos, *Biochim. Biophys. Acta 640*:66 (1981).
62. T. D. Heath, D. Robertson, M. S. C. Birbeck, and A. J. S. Davies, *Biochim. Biophys. Acta 599*:42 (1980).
63. M.-M. Chua, S.-T. Fan, and F. Karush, *Biochim. Biophys. Acta 800*:291 (1984).
64. R. L. Lundblad and C. M. Noyes, *Chemical Reagents for Protein Modification,* CRC Press, Boca Raton, Florida, 1984.

65. H. Endoh, Y. Suzuki, and Y. Hashimoto, *J. Immunol. Methods 44*:79 (1981).
66. V. K. Jansons and P. L. Mallet, *Anal. Biochem. 111*:54 (1981).
67. S. L. Snyder and W. E. Vannier, *Biochim. Biophys. Acta 772*:288 (1984).
68. N. Garcon, J. Senior, and G. Gregoriadis, *Biochem. Soc. Trans. 14*:1036 (1986).
69. J. T. P. Derksen and G. L. Scherphof, *Biochim. Biophys. Acta 814*:151 (1985).
70. L. D. Leserman, in *Liposomes, Drugs and Immunocompetent Cell Functions* (C. Nicolau and A. Paraf, eds.), Academic Press, London, 1981, p. 109.
71. J. Lasch, G. Niedermann, A. A. Bogdanov, and V. P. Torchilin, *FEBS Lett. 214*:13 (1987).
72. D. Sinha and F. Karush, *Biochem. Biophys. Res. Commun. 90*:554 (1979).
73. F. J. Martin, W. L. Hubbell, and D. Papahadjopoulos, *Biochemistry 20*: 4229 (1981).
74. L. D. Leserman, J. Barbet, F. Kourilsky, and J. N. Weinstein, *Nature 288*: 602 (1980).
75. A. Goundalkar, T. Ghose, and M. Mezei, *J. Pharm. Pharmacol. 36*:465 (1984).
76. F. J. Martin and D. Papahadjopoulos, *J. Biol. Chem. 257*:286 (1982).
77. T. D. Heath and F. J. Martin, *Chem. Phys. Lipids 40*:347 (1986).
78. L. D. Meyer, M. J. Hope, P. R. Cullis, and A. S. Janoff, *Biochim. Biophys. Acta 817*:193 (1985).
79. C. J. Kirby and G. Gregoriadis, in *Liposome Technology*, Vol. 1 (G. Gregoriadis, ed.), CRC Press, Boca Raton, Fla, 1984, p. 221.
80. J. H. Senior, *CRC Crit. Rev. Ther. Drug Carrier Syst. 3*:123 (1987).
81. F. Bonte and R. J. Juliano, *Chem. Phys. Lipids 40*:359 (1986).
82. F. J. Martin and V. T. Kung, *Ann. N.Y. Acad. Sci. 446*:443 (1985).
83. R. Bredehorst, F. S. Ligler, A. W. Kusterbeck, E. L. Chang, B. P. Gaber, and C. W. Vogel, *Biochemistry 25*:5693 (1986).
84. T. D. Madden, M. B. Bally, M. J. Hope, P. R. Cullis, H. P. Schieren, and A. S. Janoff, *Biochim. Biophys. Acta 817*:67 (1985).
85. L. M. Crowe, R. Mouradian, J. M. Crowe, S. A. Jackson, and C. Womersley, *Biochim. Biophys. Acta 169*:141 (1984).
86. M. Schneider and B. Lamy, U.S. Patent No. 4,229,360 (1980).
87. T. D. Heath, R. T. Fraley, J. Bentz, E. W. Voss, J. N. Herron, and D. Papahadjopoulos, *Biochim. Biophys. Acta 770*:148 (1984).
88. B. Wolff and G. Gregoriadis, *Biochim. Biophys. Acta 802*:259 (1984).
89. J. Barbet, P. Macny, and L. D. Leserman, *J. Supramol. Struct. Cell Biochem. 16*:243 (1981).
90. H. P. Weckenmann, S. Matzku, and H. Stricker, *Eur. J. Cell Biol. 42*(S15): 52 (1986).
91. K. S. Houck and L. Huang, *Biochem. Biophys. Res. Commun. 145*:1205 (1987).
92. L. D. Leserman, P. Machy, and J. Barbet, *Nature 293*:226 (1981).
93. P. Machy and L. D. Leserman, *Biochim. Biophys. Acta 730*:313 (1983).
94. S. M. Sullivan and L. Huang, *Biochim. Biophys. Acta 812*:116 (1985).

95. R. Schwendener, H. Schott, B. Leitner, and H. Hengartner, *Experientia 43*:709 (1987).
96. A. Singhal, A. Bali, and C. M. Gupta, *Biochim. Biophys. Acta 880*:72 (1986).
97. P. A. M. Peeters, C. A. M. Claessens, W. M. C. Eling, and D. J. A. Crommelin, *Biochem. Pharmacol. 37*:2215 (1988).
98. P. Machy, M. Pierres, J. Barbet, and L. D. Leserman, *J. Immunol. 129*: 2098 (1982).
99. P. Machy, J. Barbet, and L. S. Leserman, *Proc. Natl. Acad. Sci. USA 79*:4148 (1982).
100. N. Berinstein, K. K. Matthay, D. Papahadjopoulos, R. Levy, and B. I. Sikic, *Cancer Res. 47*:5954 (1987).
101. C. L. Berger, S. Yemul, G. Goldstein, A. Estabrook, R. L. Edelson, and H. Bayley, *Clin. Res. 35*:A669 (1987).
102. P. Machy and L. D. Leserman, *EMBO J. 3*:1971 (1984).
103. Y. Watanabe and T. Osawa, *Chem. Pharm. Bull. 35*:740 (1987).
104. R. Staubinger, K. Hong, D. S. Friend, and D. Papahadjopoulos, *Cell 32*: 1069 (1983).
105. A. Truneh, Z. Mishal, J. Barbet, P. Machy, and L. D. Leserman, *Biochem. J. 214*:189 (1983).
106. T. D. Heath, N. G. Lopez, W. H. Stern, and D. Papahadjopoulos, *FEBS Lett. 187*:73 (1985).
107. J. T. P. Derksen, H. W. M. Morselt, and G. L. Scherphof, *Biochim. Biophys. Acta 931*:33 (1987).
108. J. T. P. Derksen, H. W. M. Morselt, D. Kalicharan, C. E. Hulstaert, and G. L. Scherphof, *Exp. Cell Res. 168*:105 (1987).
109. L. Huang, in *Liposomes* (M. J. Ostro, ed.), Marcel Dekker, New York, 1983, p. 87.
110. M. Onuma, T. Odawara, S. Watarai, Y. Aida, K. Ochiai, B. Syuto, K. Matsumoto, T. Yasuda, Y. Fujimoto, and H. Izawa, *Jap. J. Cancer Res. 77*:1161 (1986).
111. R. M. Straubinger, N. G. Lopez, R. Debs, K. Hong, and D. Papahadjopoulos, *J. Cell Biol. 103*:A447 (1986).
112. J. M. Boggs, A. Goundalkar, F. Doganoglu, N. Samji, J. Kurantsin-Mills, and K. M. Koshy, *J. Neuroimm. 17*:35 (1987).
113. D. R. Hospehthal, A. L. Rogers, and G. L. Mills, *Mycopathology 101*:37 (1988).
114. J. A. Haxby, C. B. Kinsky, and S. C. Kinsky, *Proc. Natl. Acad. Sci. USA 61*:300 (1968).
115. C. R. Alving, S. C. Kinsky, J. A. Haxby, and C. B. Kinsky, *Biochemistry 8*:1582 (1969).
116. H. R. Six, K. Uemura, and S. C. Kinsky, *Biochemistry 12*:4003 (1973).
117. H. R. Six, W. W. Young, Jr., K. Uemura, and S. C. Kinsky, *Biochemistry 13*:4050 (1974).
118. T. Kataoka, J. R. Williamson, and S. C. Kinsky, *Biochim. Biophys. Acta 298*:158 (1973).

119. S. C. Kinsky, *Ann. N.Y. Acad. Sci. 195*:429 (1972).
120. J. A. Haxby, O. Gotze, H. J. Muller-Eberhard, and S. C. Kinsky, *Proc Natl. Acad. Sci. USA 64*:290 (1969).
121. S. C. Kinsky, in *Cell Membranes: Biochemistry, Cell Biology and Pathology* (G. Weissmann and R. Claiborne, eds.), H. P. Publications, New York, 1975, p. 231.
122. D. Monroe, *Am. Biotech, Lab. 5*:10 (1987).
123. B. S. Yu, Y. K. Choi, and H. Chung, *Biotech, Appl. Biochem. 9*:209 (1987).
124. F. S. Ligler, R. Bredehorst, A. Talebian, L. C. Shriver, C. F. Hammer, J. P. Sheridan, C. W. Vogel, and B. P. Gaber, *Anal. Biochem. 163*:369 (1987).
125. E. Canova-Davis, C. T. Redemann, Y. P. Vollmer, and V. T. Kung, *Clin. Chem. 32*:1687 (1986).
126. E. F. Ullman, T. Tarnowski, P. Felgner, and I. Gibbons, *Clin. Chem. 33*: 1579 (1987).
127. M. Umeda, Y. Ishimori, K. Yoshikawa, M. Takada, and T. Yasuda, *J. Immunol. 95*:15 (1986).
128. T. Yasuda, in *Medical Applications of Liposomes* (K. Yagi, ed.), Japan Scientific Societies Press, Tokyo, 1986, p. 155.
129. A. M. Butt, H. Rutner, M. Cobianch, M. Baran, N. Y. Oraivej, and J. Readio, *Clin. Chem. 34*:1255 (1988).
130. L. Pollack, M. Sylvestre, R. Czarnecky, L. Kelley, J. Readio, H. Rutner, and M. Baran, *Clin. Chem. 33*:1015 (1987).
131. H. Kumar, M. Baran, J. Costello, J. Readio, and H. Rutner, *Clin. Chem. 33*:883 (1987).
132. M. Baran, H. Mindicino, J. Readio, and N. Leonard, *Clin. Chem. 33*:882 (1987).
133. A. M. Butt, B. Monfiston, H. Rutner, J. Readio, and M. Baran, *Clin. Chem. 33*:1023 (1987).
134. L. Guo, A. S. Michaels, F. J. Martin, N. M. Weinshenker, and P. Huertas, Patent Cooperation Treaty, Publication number WO87/02778 (1987).
135. A. S. Janoff, S. S. Carpenter-Green, A. L. Weiner, J. Seibold, G. Weissmann, and M. J. Ostro, *Clin. Chem. 29*:1587 (1983).
136. R. A. Sells, R. R. Owen, R. R. C. New, and I. T. Gilmore, *Lancet 2*:624 (1987).
137. A. Gabizon, T. Peretz, R. Ben-Yosef, R. Catane, S. Biran, and Y. Barenholz, *Proc. Am. Soc. Clin. Oncol. 4*:789 (1987).
138. R. T. Proffitt, L. E. Williams, C. A. Presant, G. W. Tin, J. A. Uliana, R. C. Gamble, and J. D. Baldeschwieler, *Science 220*:502 (1983).
139. J. D. Baldeschwieler, R. C. Gamble, M. R. Mauk, T-Y. Shen, and M. M. Ponipom, U.S. Patent No. 4,310,505 (1982).
140. Vestar Corp., Annual report, 1988.
141. A. Gabizon, J. Huberty, N. Lopez, R. M. Strabinger, D. Price, and D. Papahadjopoulos, *Proc. Am. Assoc. Cancer Res. 28*:420 (1987).
142. R. J. Debs, T. D. Heath, and D. Papahadjopoulos, *Biochim. Biophys. Acta 901*:183 (1987).

143. M. Udayachander, A. Meenakshi, R. Muthiah, and M. Sivanadham, *Int. J. Radiat. Oncol. Biol. Phys. 13*:1713 (1987).
144. J. T. P. Derksen, H. W. M. Morselt, F. H. Roerdink, and G. L. Scherphof, *J. Leuk. Biol. 40*:316 (1986).
145. G. Gregoriadis, J. Senior, and B. Wolff, in *Medical Applications of Liposomes* (K. Yagi, ed.), Japan Scientific Societies Press, Tokyo, 1986, p. 67.
146. H. Konno, H. Suzuki, T. Tadakuma, K. Kumai, T. Yasuda, T. Kubota, S. Ohta, K. Nagaike, S. Hosokawa, and K. Ishibiki, *Cancer Res. 47*:4471 (1987).
147. T. Tadakuma, in *Medical Applications of Liposomes* (K. Yagi, ed.), Japan Scientific Societies Press, Tokyo, 1986, p. 113.
148. Y. Hashimoto, H. Hojo, T. Kamiya, M. Sugawara, in *Medical Applications of Liposomes*, Japan Scientific Societies Press, Tokyo, 1986, p. 131.
149. A. Singhal and C. M. Gupta, *FEBS Lett. 201*:321 (1986).
150. A. K. Agrawal, A. Singhal, and C. M. Gupta, *Biochem. Biophys. Res. Commun. 148*:357 (1987).

13
Liposomes as Carriers for Antibiotics

MARCELO C. NACUCCHIO, PEDRO H. DI ROCCO, and
DANIEL O. SORDELLI
University of Buenos Aires, Buenos Aires, Argentina

I. INTRODUCTION

Much research has been aimed at prevention and treatment of infectious diseases in the last few decades, with a significant degree of success. This effort, however, has not provided solutions to all the problems posed by human and animal bacterial pathogens, which may hamper direct administration of antibacterial agents by a variety of mechanisms. These include development of drug resistance by the microorganism; uptake of the antibacterial agent by nondiseased tissues, which often leads to serious side effects; and even the inability of some drugs to penetrate target cells or reach the infected microenvironment in appropriate concentrations.

While the search for new antibiotics continues (an extremely laborious and expensive process), several groups have developed new strategies to improve the activity and widen the therapeutic spectrum of antibacterial agents used in clinical practice. New carrier systems have been designed to procure either site-specific pharmacological action or controlled release of the drug, thus enhancing efficacy while diminishing undesirable side effects. A suitable carrier by which drugs could be entrapped and directed to target tissues might help to eliminate some of these problems. It has been suggested that the need for such a carrier could be satisfied by liposomes [1].

Liposomes may be well suited to carry antibiotics since they provide any or all of the following: (1) a means for entrapment and delivery of a variety of agents; (2) natural targeting to cells of the reticuloendothelial system (RES) as well as adaptability to specific targeting to other cells; (3) protection of the drug from the biological environment; (4) controlled release of the liposome-associated agent; and (5) a means to diminish drug toxicity. The use of liposomes as a drug delivery system for the administration of antibiotics is a modern approach to the treatment of infections due to organisms that are difficult or impossible to eradicate by conventional therapy. The liposome is a promising pharmaceutical form that permits modification of antibiotic bioavailability. High concentrations of the drug, therefore, may be attained in organs of the RES, whereas the drug would reach nontarget sites in low concentrations. Entrapment within liposomes may overcome the clinical problems related to bacterial resistance or inaccessibility of antibiotics to the site of infection.

The usefulness of antibiotic encapsulation in liposomes for treatment of bacterial infections has been widely discussed by many authors [2–4]. Different experimental systems have been utilized to ascertain whether antibiotic encapsulation in liposomes enhanced the therapeutic efficacy of the drug. Using these systems, researchers have studied: (1) the interaction of the encapsulated antibiotic with the target microorganism in broth cultures; (2) the interaction of the liposomal antibiotic and cultured cells infected with bacteria; and (3) the interaction of the liposomal antibiotic and the target bacteria in animal models of infectious diseases. Each of these experimental systems examines different aspects by which liposomes may modify the activity of the encapsulated drug. These three types of interactions will be addressed further in the following text. The overall conclusion from these studies is that liposome encapsulation may provide a significant therapeutic benefit in the treatment of certain bacterial infections.

II. PHARMACEUTICAL ASPECTS

A. Liposome Preparation and Pharmaceutical Properties

Phospholipid vesicles (or liposomes) were described nearly 20 years ago by Bangham et al. [5]. Liposomes can be prepared from many amphiphilic lipids or lipid mixtures. By far the most frequently utilized component is a mixture of phospholipids and cholesterol. Upon hydration, most phospholipids tend to adopt a bilayer organization and form closed, bilayer "liposomal" structures. According to their size, lipid vesicles have been defined as large or small liposomes. With regard to the number of layers, they have been defined as multilamellar or unilamellar.

The method used to prepare liposomes defines the kind of structure exhibited by the vesicle. The easiest way to prepare a vesicle consists of dispersing phospholipids in water. This procedure yields multilamellar vesicles (MLVs) characterized by concentric bilayers that entrap between them the solution of the active molecule. Unilamellar vesicles, which are either small (SUVs) or large (LUVs), can be obtained either directly from MLV preparations or by several other independent procedures. The definition of SUV and LUV is rather arbitrary. Vesicles with a diameter of approximately 0.1 μm or greater are usually considered LUVs, whereas those that are smaller are referred to as SUVs.

Water-soluble antibiotics are carried in liposomes in the aqueous phase located either inside SUVs and LUVs or in between lipid layers of MLVs. Water-insoluble antibiotics are, conversely, dissolved in the lipid bilayer itself. Authors have used the terms encapsulated, entrapped, and liposomal indistinctly to refer to an agent associated to a lipid vesicle. The term encapsulated, however, may not be applicable to a water-insoluble compound. Because most of the experience in liposome encapsulation comes from studies using water-soluble antibiotics, we will not consider the nomenclature issue and will use these terms synonymously.

A description of the wide variety of methods for preparing lipid vesicles exceeds the scope of this review. Additional information in this regard can be found in several recent reviews [6–8]. Most of the available information on the encapsulation of antibiotics in liposomes can be obtained from the references detailed in Table 1.

From a pharmaceutical point of view, it is desirable that a liposome formulation: (1) exhibit high trapping efficiency; (2) contain reproducible vesicles with regard to composition and size; (3) should not contain impurities, derived from the method of preparation; that would affect stability and/or in vivo efficacy; (4) can be prepared by a method that is economically feasible, with an easy transfer to industrial-scale production; and (5) exhibit physical and chemical stability for extended periods of time (at least 1 year). It should be pointed out that stability of a liposome formulation refers not only to the intrinsic stability of the encapsulated antibiotic but also to the stability of the carrier system; i.e., the liposome should neither permit leakage of the entrapped agent nor change its physical and chemical properties.

B. Fate of Liposomes In Vivo

It is well established that after intravenous injection liposomes and entrapped drugs are sooner or later taken up by the RES. Liposomes, consequently, are captured by phagocytic cells of the liver, spleen, and bone marrow [9]. The rate of uptake by these tissues depends on, among other factors, vesicle size, surface charge, amount of liposomal lipid given, animal species, and physiological state

Table 1 Studies Addressing Encapsulation of Antibacterial Agents

Antibiotic	References
Amikacin	32,37
Ampicillin	4,32,42,57
Cefazolin	54,57
Cefoperazone	This report
Cephalotin	16,17
Cephaloridin	57
Chloramphenicol	16,17
Dihydrostreptomycin	30,52
Gentamicin	24,31–33,44,51,53,56,59
Kanamycin	31,32,59
Methicillin	13
Neamine	59
Neomycin	15,59
Netilmicin	59
Oxytetracycline	40
Penicillin G	1,13,58
Piperacillin	24,25
Sisomycin	34
Spectinomycin	59
Streptomycin	17,31,38–40
Tobramycin	31,59

of the host. The rate of liposome deposition can be modified, for instance, by introducing changes in vesicle size or surface charge; i.e., large size and/or negative charge promote rapid uptake by tissues. This phenomenon has been defined as passive targeting.

Gregoriadis et al. have discussed the influence of lipid composition on degradation of lipid vesicles in vivo by plasma high-density lipoproteins. Indeed, these lipoproteins are able to remove phospholipid molecules, thus destabilizing vesicle structure and potentially leading to the release of the entrapped drug into the circulation. Tissues, therefore, may interact with degrading vesicles when a portion of the entrapped agent has already been lost [10]. A liposome can be stabilized by addition of cholesterol to the vesicle bilayer in order to inhibit high-density lipoprotein activity on the vesicles.

The fate of liposomes delivered to a host will obviously be modified by the route of administration. For instance, the intramuscular route is chosen when the encapsulated agent should be released slowly. For slow release of the agent the vesicle of choice is the MLV, because it has been shown that the larger the liposome, the lower the rate of disappearance from the site of injection [11,12]. The uptake of liposomes by the RES is a major drawback to targeting an encapsulated agent to non-RES cells. Many researchers, however, have taken advantage of this characteristic of the liposome-host interaction to deliver drugs to phagocytic cells. Addition of different ligands to the liposome lipid layer may result in active targeting of the entrapped drug. These ligands include monoclonal antibodies, sugars, glycolipids, proteins, and others. The advantages of site-specific delivery (ligand targeting) of liposomal antibiotics is addressed in Section VI.

III. INTERACTIONS BETWEEN LIPOSOMES AND BACTERIA

Different authors have reported that encapsulation in liposomes enhances the activity of certain antibiotics against bacteria in broth cultures. Chowdhury et al. [13] have shown that encapsulation of methicillin, penicillin, and cloxacillin in liposomes of different composition enhanced the activity of these agents against *Staphylococcus aureus, Bacillus licheniformis*, and *Escherichia coli*. The authors suggested that this effect may have been due to an electrostatic attraction between the liposome and the bacterial membrane, facilitating fusion between the two. This hypothesis may be correct in experiments where liposomes are prepared with lipids that confer charges on the surface, but it does not answer the question of how drug activity can be enhanced by encapsulation in electrostatically neutral liposomes. An alternative explanation might be leakage of the entrapped penicillins, as hypothesized by other authors [1,14]. Hodges et al. [15] have also reported that liposomal entrapment of neomycin enhanced its activity against *E. coli* in broth culture. These authors, unfortunately, did not suggest a mechanism for this effect.

The effect of free or encapsulated chloramphenicol [16] and streptomycin [17] on *E. coli* in broth culture was investigated by Stevenson et al. These authors found that washing rendered the liposomes containing the antibiotics inactive against the bacteria, in spite of the fact that the vesicles carried an amount of antibiotic that, if free, would have inhibited bacterial growth. Likewise, Bakker-Woudenberg et al. reported that washed liposomes made of cholesterol, sphingomyelin, and L-phosphatidylserine containing ampicillin did not inhibit growth of *Listeria monocytogenes* in culture [18]. Taken together, the results from these two laboratories suggest that the entrapped drug may not be available to interact with the target organism.

Popescu et al. have recently discussed the fact that liposome-entrapped antibiotics kill bacteria in culture more efficiently than free drugs do. These authors think that particles in suspension (liposomes and bacteria) tend to sediment, causing local concentration effects that result in increased inhibition of bacterial cell growth. The authors suggest that a leaky liposome may be more likely to show an effect than a nonleaky one [19], and conclude that water-soluble drug encapsulated in washed, nonleaky liposome would not be available to interact with the target bacteria. Conversely, a water-soluble drug in a leaky liposome may show activity. The fact that an entrapped agent can leak from a lipid vesicle has been proved by Szoka et al. [20]. Sedimentation of liposomes and bacteria in broth culture under vigorous shaking, however, appears unlikely.

The rate of antibiotic leakage from a liposome in the internal medium, as explained before, can be modulated by modifying the vesicle lipid composition. It must be pointed out that it is of paramount importance to know which exoproducts the microorganism secretes at the time the liposome-bacteria interaction takes place. Most strains of *Pseudomonas aeruginosa* and other bacterial strains, for instance, are able to produce a phospholipase [21] that would disrupt the bilayer structure, thus releasing the encapsulated drug at a site of high bacterial density in vivo. Whether advantage can be taken of this kind of interaction to achieve site-specific delivery of an encapsulated agent requires further investigation.

In 1977, Jones and Osborn demonstrated that phospholipids can be transferred from vesicles to intact bacterial cells [22]. This finding led other researchers to postulate that fusion of liposomes and bacteria may be a mechanism through which liposomes could specifically deliver their entrapped drug into target cells [23]. In a recent report, we presented evidence that liposome encapsulation enhances the in vitro antibacterial activity of gentamicin and piperacillin against strains of *E. coli* and *P. aeruginosa* that are resistant to those antibiotics [24]. At least three mechanisms may be involved in the liposome-mediated enhancement of the antibacterial activity. Two of them are related to the protection of the entrapped drug from the action of hydrolytic enzymes.

The results from a study performed at our laboratory have shown that intrinsic protection of piperacillin may be obtained by liposome encapsulation of the drug, thus preventing its hydrolysis by either *S. aureus* beta-lactamases or the exogenous enzyme (from *Bacillus cereus*) added to the culture media [25]. In the same report we showed that not only encapsulation within but also adsorption to liposomes produces enhancement in the antibiotic activity. A likely explanation for this phenomenon is that the liposome produces steric hindrance to the action of the beta-lactamase. Protection of the antibiotic molecule, however, may not explain satisfactorily the enhancement of gentamicin activity against *E. coli* and *P. aeruginosa*. Changes in the cell-envelope permeability rather than enzymatic inactivation is the resistance mechanism more often displayed by gram negatives. In this regard, it has been suggested that all gentamicin resistance may ultimately be due to impermeability of the bacterial envelope to the antibiotic [26]. Furthermore, Zimmerman et al. have demonstrated that changes in the permeability of the outer membrane play in important role in the resistance of *E. coli* to several beta-lactam antibiotics [27].

Our findings suggest that association of antibiotics to liposomes may facilitate the diffusion of the drug across the bacterial envelope. This hypothesis is in accordance with the findings of Sekeri-Pataryas et al. [28], who demonstrated that nonpermeable substances can permeate through *P. aeruginosa* external envelopes by the use of liposomes. In conclusion, we suggest that the enhancement of the antibacterial activity by any of the mechanisms explained are operating concomitantly. The extent to which each mechanism is participating in each particular bacteria-liposomal antibiotic interaction remains to be elucidated.

IV. INTERACTIONS BETWEEN LIPOSOMES AND PHAGOCYTIC CELLS

Phagocytic cells are usual hosts of the microorganisms in infections caused by intracellular bacteria. Although there are several antibiotics that can act in the intracellular milieu of phagocytic cells, intracellular infections still remain a significant problem in medical practice. In spite of the fact that most antibiotics exhibit high sensitivity in vitro, many of them cannot be used in these infections because they do not reach the intracellular milieu, where bacteria multiply and remain. The use of lipid vesicles to transport antimicrobial drugs to the intracellular medium have been previously hypothesized [29]. Therefore, the possibility that liposome encapsulation of an antibiotic would increase the phagocyte capacity to kill intracellular bacteria has been examined by several groups.

Bonventre and Gregoriadis [30] determined the efficacy of liposomal dihydrostreptomycin in SUV (egg phosphatidylcholine, cholesterol and phosphatidic acid in a 7:2:1 molar ratio) against intracellular *S. aureus*. They found that, after incubating infected macrophages for 16 h, the intracellular killing activity of the liposomal antibiotic was 40 times higher than that of the free drug.

In another study [17], streptomycin and chloramphenicol were entrapped within large neutral (egg phosphatidylcholine and cholesterol) or anionic unilamellar vesicles of the same lipid composition, with the addition of either phosphatidic acid or phosphatidylserine. The activities of both antibiotics against *E. coli* located within murine macrophages were measured. The results of this study showed that liposomal streptomycin was 12 times, and chloramphenicol 10 times, more effective than the free drug in killing intraphagocytic bacteria. The authors suggested that the enhanced activity of chloramphenicol was due to increased uptake by phagocytosis with the consequent modification of the intracellular distribution of the drug; i.e., delivery of the antibiotic to phagocytic vacuoles occurred, followed by delayed partition of the drug toward lipid-rich regions, such as the cell membrane [17].

Antibiotic entrapment in liposomes to strengthen its activity against different intracellular pathogens has been investigated by several authors. In a series of studies using infected monocyte cultures, Fountain et al. demonstrated that encapsulated streptomycin sulfate, amikacin, gentamicin, kanamycin, and tobramycin were significantly more active than the free drug in inactivating intracellular *Brucella spp.* [31-33]. Sunamoto et al. [34] have studied the interactions of liposomal sisomycin and intracellular *Legionella pneumophila* in vitro. They encapsulated the antibiotic in LUVs containing, in addition to the lipids, a specific ligand to be used in subsequent targeting experiments in vivo (see Section VI.A). Guinea pig alveolar macrophages and human blood monocytes in culture were used in the experiments. The authors found that even 50 μg/ml of free sisomycin was unable to kill intracellular *L. pneumophila*, whereas the liposome-encapsulated drug produced a drastic decrease in the number of intracellular colony-forming units.

Desiderio and Campbell [35] determined the effect of encapsulated cephalotin in MLVs (composed of phosphatidylcholine, cholesterol, and phosphatidylserine, molar ratio 6:3:1) on the viability of *Salmonella typhimurium* residing intracellularly in murine peritoneal macrophages. They found that treatment of infected macrophages with the liposomal antibiotic was more effective than the free drug in enhancing the intraphagocytic killing of the microorganism.

The use of liposomal ampicillin to inactivate intracellular *Listeria monocytogenes* has been analyzed by Bakker-Woudenberg et al. [18]. The model system utilized this time was a culture of mouse peritoneal macrophages. Lipid vesicles contained cholesterol, sphingomyelin, and phosphatidylserine in a 5:4:1 molar ratio. The authors found that 150 μg/ml of ampicillin (free drug) plus empty liposomes inhibited the growth of intracellular *L. monocytogenes* but did not reduce the number of viable organisms. In the presence of 50 μg/ml ampicillin, bacterial intracellular growth was still observed, although less as compared with intracellular growth in the absence of ampicillin. At a concentration of 50 μg/ml of free ampicillin plus 100 μg/ml of liposomal ampicillin, 99% of the intracellular bacteria were killed. The fact that 100 μg/ml of liposomal ampicillin in broth culture did not affect *L. monocytogenes* growth indicated that significant leakage of ampicillin from the liposomes with subsequent killing of the bacteria by the free drug did not occur. According to the results, leakage was less than 0.2% of ampicillin from the lipid vesicles within the first 2 h.

The interaction of a liposomal beta-lactam antibiotic with *L. monocytogenes* was also studied by Ito and co-workers. They reported that administration of penicillin G encapsulated in liposomes (phosphatidylcholine, cholesterol, and phosphatidylserine, molar ratio 6:3:1, respectively) inhibited the proliferation of *L. monocytogenes* within murine intraperitoneal macrophages, whereas penicillin G plus liposomes did not [36]. These results are in agreement with those of Bakker-Woudenberg described above.

Bermúdez and co-workers studied the effect of encapsulated amikacin on intracellular *Mycobacterium avium*, either alone or in combination with free rifapentine [37]. Encapsulation of amikacin in liposomes conferred a significant increase in its antibacterial activity against three microorganisms within human macrophages in culture. The authors hypothesized that the enhanced intracellular killing was due to increased uptake of liposomal amikacin by macrophages when compared with that of the free drug. Although liposomal amikacin activity was not enhanced by addition of rifapentine to the culture medium, the authors suggested the use of the liposomal amikacin-free rifapentine combination to reduce the emergence of resistance in cases of chronic infection.

From the experiments cited above, there is no evidence that liposomes are toxic to cells in culture. Phagocytic and intracellular lytic activity seemed to be unaffected during cell-liposome interaction. The interaction between a phagocyte and a particulate agent (liposomes), however, may cause cell activation with subsequent increase in bacterial killing activity. This activation could counterbalance a putative inhibitory effect. The results observed in all studies cited above, therefore, do not discard the fact that phagocyte activation and toxic effects induced by liposomes may be present at the same time during phagocyte-lipid vesicle interaction. More studies designed specifically to address this subject are required to satisfactorily explain the effect of liposomes on phagocytic cells.

V. TESTING OF LIPOSOME PREPARATIONS IN VIVO

The ultimate evidence as to whether encapsulation of antibiotics in lipid vesicles will be therapeutically useful is the actual demonstration of their effect in a model of infectious disease. Several studies in this regard have been published, most of which addressed the effect of a liposomal antibiotic on facultative intracellular bacteria in vivo.

Guinea pigs experimentally infected by intratracheal instillation of *L. pneumophila* were treated with liposomes containing sisomycin, either intravenously or intramuscularly [34]. The antibiotic-liposome complex consisted of LUV coated with a polysaccharide ligand, with 36% of the antibiotic actually entrapped in the lipid vesicle. Control animals were treated with free sisomycin intramuscularly in a similar dose. Although sample size in this experiment appeared to be somewhat small, results demonstrated that the experiment was successful indeed. One hundred percent of the animals treated intramuscularly with free antibiotic died within 6 days of infection, 40% survived when treated intramuscularly with the liposomal sisomycin, and 100% of those treated intravenously with liposomal sisomycin survived for 10 days. The higher efficacy of intravenous over intramuscular treatment with liposomal sisomycin may have been related to rapid delivery of the agent to the target tissue. The authors, however, did not elaborate further to explain this finding. The half-life of sisomycin is relatively short, and increased animal survival after intramuscular treatment with liposomal antibiotic may have been due to a sustained-release effect.

The feasibility of treating *Salmonella typhimurium* infections in mice with liposome-entrapped antibiotics was investigated by Tadakuma et al. [38], who used encapsulated streptomycin sulfate, and by Desiderio and Campbell [39], who used encapsulated cephalotin. Both research teams reported that antibiotic activity in blood was decreased in liposomal antibiotic-treated mice when compared with mice treated with the free drug. However, antibiotic activity in the liver and spleen, two major sites of bacterial localization in *Salmonella* infections, was 4 to 100 times higher in liposomal antibiotic-treated mice when compared to those receiving free drug. In both studies, the liposomal preparations were significantly more effective than the free drug in reducing the number of bacteria in spleen homogenates 7–10 days after infection. Treatment of mice with liposomal streptomycin, at a fraction of the amount of free drug necessary to produce any effect, prolonged mouse survival significantly. Furthermore, treatment with the entrapped antibiotic caused less acute toxicity than treatment with the free drug, which would permit the use of higher antibiotic dosages.

Fountain et al. [31] evaluated the effect of liposomal aminoglycosides in the treatment of experimental *Brucella canis* infections in mice and guinea pigs. A single intraperitoneal dose of either free or encapsulated streptomycin after intraperitoneal chllenge did not modify the number of colony-forming units recovered from the spleens of infected mice. No viable organisms were isolated from the spleen, heart, liver, lungs, kidneys, or testes on days 7 and 10 after infection from animals that received two injections of liposome-entrapped streptomycin. Large numbers of colony-forming units of *B. canis* were isolated from organs of animals treated with free streptomycin. Treatment with liposomal dihydrostreptomycin, gentamicin, and kanamycin were also highly effective when compared to treatment with the free aminoglycosides.

Milward et al. [40] determined the effect of treatment with oxytetracycline and/or streptomycin under various regimes, including liposomal preparations, on

Brucella abortus infection in cows. The authors found no evidence that liposome antibiotics were more effective than free drug. In these studies, however, the percent of antibiotic encapsulation was very low, to the extent that less than 1% of the administered antibiotic in the preparation was actually encapsulated.

In addition to those experiments performed with phagocytic cells in vitro, Bakker-Woudenberg et al. [41] compared the effect of free and encapsulated ampicillin on the course of *L. monocytogenes* infections in normal and athymic nude mice. These authors found that the therapeutic activity of liposomal ampicillin increased 80 times over that of the free drug in normal mice. Treatment with either free or liposomal ampicillin could not eradicate bacteria from the liver and spleen of athymic mice. Similar results were obtained by the same authors using MLV (cholesterol, sphyngomyelin, and phosphatidylserine, molar ratio 5:4:1) using a specific pathogen-free mouse model of *L. monocytogenes* infection [42].

Vladimirsky and Ladigina [43] investigated the effect of treatment with a liposomal preparation of streptomycin in mice infected with *Mycobacterium tuberculosis*. They reported that treatment with the liposomal preparation was more effective than the free drug in prolonging mouse survival and reducing the number of microorganisms in the spleen, but not in the lungs.

In their recent review, Swenson et al. described some results from their experiments in which they determined the effect of encapsulation on the efficacy of gentamicin in an experimental murine model of *Klebsiella pneumoniae* lung infection [44]. In these experiments, mice were treated with a single 20-mg/kg dose of either free or liposomal gentamicin 32 h after bacterial challenge. Seventy-two hours after treatment, the lungs from animals treated with the liposomal preparation had 1000-fold fewer viable bacteria than those given the free drug. The mean survival time for untreated animals was 4 days, whereas the mean survival time was 7.8 days for mice treated with the free drug and more than 15 days for the liposomal gentamicin-treated group.

A model of acute pneumonia was utilized by Nacucchio and co-workers to determine whether treatment with liposome-encapsulated cefoperazone had advantages over treatment with the free antibiotic. Mice were rendered granulocytopenic by a single I.P. 200-mg/kg cyclophosphamide injection. Four days after injection, the animals had less than 50 blood polymorphonuclear leukocytes per mm^3. This model provided a short period during which the animals become sensitive to *P. aeruginosa* infection. Granulocytopenic mice were challenged by exposure to an aerosol containing the bacteria. The LD_{50} after a single administration of 200 mg/kg of cyclophosphamide given 96 h before challenge was ca. 2000 intrapulmonary colony-forming units of *P. aeruginosa*. The results showed that encapsulation of cefoperazone in multilamellar liposomes of a defined composition (phosphadidylcholine, cholesterol, and phosphatidylserine 7:4:1 molar ratio, respectively) enhanced the antibacterial activity of the agent against *P. aeruginosa* in vivo (Nacucchio et al., unpublished data). The mechanism by which liposome encapsulation increased the efficacy of cefoperazone against *P. aeruginosa* lung infection is not understood, but may involve mechanical localization of liposomes in the capillary beds of the lung or uptake of antibiotic by blood phagocytes that subsequently migrate into the alveoli. Either mechanism might increase the amount of drug in the target organ.

The results from studies concerning the interaction of a bacterial pathogen and a liposome-entrapped antibiotic in vivo generally agree with the hypothesis that liposomes may be a suitable carrier system to deliver the drug to tissues harboring facultative intracellular bacteria. The actual mechanism by which liposomal encapsulation confers enhanced antibacterial activity within cells has not been completely elucidated yet. Several hypotheses have been put forward to explain this increased activity.

Macrophages within the major RES organs sequester particles from circulation, thus having the capacity to capture liposomes. This accumulation of liposomes in the RES organs leads to increased concentrations of the antibiotic in the target site. Conversely, the free antibiotic is normally distributed according to a standard biodistribution pattern characteristic for each agent and for a given host pathophysiological status. Antibiotic leakage from liposomes trapped within a RES organ, therefore, may result in increased local concentration of antibiotic, which in turn would be responsible for the beneficial effects of treatment with the encapsulated antibiotic. Another hypothesis to explain the advantages of liposomal antibiotic treatment is related to nonspecific activation of phagocytic cells after ingestion of a significant number of lipid vesicles. Whether this mechanism may play an important role in vivo cannot be demonstrated from the experiments cited above.

The potential positive tropism of liposomes toward lisosomes has been discussed by several authors [45-47]. The actual subcellular site of delivery of liposomal antibiotics within infected cells, however, has not been demonstrated in any of the studies discussed thus far. Because certain intracellular organisms can avoid intraphagocyte inactivation by preventing phagolysosomal fusion, it seems unlikely that a liposome-encapsulated antibiotic would be delivered directly to the intracellular pathogen. Perhaps just the increase in the overall intracellular concentration of the antibiotic is sufficient to increase efficacy in certain cases [19].

It seems reasonable to conclude from the experiments discussed so far that the liposome is a promising carrier system to deliver antibiotics to RES organs. The immediate usefulness of liposomal antibiotics appears to be restricted to the treatment of infections due to bacteria capable of persisting in the intracellular milieu. Except where lipid-based delivery systems diminish toxicity of the drug or provide sustained release of the antimicrobial agent, the potential for liposomal antibiotic therapy in bacterial infections outside RES organs remains unclear. Coating the liposome with a ligand with distinctive affinity for the target organ has been suggested as a method to curtail trapping of circulating lipid vesicles by the RES.

VI. ADVANTAGES OF ANTIBIOTIC ENTRAPMENT IN LIPOSOMES

A. Targeting

Before discussing the importance of targeting, the definition of this term deserves some clarification. In general, targeting is the modification of the natural biodistribution of an agent by any means to achieve its routing to a given site. There are different approaches to reach this goal. One of them is the pharmaceutical approach, which achieves targeting by site-specific delivery of the drug. The physical approach uses carriers as drug delivery systems. Liposomes, microspheres, and nanoparticles, among others, are examples of these systems. The use of any of these carriers modifies the normal fate of the drug and, therefore, may be regarded as active targeting. This is true if we considered as the free drug reference point. The fate of the agent would be the natural fate of the carrier structure. Although the liposome would reach its fate through a passive process, the active agent would still be targeted in an active fashion.

Active targeting may be accomplished by different methods, which fall into two major groups. One of them involves the change of either the physical characteristics of the liposome or its composition, without the addition of a specific ligand. The other involves the addition on the surface of the vesicle of a molecule, the specific ligand, which recognizes a receptor on the surface of the target cell.

Research concerning the delivery of liposomal antibiotics to different hosts has been detailed above. In a way, all these experiments are examples of targeting without the use of a specific ligand. Although ligand targeting of antineoplastic agents in vivo has attracted the attention of researchers [48], there is little experience with ligand targeting of antibiotics. The use of the ligand O-palmitoylamylopectin to target liposomal sysomicin to intracellular sites in the lung has been attempted by Sunamoto et al. [34]. They have shown that the ligand-coated liposomes accumulated to a greater extent in the lungs when compared with liposomes of similar size and composition without the ligand. This is a very interesting and promising approach to achieving antibiotic targeting for the treatment of pulmonary infections.

B. Reduction of Drug Toxicity

Toxicity of an antimicrobial agent can be reduced by entrapment in liposomes. There is wide experience in the prevention of toxicity by encapsulation of highly toxic antiparasitic and antifungal agents. There are few reports concerning the decrease in antibiotic toxicity by entrapment in liposomes due, in part, to the availability of a good number of antibiotics with low toxicity for human beings. It is not surprising to find, therefore, that the few reports addressing this subject are related to aminoglycosides. Using an experimental model system, Vladimirsky and Ladigina [43] and Tadakuma et al. [38] showed that encapsulated streptomycin was less toxic than the free drug when these preparations were injected intravenously. Similar results were obtained by other authors working with gentamicin [44]. The mechanism through which liposome entrapment prevents antibiotic toxicity remains to be elucidated. To our knowledge, there are no studies addressing the effects of liposome encapsulation on oto- or nephrotoxicity induced by aminoglycosides.

C. Controlled Release

Antineoplastic drugs [49] and hormones [50] have been encapsulated in liposomes with the aim of preparing a formulation to accomplish sustained release. Although this approach has potential usefulness for controlled delivery of antibiotics, there are few reports on the subject. Barza et al. [51] injected gentamicin entrapped in liposomes subconjunctivally to rabbits and found that this preparation produced significantly higher antibiotic levels in the sclera and the cornea, when compared with injection of the free drug. It was presumed that the higher concentration of the drug was the result of retarded antibiotic release from the subconjuntival depot. Singh and Mezei, however, found that topical administration of encapsulated dihydrostreptomycin sulfate to the eye did not confer any advantage over the free drug in a rabbit model [52].

Schreier et al. have shown that large-size liposomes containing gentamicin appear to reside at the site of intramuscular injection for several days, providing a depot for short- and medium-term sustained systemic drug delivery. They have also found that erosion of liposomes and release of the encapsulated drug seems to depend on the fluidity of the lipid layer. The authors suggested that varying drug absorption rates may be achieved by mixing lipids with different degrees of fluidity [53,54]. Other authors have also shown slow release of liposomal cefazolin sodium from an intramuscular injection site [55].

VII. FUTURE

Among other vehicles for drug delivery that are under investigation, liposomes are an innovative and sophisticated drug carrier system. There are a number of biodegradable or bioadsorbable materials that can potentially be used to generate microspheres and microcapsules to entrap antibiotics. These carriers can, similarly to what occurs with liposomes, be targeted to RES organs. There is, unfortunately, little information on the comparative advantages of the former carrier systems over liposomes. Although there are many studies addressing the use of liposomes for treatment of infectious diseases, it remains difficult to evaluate accurately the potential usefulness of this novel pharmaceutical form. The reason for this obstacle is the lack of standardized procedures to obtain the vesicles and the variety of systems utilized to evaluate their efficacy. In any event, there are certain specific cases where the value of antibiotic encapsulation in liposomes has been clearly established.

Because liposomes entering the circulation will eventually be trapped by organs of the RES, encapsulation of antibiotics in lipid vesicles offers a suitable means to treat infections of these organs. Treatment of infections in sites other than the RES organs, however, requires a different approach to the problem. Although the solution to this problem remains elusive, the concept of targeting is a promising idea for developing an effective carrier. Several research groups have already begun working in this field, but much of this research is still in the initial stages.

If an appropriate experimental formulation is obtained, there remains to be determined the level of toxicity of such liposomal preparations. This requires additional preclinical studies that would involve determination of short- and long-term toxicity, teratogenicity, and mutagenicity. It must be noted that targeting not only concentrates the active agent in a given organ but also accumulates other liposomal components therein with potential toxic effect. More information is required in each particular case to determine whether the specific interaction between lipid vesicles and the target cells in a given organ may cause undesirable side effects.

There is evidence that liposome encapsulation can provide sustained release of an antibiotic under appropriate circumstances. The actual use of liposomes for this purpose, however, will depend on their competitiveness when compared with other available technologies.

From the industrial viewpoint there are still requirements to be met by the final pharmaceutical form containing liposomes to become a successful product. First, parenteral preparations should be sterile and nonpyrogenic. To the present time, preparation of liposomes on a large scale that fulfill these properties remains a challenge to be solved. Second, the liposomal preparation should be stable for a significant period of time. There is only scant information about long-term stability, and little is known about interactions between lipid vesicle components and other components of the pharmaceutical form. More information is required to determine whether a standardized and reproducible liposome formulation can be stable enough to become a successful carrier for antibiotics on an industrial scale.

In light of all available information, we believe that encapsulation of antibiotics within liposomes may become a successful way not only to overcome the problems caused by infections that are refractory to conventional treatment, but also widen the therapeutic margin of antibiotics currently used in clinical practice.

REFERENCES

1. G. Gregoriadis, *FEBS Lett. 36*:292 (1973).
2. V. J. Richardson, *J. Antimicrob. Chemother. 12*:532 (1983).
3. G. Gregoriadis, *Trends Biotech. 3*:235 (1985).
4. I. A. J. M. Bakker-Woundenberg and F. H. Roerdink, *J. Antimicrob. Chemother. 17*:547 (1986).
5. A. D. Bangham, M. M. Standish, and J. C. Watkins, *J. Mol. Biol. 13*:238 (1965).
6. L. D. Mayer, M. B. Bally, M. J. Hope, and P. R. Cullis, *Chem. Phys. Lipids 40*:333 (1986).
7. F. C. Szoka and D. Papahadjopoulos, *Ann. Rev. Biophys. Bioeng. 9*:467 (1980).
8. G. Gregoriadis (ed.), *Liposome Technology*, CRC Press, Boca Raton, Fla., 1983.
9. G. Poste, *Biol. Cell 47*:19 (1983).
10. G. Gregoriadis, J. Senior, B. Wolff, and C. Kirby, in *Receptor-Mediated Targeting of Drugs* (G. Gregoriadis, G. Poste, J. Senior, and A. Trouet, eds.), Plenum Press, New York, 1984, p. 243.
11. E. Arakawa, Y. Imai, H. Kobayashi, K. Okumura, and H. Sezaki, *Chem. Pharm. Bull. 23*:2218 (1975).
12. A. J. Jackson, *Drug. Metabol. Dispos. 9*:535 (1981).
13. M. K. R. Chowdhury, R. Goswami, and P. Chakrabarti, *J. Appl. Bacteriol. 51*:223 (1981).
14. H. E. Schaeffer and D. L. Krohn, *Invest. Ophthalmol. Visual Sci. 21*:220 (1982).
15. N. A. Hodges, R. Mounajed, C. J. Olliff, and J. M. Padfield, *J. Pharm. Pharmacol. 31 (Suppl.)*:85P (1979).
16. M. Stevenson, A. J. Baillie, and R. M. E. Richards, *J. Pharm. Pharmacol. 33 (Suppl.)*:31P (1983).
17. M. Stevenson, A. J. Baillie, and R. M. E. Richards, *Antimicrob. Agents Chemother. 24*:742 (1983).
18. I. A. J. M. Bakker-Woudenberg, A. F. Lokerse, J. C. Vink-van den Berg, F. H. Roerdink, and M. F. Michel, *Antimicrob. Agents Chemother. 30*:295 (1986).

19. M. C. Popescu, C. E. Swenson, and R. S. Ginsgerg, in *Liposomes. From Biophysics to Therapeutics* (M. J. Ostro, ed.), Marcel Dekker, New York, p. 219, 1987.
20. F. C. Szoka, K. Jacobson, and D. Papahadjopoulos, *Biophys. Acta 551*:295 (1979).
21. P. V. Liu, in *Pseudomonas Aeruginosa. Clinical Manifestations of Infection and Current Therapy* (R. G. Dogget, ed.), Academic Press, New York, 1979, p. 63.
22. N. C. Jones and M. J. Osborn, *J. Biol. Chem. 252*:7398 (1977).
23. L. Huang, in *Liposomes* (M. J. Ostro, ed.), Marcel Dekker, New York, 1983, p. 87.
24. M. C. Nacucchio, M. J. Gatto-Bellora, D. O. Sordelli, and M. D'Aquino, *J. Microencap. 5*:303 (1988).
25. M. C. Nacucchio, G. J. Gatto, M. J. Bellora, D. O. Sordelli, and M. D'Aquino *Antimicrob. Agents Chemother. 27*:137 (1985).
26. L. E. Brian, in *Pseudomonas aeruginosa. Clinical Manifestations of Infection and Current Therapy* (R. G. Dogget, ed.), Academic Press, New York, 1979, p. 219.
27. W. Zimmerman and A. Rosselet, *Antimicrob. Agents Chemother. 12*:368 (1977).
28. K. H. Sekery-Pataryas, C. Vakirtzi-Lemonias, H. A. Pataryas, and J. N. Legakis, *Int. J. Biol. Macromol. 7*:3791 (1985).
29. C. R. Alving, in *Targeting of Drugs* (G. Gregoriadis, J. Senior, and A. Trouet, eds.), Plenum Press, 1982, p. 337.
30. P. F. Bonventre and G. Gregoriadis, *Antimicrob. Agents Chemother. 13*: 1049 (1978).
31. M. W. Fountain, D. K. Weiss, A. G. Fountain, A. Shen, and R. P. Lenk, *J. Infect. Dis. 152*:529 (1985).
32. M. W. Fountain, C. Dees, and R. D. Schultz, *Curr. Microbiol. 6*:373 (1981).
33. C. Dees, M. W. Fountain, J. R. Taylor, and R. D. Schultz, *Vet. Immunol. Immunopathol. 8*:171 (1985).
34. J. Sunamoto, M. Goto, T. Ilida, K. Hara, A. Saito, and A. Tomonaga, in *Receptor-Mediated Targeting of Drugs* (G. Gregoriadis, G. Poste, J. Senior, and A. Trouet, eds.), Plenum Press, New York, 1984, p. 359.
35. J. B. Desiderio and S. G. Campbell, *J. Infect. Dis. 148*:563 (1983).
36. M. Ito, E. Ishida, F. Tanabe, N. Mori, and S. Shigeta, *Tohoku J. Exp. Med. 150*:281 (1986).
37. L. E. M. Bermúdez, M. Wu, and L. E. S. Young, *J. Infect. Dis. 156*:510 (1987).
38. T. Tadakuma, N. Ikmewaki, T. Yasuda, M. Tsutsumi, S. Saito, and K. Saito, *Antimicrob. Agents Chemother. 28*:28 (1985).
39. J. V. Desiderio and S. G. Campbell, *J. Reticuloendothel. Soc. 34*:279 (1983).
40. F. W. Milward, P. Nicoletti, and E. Hoffmann, *Am. J. Vet. Res. 45*:1825 (1984).

41. I. A. J. M. Bakker-Woudenberg, A. F. Lokerse, F. H. Roerdink, D. Regts, and M. F. Michel, *J. Infect. Dis. 131*:917 (1985).
42. I. A. J. M. Bakker-Woudenberg, A. F. Lokerse, F. H. Roerdink, D. Regts, and M. F. Michel, *Cells of the Hepatic Sinusoid* (A. Kirn, D. L. Knook, and E. Wisse, eds.), The Kupffer Cell Foundation, Rijswijk, The Netherlands, 1986, p. 139.
43. M. A. Vladimirsky and G. A. Ladigina, *Biomedicine* (Paris) *36*:375 (1982).
44. C. E. Swenson, M. C. Popescu, and R. S. Ginsberg, *Crit. Rev. Microbiol. 15*: S1 (1988).
45. G. Gregoriadis, *N. Engl. J. Med. 295*:705 (1976).
46. C. R. Alving, J. S. Vledon, J. F. Munell, and W. L. Hanson, in *Receptor-Mediated Targeting of Drugs* (G. Gregoriadis, G. Poste, J. Senior, and A. Trouet, eds.), Plenum Press, New York, 1984, p. 317.
47. C. De Duve, T. De Barsy, B. Poole, A. Trouet, P. Tulkens, and F. Van Hoot, *Biochem. Pharmacol. 23*:2495 (1974).
48. E. Mayhew and D. Papahadjopoulos, in *Liposomes* (M. J. Ostro, ed.), Marcel Dekker, New York, 1983, p. 289.
49. H. Sasaki, T. Kakutani, M. Hashida, and H. Sezaki, *Clin. Sci. Mol. Med. 49*: 99 (1985).
50. A. L. Weiner, S. S. Carpenter Green, E. C. Soehngen, L. P. Lenk, and M. C. Popescu, *J. Pharmacol. Sci. 74*:922 (1985).
51. M. Barza, J. Baum, and F. C. Szoka, *Invest. Ophthamol. Vis. Sci. 25*:486 (1984).
52. K. Singh and M. Mezei, *Int. J. Pharm. 19*:263 (1984).
53. H. Schreier and P. Mihalko, *Pharm. Res. 3*:90S (1986).
54. H. Schreier, M. Levy, and P. Mihalko, *J. Control. Release 5*:187 (1987).
55. Fujisawa Pharm. Ind. Ltd., Japanese Patent JA74110818 (1974).
56. J. R. Morgan and K. E. Williams, *Antimicrob. Agents Chemother. 17*:544 (1980).
57. A. Yamaguchi, R. Hiruma, and T. Sawai, *J. Antibiotics 25*:1692 (1982).
58. P. deHaan, E. Claassen, and N. van Rooijen, *Int. Arch. Allergy Appl. Immunol. 81*:186 (1986).
59. S. Au, N. D. Weiner, and J. Schacht, *Biochim. Biophys. Acta 902*:80 (1987).

14

LDL as a Carrier in Site-Specific Drug Delivery

PASSCHIER C. DE SMIDT, MARTIN K. BIJSTERBOSCH, and THEO J. C. VAN BERKEL
University of Leiden, Leiden, The Netherlands

I. INTRODUCTION

One of the major drawbacks in systemic drug therapy is the existence of side effects. Sometimes, adverse effects originate from the target organ or tissue itself, as is the case with toxic plasma levels of digoxine. More often, side effects are caused by tissues other than the target tissues or cells. For instance, levodopa, widely used in the treatment of Parkinson'sdisease, is ideally processed only in the brain, leading to an increase in brain dopamine; however, high systemic blood levels can elicit unwanted effects such as diarrhea, nausea, hemolytic anemia, and cardiac arrhythmias.

Obviously, the need for site-specific drug delivery is highest in pathological conditions such as cancer and severe viral infections. Side effects of the drugs utilized in the treatment of neoplastic diseases often are quite severe. Apart from the target cancer cells, many fast-growing tissues such as bone marrow and gastrointestinal border cells are affected by the antineoplastic drug, and therapy will often have to be stopped.

In some cases, chemical modification of a drug can enhance its target organ specificity. Brain-targeted delivery systems based on the dihydropyridine-pyridinium salt redox interconversion have been developed for compounds such as estradiol and ethinylestradiol [1]. Covalent coupling of some sugars to drugs can enhance their uptake by the liver [2].

In most cases, however, a specific carrier is needed that is designed for transport and delivery of the drug to its target tissue. Distribution characteristics of the drug itself will then be irrelevant, since the carrier will determine its fate. However, once delivered to the target cells, it still will be necessary to consider the delivery of the drug to its intracellular effective site.

A number of particles have been proposed as drug carrier systems. Recent interest has focused on monoclonal antibodies and colloidal delivery systems [3] such as liposomes and polymeric microspheres.

In addition, soluble molecules have been proposed as drug carriers, including DNA, lectins, poly-L-lysine, virosomes, insulin, dextran, HCG, dipeptides, and even cellular systems such as erythrocytes and fibroblasts. A more extensive review of the so-called biological drug carrier systems is provided by Poznansky and Juliano [4].

In our group recently, the possible use of lipoproteins as drug carriers has been investigated. Up to now, the characteristics of the low-density lipoprotein (LDL) seems especially to fit its possible use as a drug carrier [5,6]. LDL is a well-defined particle, and its fate in the body has been thoroughly investigated. Interest in LDL as a drug carrier has been triggered by the discovery that many cancer cells show a high receptor-mediated uptake of LDL [7,8]. Cancer cells presumably need much cholesterol for the formation of new membrane material, and the increased LDL uptake from the blood circulation may meet this requirement. The rationale for an LDL-mediated antineoplastic therapy can be summarized as follows:

1. Cancer cells show an increased uptake of LDL. This increased uptake is caused by an elevated LDL receptor expression on the cellular surface. This receptor upregulation cannot be down-regulated by LDL or other lipoproteins; in other words, in this respect cancer cells have become autonomous.
2. In tissues that are responsible for the uptake of LDL under normal conditions, the amount of LDL receptors can be effectively down-regulated: in the liver, for example, by means of dietary fats and cholesterol [9].

In many respects, LDL can be regarded as a professional drug carrier. Furthermore, it lacks immunogenicity, which can be considered an advantage over many foreign particles. Its native structure comprises a lipid droplet that may be used to store drugs, shielded by a coat of more polar phospholipids. A surface protein, apo B-100, will recognize LDL receptors and mediate the cellular uptake.

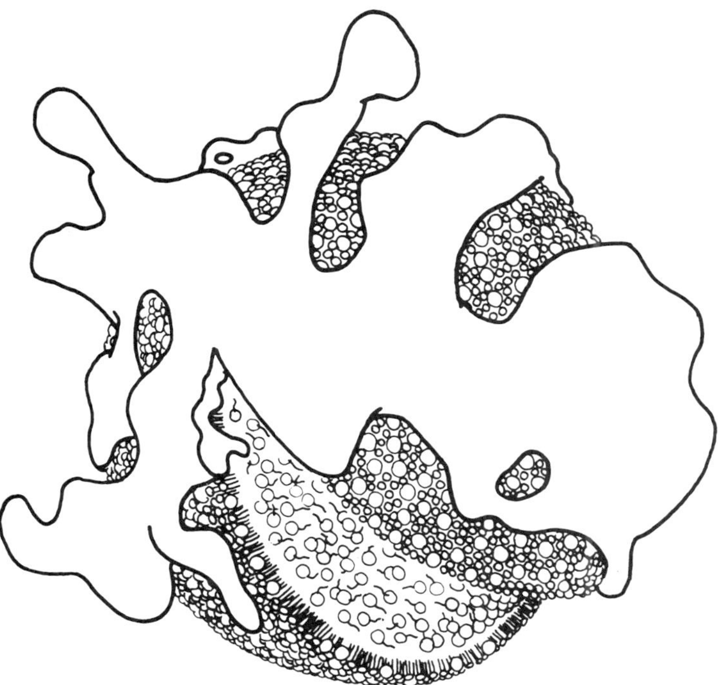

Figure 1 LDL particle: structure of low-density lipoprotein. The inner core of LDL, a droplet of cholesteryl ester molecules, is shielded by a coat of phospholipid and free cholesterol. Covering 60–70% of the surface of the spherical particle, the 514,000-dalton apo B will be mainly responsible for receptor recognition. (Based on Ref. 10.)

II. LDL AS A CARRIER IN SITE-SPECIFIC DRUG DELIVERY

A. Native LDL

During the last decade, knowledge of lipoproteins and their fate in vivo has grown exponentially, with a major emphasis on LDL.

LDL is a spherical particle with a diameter of approximately 23 nm and a mass of 3.10^6 daltons (Figure 1). LDL can be regarded as a lipid droplet that contains about 1500 molecules of cholesteryl ester. It is shielded from the aqueous blood by an amphiphilic shell that consists of 800 molecules of phospholipid and 500 molecules of unesterified cholesterol. This unesterified cholesterol can exchange rapidly with other structures such as cellular mem-

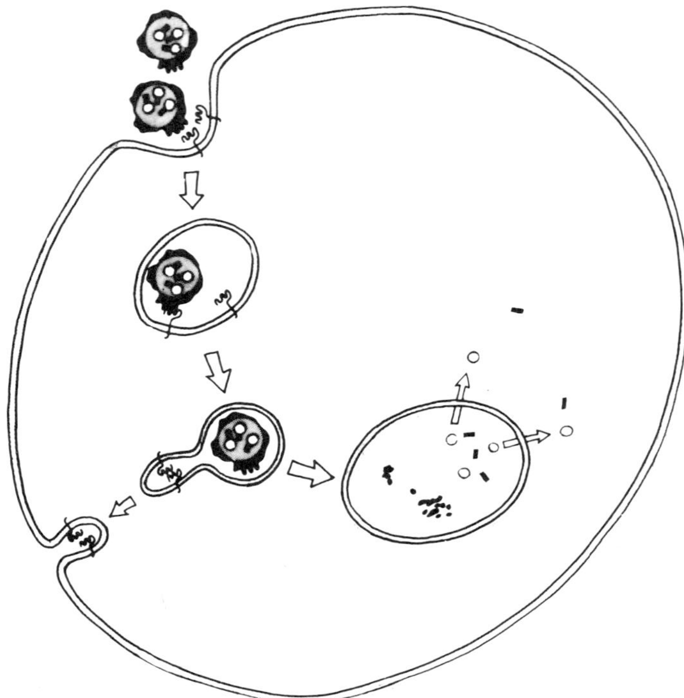

Figure 2 Fate of LDL particles and LDL receptors after endocytosis. After an LDL particle binds to an LDL receptor, the receptor-ligand complex is internalized in a coated pit that pinches off to become a vesicle. In a subsequent step, the LDL dissociates from the receptor and the receptor is transferred back to the plasma membrane. The endosome containing LDL now fuses with lysosomes, forming a large secondary lysosome. In this lysosome the apo-B protein is degraded to amino acids, and the cholesteryl esters are hydrolyzed to fatty acids and cholesterol, which can enter the cytosol through the lysosomal membrane.

branes, whereas the cholesteryl ester molecules remain more firmly trapped in the oily core.

Also present on the shell of LDL is a single 514,000-dalton protein called apo B-100. Cellular LDL receptors bind apo B-100 and carry LDL into the cell by receptor-mediated endocytosis.

After binding of LDL to the LDL receptor, the lipoprotein is internalized and delivered to the lysosomes, where the cholesteryl esters can be hydrolyzed.

Liberated cholesterol can pass the lysosomal membrane and is available for cellular metabolism, such as formation of membranes or steroids (Figure 2).

The number of LDL receptors on many cells is not constant and can be up- or down-regulated. For instance, with HepG2 cells, preincubation with HDL or lipoprotein-depleted serum will increase the number of LDL receptors. On the other hand, preincubation with high LDL concentrations will lead to down-regulation of the LDL receptors [11].

Many tissues in the human body have a low LDL receptor expression, which is probably due to high LDL blood levels in the human species compared to other animal species. In major tissues such as the central nervous system (CNS), heart, skeletal muscle, and skin, it is impossible to detect any receptor-mediated LDL uptake [12]. A limited number of tissues, however, show substantial LDL uptake. In the hamster, which species probably resembles the human situation best, receptor-mediated uptake is seen in the liver, endocrine glands, spleen, and small intestine. It can be calculated that at least 75% of LDL turnover from the plasms compartment is accounted for by uptake in the liver. In addition, the small intestine may account for about 7% of the total LDL uptake [12].

It must be mentioned that in addition to the receptor-mediated uptake, cells can internalize LDL nonspecifically by means of pinocytosis. This process exhibits low-affinity, high-capacity characteristics.

Replicating cells such as cancer cells require large amounts of cholesterol for membrane synthesis. In addition to direct endogenous synthesis, uptake of LDL-cholesterol from the bloodstream may meet such requirements. This need of neoplastic tissues for LDL forms the rationale for LDL-mediated antineoplastic drug therapy. In vitro, several authors have provided evidence for high LDL receptor expression in neoplastic cells. Gal and others [13] found that neoplastic cells of gynecological origin degraded LDL in a similar way to other human cells such as skin fibroblasts, although at a higher rate.

In blood mononuclear cells from patients with acute myelogenic leukemia, the rate of receptor-mediated uptake and the degradation of LDL was shown to be threefold to 100-fold higher as compared to cells from healthy persons [7].

In vivo studies are consistent with these in vitro data. Norata et al. [14] have shown that MS-2 tumor cells, inoculated in mice, do degrade LDL at a high rate via both a receptor-dependent and a receptor-independent mechanism. Hynds and colleagues examined LDL uptake in mice inoculated with MAC13 tumors and found that the neoplastic lesions were second only to liver in their uptake of LDL [15]. A study of 59 patients with acute leukemia showed that hypocholesterolemia—commonly found in acute leukemia—was due to the high low-density lipoprotein receptor activity of the leukemic cells [16].

It must be stressed that most cancer tissues and cell lines have not yet been studied with regard to their LDL receptor expression and LDL uptake. Initial work, however, seems to justify research on LDL as a drug carrier.

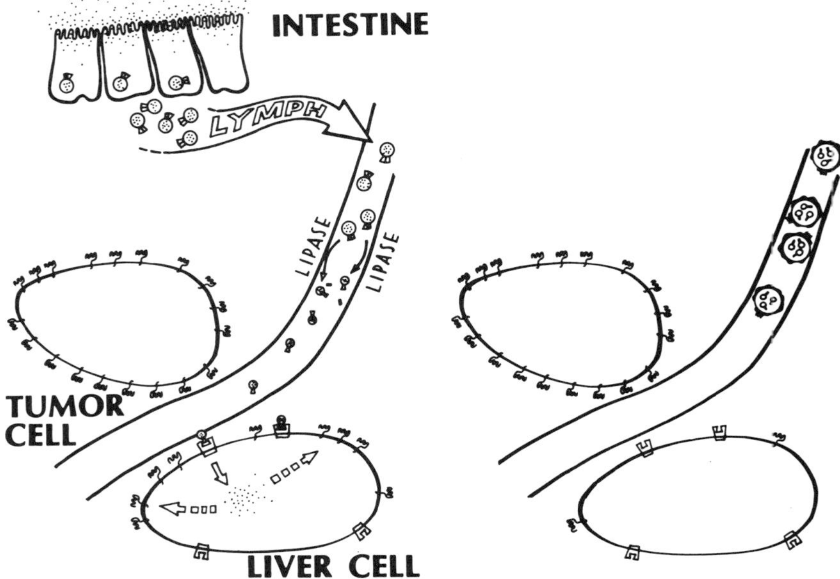

Figure 3 Influence of dietary fat on LDL receptor regulation in liver and tumor cells. When taking special diets, cholesterol and fatty acids are absorbed and processed in the intestine to become chylomicrons. After hydrolysis of triglycerides in the capillary endothelial, the chylomicrons are converted to remnants that are internalized only by liver cells. The liberated cholesterol and fatty acids will down-regulate LDL receptor synthesis in these liver cells. Tumor cells will not be influenced by dietary fat (left). After this pretreatment, a high degree of tumor specificity can be expected when cytotoxic LDL is injected into the bloodstream (right).

The neoplastic state is frequently characterized by a loss of regulation in cellular metabolism. Also for LDL it has been shown that LDL receptor regulation is lost in cancer cells, and therefore cancer cells will keep a high level of LDL receptors even under conditions where receptors in other cell types are down-regulated (high LDL concentrations).

It can be rationalized that conditions can be achieved where the LDL uptake in normal tissues can indeed be down-regulated while high LDL receptor expression in cancer cells is not influenced. Healthy tissues are then protected against the harmful action of a toxic LDL-drug particle, while cancer cells actively internalize these suicide particles. Evidence for the down regulation of LDL receptors inside the liver is provided by studies with hamsters.

Table 1 Distribution of Radioactivity Between Liver and Serum 10 and 30 min After Intravenous Injection of Acetyl-LDL

Time after injection (min)	Radioactivity distribution (% of injected dose)	
	Liver	Serum
10	83.4 ± 1.7	2.2 ± 0.2
30	18.0 ± 1.2	8.4 ± 0.3

Values are means of three different experiments.

When hamsters were fed a diet with 0.12% cholesterol and 20% saturated triglyceride, Spady and Dietschy showed that the receptor-dependent LDL uptake in the liver was suppressed for more than 90% [17]. This effect is exerted presumably by chylomicron remnants, which are supposed to transport dietary fat from the small intestine to the liver. If adrenal LDL receptor expression can be reduced by pretreatment with taurocholate and hydrocortisone [15], two major organs are protected from subsequent LDL-mediated antineoplastic therapy.

Figure 3 visualizes the possible effect of pretreatment of animals with dietary fats on LDL receptor expression in the liver as compared to human cells. Further in vivo studies are, however, necessary in order to prove that such a mechanism indeed improves the specific therapeutic efficiency on tumor cells.

B. Modified LDL

In the previous section, the possibility of targeting native LDL specifically to tumor cells was discussed. The oily core of LDL was utilized as a drug depot, and the apoprotein was still executing its receptor binding task.

However, when the apo B function can be annihilated, it will be possible to store drugs in the LDL core while taking other routes than the LDL pathway. For instance, acetylation of LDL leads to a rapid removal from serum. The uptake is mediated by a highly active scavenger receptor on the liver endothelial cell [18]. Thus, acetylated LDL negates the LDL pathway and follows other physiological routes.

Addition of recognition markers for carbohydrate receptors might be used to target the lipoproteins toward such receptors—for instance, the asialoglycoprotein receptor [19] on parenchymal liver cells.

In this section we will discuss the targeting possibilities of (1) acetylated LDL, (2) tris-gal-chol-associated lipoproteins, and (3) lactosylated LDL.

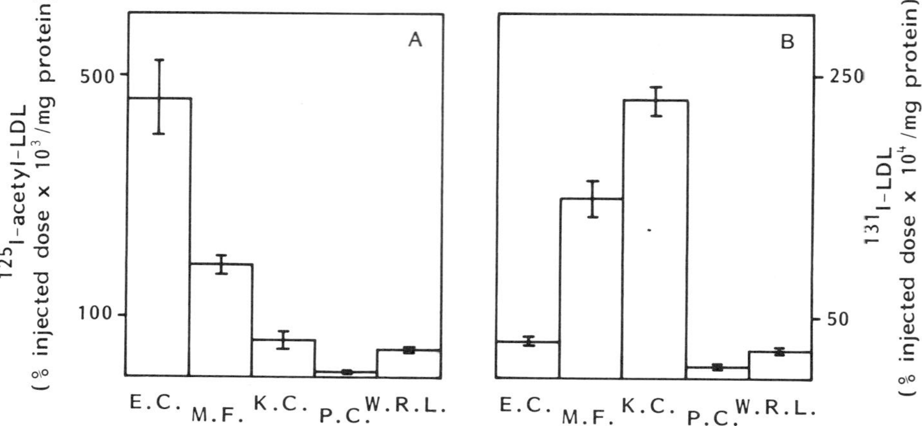

Figure 4 Distribution of acetylated LDL compared to LDL between the various liver cell types at 10 min after injection. P.C., parenchymal cells; E.C., endothelial cells; K.C. Kupffer cells; M.F., mixed fraction Kupffer-endothelial cells; W.R.L., whole rat liver.

Acetylated LDL

The fate of the lipoproteins can be changed upon a chemical modification of the apolipoproteins. For instance, acetylation of LDL leads to a rapid removal of LDL from serum by the liver (Table 1).

Isolated endothelial cells contain five times more acetyl-LDL per milligram of cell protein than the Kupffer cells, and 31 times more than the hepatocytes (Figure 4). This uptake is mediated by a highly active receptor (scavenger receptor) on the liver endothelial cells [18]. Recent morphological studies on the interaction of acetyl-LDL conjugated to 20-nm colloidal gold illustrate that the liver endothelial cells bind acetyl-LDL in coated pits, which is followed by rapid uptake [20]. Uptake proceeds through small coated vesicles, and degradation of the apoprotein occurs in the lysosomes. According to these data, acetylated LDL seems an appropriate carrier for endothelial liver cell-directed targeting.

Tris-gal-chol-Associated Lipoproteins

Because the liver endothelial lining contains "sieve plates" with a diameter of about 100 nm [21], targeting to underlying parenchymal liver cells can be achieved only with relatively small particles. Recently we synthesized a com-

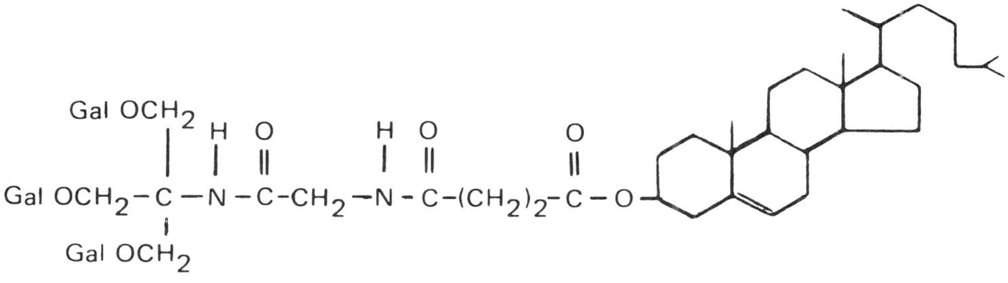

Diagram I The structure of tris-gal-chol.

pound, tris-gal-chol [22] , which enables us to utilize the active receptors for galactose-terminated macromolecules as a trigger for the uptake of lipoproteins (Diagram I). The triantennary galactose-terminated cholesterol derivative (tris-gal-chol) dissolves easily in water, and upon mixing with lipoproteins it is immediately incorporated into these particles.

In Figure 5 the decay of LDL in serum and uptake in liver is plotted after loading the particles with varying amounts of tris-gal-chol. Incorporation of tris-gal-chol into LDL leads to a markedly increased removal from serum paralleled with a quantitative recovery of the label in liver (Figure 5). Similar experiments performed with HDL apparently show a similar induction of liver uptake by tris-gal-chol (Figure 6).

The increased decay of tris-gal-chol LDL (20:13) in serum is nearly completely blocked by preinjection (1 min before the tris-gal-chol LLD) of 0.5 mmol/rat of N-acetylgalactosamine (GalNAc, Figure 7, final concentration in the blood approximately 50 mM), while preinjection of the same dose of N-acetylglucosamine (GlcNAc) has no effect at all (not shown). Compared to GalNAc, the preinjection of 5 or 25 mg of asialofetuin influences the serum radioactivity and the liver association of tris-gal-chol LDL (20:13) to only a small extent.

The increased decay of tris-gal-chol HDL in the serum and liver association is completely blocked by preinjection (1 min before tris-gal-chol HDL) of 0.5 mmol/rat of N-acetylgalactosamine (GalNAc, final concentration in the blood approximately 50 mM), while preinjection of N-acetylglucosamine (GlcNAc) has hardly any effect. Preinjection of 5 mg of asialofetuin/rat leads to a comp-

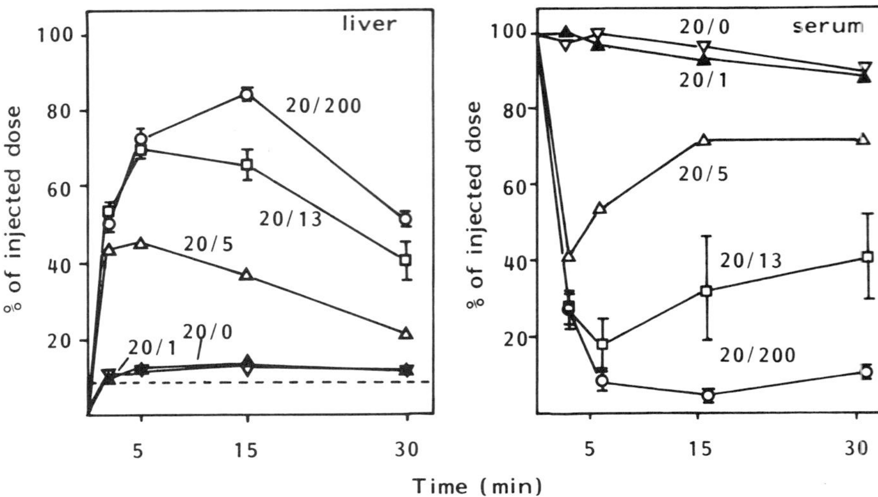

Figure 5 Effect of tris-gal-chol on the liver association and serum decay of LDL. ^{125}I-LDL (20 μg of apolipoproteins) was mixed with 0 (▽), 1 (▲), 5 (△), 13 (□), or 200 (○) μg of tris-gal-chol. The mixture was injected into anesthetized rats, and the liver association and serum decay were determined. When indicated, bars represent S.E. for three animals. The livers were not perfused, and the dashed line represents the maximal contribution of the serum to the liver uptake (determined with ^{3}H-albumin).

plete blockade of the increased liver association of tris-gal-chol HDL, similar to GalNAc (Figure 7).

It can be concluded that lipoproteins, when associated with tris-gal-chol, can be used as carriers for liver targeting. The highly active galactose receptor internalizes the particle, thereby ignoring apo B receptor recognition.

Lactosylated LDL

The idea of sugar-mediated lipoprotein targeting was extended into further research by covalent coupling of D-galactose to LDL. Incubation of LDL with lactose and sodium cyanoborohydrid results in the time-dependent coupling of lactose to the lipoprotein. The coupling of lactose to LDL induces rapid plasma clearance, due to galactose-specific uptake by the liver. At a low degree of lactosylation, parenchymal cells are found to be the main site of uptake; whereas at high degrees of substitution, Lac-LDL was taken up predominantly by Kupffer cells (Figure 8).

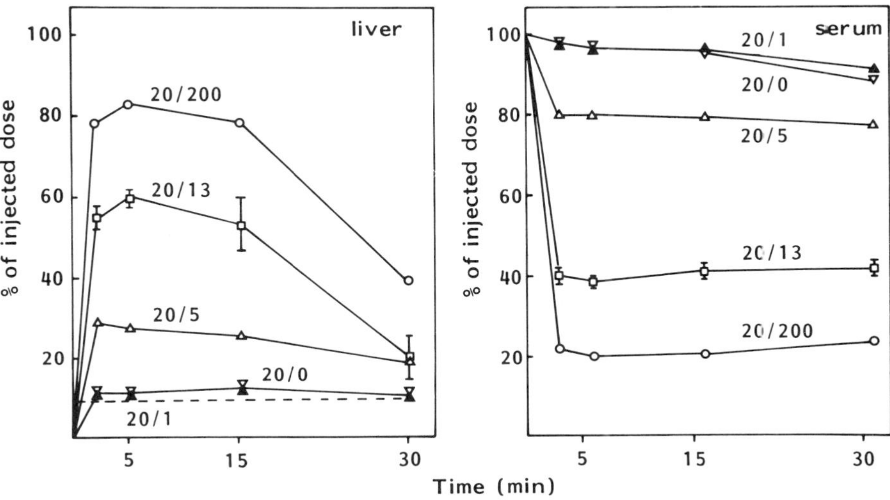

Figure 6 The effect of tris-gal-chol on the liver association and serum decay of HDL. ^{125}I-HDL (20 μg of apolipoprotein was mixed with 0 (▽), 1 (▲), 5 (△), 13 (□), or 200 (○) μg of tris-gal-chol. The mixture was injected into anesthetized rats, and the liver association and serum decay was determined. When indicated, bars represent S.E. for three animals. The livers were not perfused, and the dotted line represents the maximal contribution of the serum to the liver uptake (determined with ^{3}H-albumin).

Earlier studies showed that galactose-containing particles with a size exceeding 15 nm are taken up primarily by Kupffer cells but not by parenchymal cells [23]. Our present data indicate that the LDL particle with a size of 23 nm can be directed mainly to parenchymal cells, provided that it contains a relatively small amount of terminal galactose residues. An increase in the number of galactose residues on LDL resulted in an increase in uptake by Kupffer cells. The uptake by parenchymal cells was also increased, but to a much smaller extent than that by Kupffer cells. The ratio of the specific uptake of Kupffer cells versus that of parenchymal cells increased from 28 for 8h-LacLDL to 90 and 69 for 72h- and 168h-LacLDL, respectively. This suggests that extensive lactosylation can increase the relative targeting of LDL to the galactose-specific receptor on Kupffer cells.

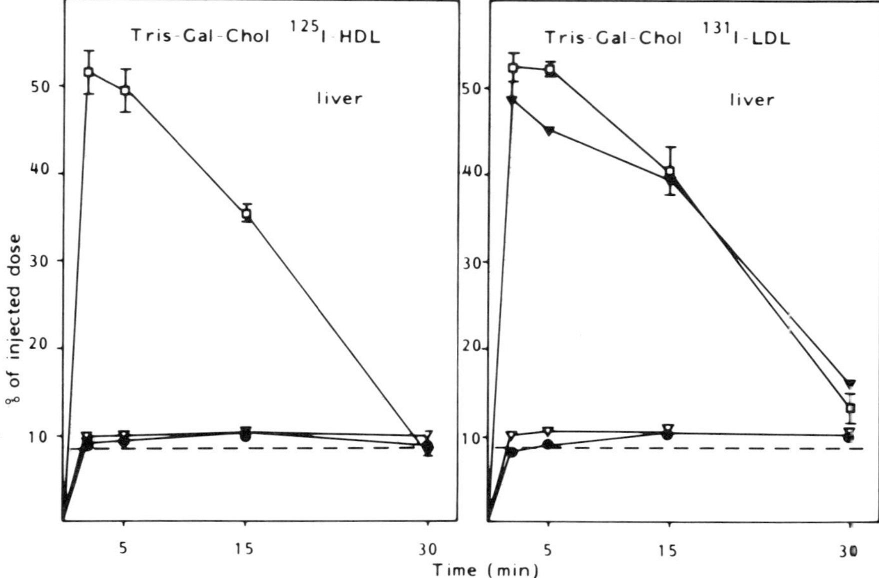

Figure 7 The effect of asialofetuin and GalNAc on the liver association of tris-gal-chol HDL and tris-gal-chol LDL. ^{125}I-HDL (20 μg of apolipoprotein) and ^{131}I-LDL (20 μg of apolipoprotein) were separately mixed with 13 μg of tris-gal-chol. One minute prior to the injection of the tris-gal-chol-loaded lipoproteins, 5 mg asialofetuin (▼-▼) or 0.5 mmol GalNAc (●-●) were injected. □-□ represent tris-gal-chol-loaded lipoproteins without preinjection and ▽-▽ represent unloaded lipoproteins.

III. METHODS—PREPARATION OF DRUG-LDL-PARTICLES

A. Prodrugs

The vast majority of drugs currently used in therapeutics are too hydrophylic for incorporation in LDL. For oral administration, it is generally accepted that a drug should possess a certain hydrophilicity/lipophilicity balance. In order to cross epithelial membranes, the compound must show some hydrophobic characteristics, but hydrophobicity should not exceed certain levels since water (serum!) solubility then becomes a problem.

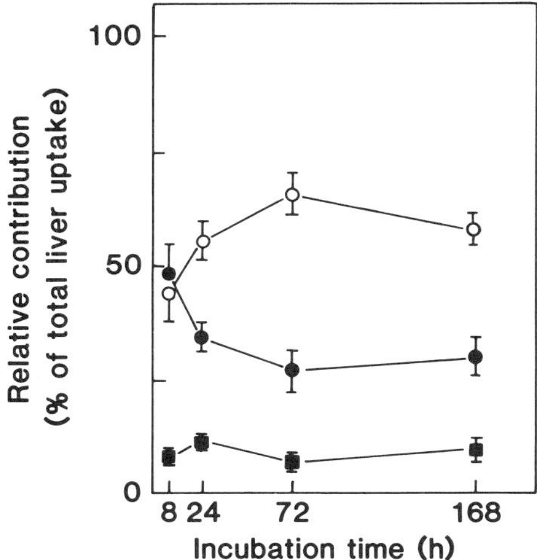

Figure 8 Effect of the extent of lactosylation of LDL on the intrahepatic distribution of lactosylated LDL. Rats were injected with ^{125}I-LDL that had been lactosylated for 8–168 h. Ten minutes later, liver cells were isolated. The intrahepatic distribution of label was calculated from the specific uptake of each cell type and their contribution to the total liver protein. Values are means ± S.E.M. of three experiments. (○), Kupffer cells; (●), parenchymal cells; (■), endothelial cells.

The core of the low-density lipoprotein is extremely hydrophobic. This core exists of about 1500 molecules of cholesteryl ester, which prohibits the inclusion of water-soluble compounds. Only very lipophilic substances such as triglycerides and retinol esters are able to remain in this oily environment. Apart from the lipid core, incorporation can also take place in the more amphiphilic phospholipid coat. Similar to cholesterol itself or to phospholipids, compounds may be more hydrophilic and thereby incorporate into the phospholipid coat. An example is the amphiphilic tris-gal-chol [19], which incorporates its cholesterol moiety in the LDL shell and has its hydrophilic galactose antennas sticking in the serum.

Successful reconstitution of LDL is facilitated if the compound to be reconstituted is provided with certain groups that render it compatible with the LDL's phospholipid coat and lipophilic core. Groups that are particularly suitable for this purpose are, for instance, oleyl, retinyl, and cholesteryl [24]. When a drug is coupled with LDL anchors such as retinyl or oleyl, the newly formed compound may be referred to as a *prodrug*. In most cases, the prodrug has lost some if not all of the pharmacological activity of its "mother compound." The LDL anchor will have to be split off to release the original drug in the target cells, which subsequently can elicit its effect.

When derivatizing a drug for LDL targeting, one should consider the following points:

1. The LDL anchor should give a pronounced increase of the drug's affinity for LDL.
2. The drug-anchor bond should be stable enough to resist hydrolysis in plasma.
3. The drug-anchor bond must be accessible for hydrolysis by lysosomal enzymes. The original drug can then diffuse from the lysosomes into the cytoplasm toward its intracellular target.

A general scheme for the fate of an LDL-drug particle is shown in Figure 2 and is essentially the same as the native LDL particle.

After binding to the LDL receptor, the toxic LDL is internalized and the endosome is delivered to lysosomes. The particle is now attacked by lysosomal enzymes. It is essential that lysosomal enzymes can split off the LDL anchor (e.g., a long fatty acid chain) and leave the released drug intact. This problem can be tackled by using a degradable peptide spacer arm between drug and LDL anchor [25]. It is equally important that the released drug is able to pass the lysosomal membrane and become available in the cytoplasm. When choosing a drug for LDL-mediated targeting, its molecular weight should not be too high since lysosomal membrane permeability may then become a problem. For instance, since nucleosides are shown to pass the lysosomal membrane [26], the choice of purine and pyrimidine antineoplastic agents seems especially relevant. Drugs with molecular weights exceeding 300 may encounter more problems passing lysosomal membranes [27].

Some tentative research has been performed in order to establish the structural requirements for effective LDL anchors. Krieger et al. found that the presence of at least one cis double bond in the fatty acyl chain was necessary for successful incorporation in LDL. Cholesteryloleate, triolein, and methyloleate could be incorporated in LDL, whereas only trace amounts of esters of saturated fatty acids (e.g., cholesterylstearate, tristearin, and methylstearate) could be incorporated into LDL [28]. However, some authors reported good results with saturated fatty acids as LDL anchors. It might also be possible to apply a single hexadecyl side chain [29], as has been done with methotrexate and ara-CMP.

A good example of a model drug for LDL-mediated tumor targeting is the so-called compound 25 (Diagram II). Its frame consists of the native molecule

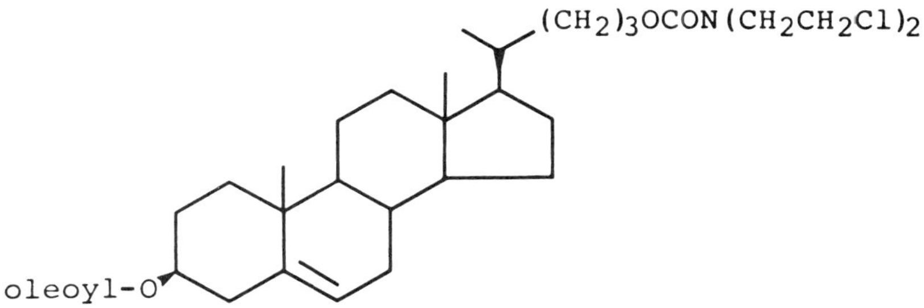

$(CH_2)_3OCON(CH_2CH_2Cl)_2$

oleoyl-O

Diagram II The structure of "compound 25."

cholesteryloleate with the antineoplastic nitrogen mustard attached to its D-ring side chain. This drug is very lipophilic and is able to kill cells in vivo via the LDL pathway [24].

B. Incorporation Methods

One of the fundamental problems in evaluating LDL-mediated targeting is the development of a satisfactory drug incorporation method. Because apo B is responsible for the delivery of the LDL particle to the cell, this protein should preferably not be changed, and it seems rational to use the lipid part of LDL as a "drug reservoir." The LDL incorporation method must meet two basic requirements:

1. It must form a stable, nonleaking drug-LDL complex.
2. It must leave apo B intact so that the recognition of LDL by cells is not influenced.

The last requirement restricts the experimental conditions that can be used for drug incorporation. Even small changes in parameters such as pH and dielectric constant may alter the three-dimensional structure of apo B in such a way that LDL will lose its receptor recognition properties. LDL is therefore a rather delicate particle, and temperature changes, oxidation (for instance, by metal ions and air), and even vortexing [30] have been reported to change its biological behavior. Therefore, new incorporation methods should always be validated with extensive physicochemical characterization and in vitro and in vivo tests on the biological behavior of the particle.

The most important drug incorporation methods are

1. Direct addition/spontaneous association
2. Contact methods

3. Reassembly methods
4. Facilitated transfer

Direct Addition/Spontaneous Association

Direct addition with spontaneous association is the simplest of the incorporation procedures. It takes place in aqueous, buffered milieu and is basically the addition of an aqueous solution of drug to the LDL. Only a very few compounds can be satisfactorily associated with LDL, since they have to show preferential affinity for LDL, which requires high lipophilicity but also water solubility.

Aqueous addition has been used to associate aclacinomycin A with LDL [31], and to conjugate adriamycin and daunomycin with HDL [29]. In photodynamic cancer therapy, the antineoplastic photofrin II is reported to become successfully associated with LDL [32,33].

Another example of simple aqueous addition is the association of tris-galchol to lipoproteins [19,22]. This compound is an amphiphilic cholesteryl-containing tris-galactoside. When dissolved in water, it forms micelles that contain approximately 35 molecules of tris-gal-chol. In the presence of lipoproteins, these micelles spontaneously dissociate and the individual tris-gal-chol molecules associate with the lipoproteins. Since the galactose moiety can be recognized by liver receptors, this presumably occurs on the outer shell.

This interaction is not specific for LDL, since a similar tris-gal-chol association to HDL and liposomes is found. Also, leakage of tris-gal-chol from LDL to other lipid structures such as erythrocyte membranes may occur [34].

Contact Methods

The basis of the contact methods is the transfer of the drug from a solid surface to the lipoprotein. The dry film method utilizes the wall of a glass tube or the surface of glass beads. The lipoprotein-drug contact can be enhanced by using celite, a very fine powdery substance, as the solid surface. An essential step is the partitioning of the solid surface to the lipid particle. Initial binding of the drug to the lipoprotein shell may be followed by diffusion into the LDL core.

The dry film procedure involves evaporation of the solvent under an inert gas to a glass surface (for instance, a glass wall or glass beads), followed by incubation with a buffered LDL solution. With the dry film method, incorporation has been reported of up to seven daunomycin molecules per LDL particle, 55 hexadecylmethotrexate molecules/LDL particle, 100 AD-32 molecules/LDL particle, and 140 muramyltripeptide phosphatidylethanolamine (MTP-PE) molecules/acetylated LDL particle [29].

Plant sterylglucosides have been successfully associated with lipoproteins by immobilizing them on the surface of celite, followed by an incubation with lipoprotein solution [35].

The contact methods are simple and clean procedures; filtration or centrifugation are usually sufficient for a final purification of the drug-lipoprotein complexes. In vitro data with LDL-drug particles prepared by these contact procedures seem promising. Some preparations show evidence of LDL-mediated cellular uptake [32,36]. Sometimes, however, compounds utilize routes other than the LDL pathway, for instance by leaking out of the LDL particle and associating with cellular membranes. Remsen has described a rapid redistribution of benzo(a)pyrene between LDL and cell membranes [37].

The in vivo behavior of the drug-LDL complexes has so far been scarcely studied, and it is not yet clear whether particles prepared by the contact procedures behave in vivo in a similar physiological way as in vitro.

Delipidation

The delipidation methods are the most radical means of incorporation. The lipid core of LDL is extracted, resulting in a particle that is enriched in protein and phospholipids. This particle is presumably more "eager" to be loaded with lipid and can be reconstituted with the new compound. A number of different ways have been described for delipidation of LDL.

Extraction by Organic Solvents. In this procedure, LDL is freeze-dried in the presence of starch [38] or sucrose [25] until all water has disappeared. In the next step, LDL core lipid is extracted by organic solvents such as heptane, petroleum ether, carbon tetrachloride, or benzene. The solvent chosen must be quite lipophilic, since phospholipids should not be extracted in this process. After delipidation, the compound that must be incorporated is added and incubated for a short time. After thorough evaporation, tricine buffer is added and the sample is incubated at 4°C for 10-20 h [39]. The sample can undergo further purification—for instance, centrifugation to discard starch or a surplus of lipid.

Delipidation by Detergents. Several detergents have been investigated for their ability to extrace lipids from the LDL core. Best results were obtained by using SDOC (sodium deoxycholate) as detergent [40]. The resulting apo B-SDOC complexes can be purified by gel filtration and then reconstituted with drug-phospholipid microemulsions. The resulting LDL particles had a size and agarose electrophoresis migration similar to native LDL.

Enzymatic Delipidation. The LDL cholesteryl ester core can be depleted by treating LDL with cholesteryl esterase (also referred to as sterol ester hydrolase). After cleavage of the sterol-fatty acid bond by this enzyme, the resulting cholesterol and fatty acid leave the LDL core and can be bound by albumin and phospholipid vesicles in the medium. The resulting cholesteryl ester-depleted particles can be purified and incubated with a drug suspension. After purification, toxic drug-LDL particles were shown to effectively kill cultured cells by utilizing the LDL receptor pathway [41]. In vivo validations have not yet been performed.

Most studies of reconstituted LDL prepared by delipidation have been carried out in vitro. Physicochemical characterization by, for instance, agarose gel electrophoresis, electron microscopy, and sepharose chromatography showed well-defined particles in most cases. Uptake of the particles by cultured cells generally occurs via the LDL receptor pathway, although sometimes additive pathways may participate [31].

More important, in vivo data are often lacking. Our recent experiments indicate that freeze-drying of LDL alone, or in the presence of starch, results in a more rapid disappearance of LDL from plasma. Treatment of LDL with organic solvents similarly increases the plasma disappearance. Up to now, the "apo B-friendly" enzymatic digestion method has not been evaluated in vivo.

Covalent Attachment

To establish a tight bond between drug and lipoprotein, it is possible to link th the drug covalently to the LDL. Especially the apoprotein moiety offers a possibility. By using carbodiimide, it has been found that about 10 molecules of methotrexate can be attached to each LDL particle, and the drug-LDL conjugate inhibited L 1210 growth in vivo in a concentration-dependent way [42]. A potential hazard of covalent attachment may be the loss of recognition properties of apo B, and this will limit the number of drug molecules that can be bound to the apoprotein.

C. Physicochemical Characterization

Drug-LDL particles can be evaluated physicochemically in a number of ways. By choosing the right method, information can be obtained about the apoprotein, the lipid part of LDL, and the quantitative incorporation of drug.

Apart from traditional tests such as cholesterol, cholesteryl ester, and phospholipid assays, a number of methods have proved their usefulness for characterization of drug-LDL particles.

Density Ultracentrifugation

The density ultracentrifugation technique yields information about the density of the LDL-drug particle and purity of the sample. If the sample contains unbound drug that has a density different than the LDL density, two separate bands will form in the centrifugation tube. The centrifugation tube can be fractionated, and different fractions can then be analyzed for density, drug content, and lipoprotein.

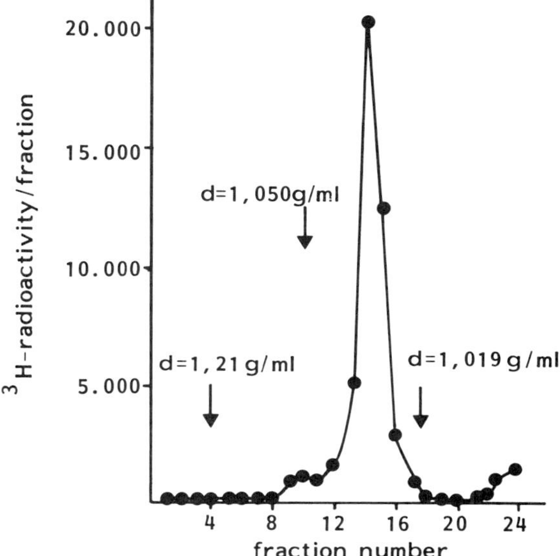

Figure 9 Density ultracentrifugation of LDL. LDL was labeled with ^{3}H-cholesteryl oleate and subsequently analyzed by density ultracentrifugation. The centrifugation tube was fractionated and analyzed for radioactivity.

A density ultracentrifugation analysis of native ^{3}H-labeled LDL is shown in Figure 9. Note that incorporation or association of compounds with LDL can alter the density of the particle, as is the case with tris-gal-chol [22].

Gel Filtration

LDL carrier complexes can be separated from non-LDL-associated drug by gel filtration chromatography. Depending on the nature of free drug (micelles, microemulsions, albumin-bound, etc.) appropriate column material is selected for separation of reconstituted LDL and unbound drug material.

Fractions of eluent can be analyzed for drug and lipoprotein, and in the case of a physicochemically stable complex, these two will co-elute to give an ideal overlapping drug-LDL peak. Figure 10 shows an elution profile of a double-labeled LDL-dioleylmethotrexate particle.

Agarose Gel Electrophoresis

In agarose gel electrophoresis, the lipid part of LDL co-migrates with the apoprotein, offering a major advantage over SDS gel electrophoresis where only the

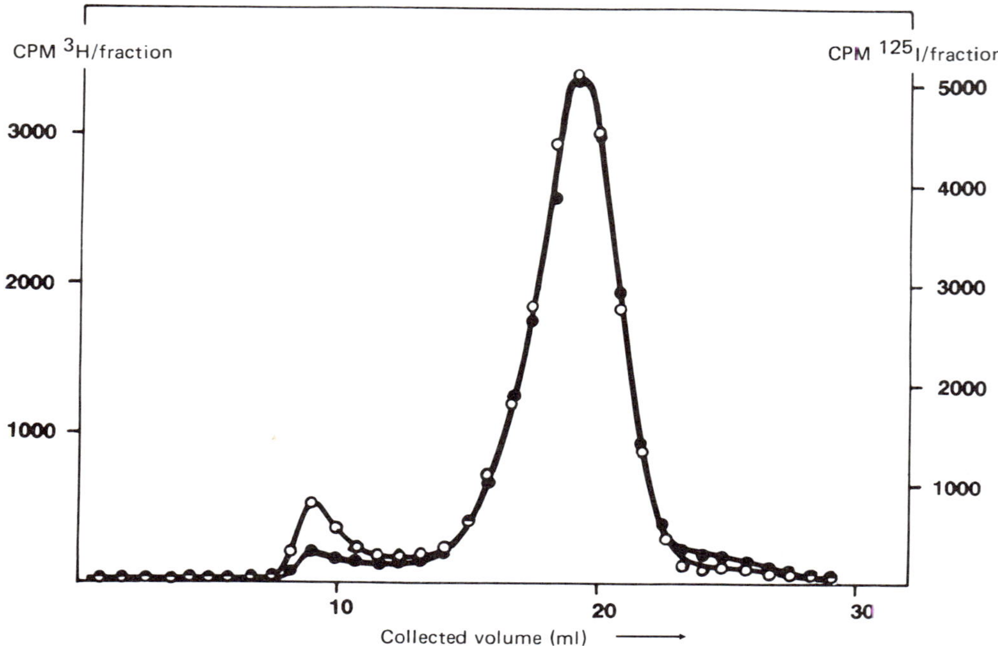

Figure 10 Gel filtration of reconstituted LDL. A sample of double-labeled ^{3}H-dioleylmethotrexate–^{125}I-LDL was prepared by following a modified starch-heptane delipidation procedure, and loaded on a 0.9 × 45 cm Biogel A 50 M column. Open circles (○), ^{3}H-radioactivity; closed circles (●), ^{125}I-radioactivity. Fractions were collected and samples were assayed for radioactivity utilizing a Packard 1500 Tri-Carb liquid scintillation analyzer programmed and validated for ^{3}H-^{125}I double-labeled samples.

apoprotein is analyzed. The complete drug-LDL particle mobilizes and, by analyzing the complete agarose slice, information can be obtained about the drug-LDL binding and the amount of unbound drug.

The agarose gel can be stained, for instance, with Red oil O, Sudan Black, or fat red B. Association of drugs with LDL can influence their cationic mobility. An LDL-photofrin II complex exhibits an increased cationic mobility, which is probably caused by insertion of photofrin II into the external leaflet of LDL, resulting in a net increase in negative net charge.

In an elegant experiment, Krieger et al. analyzed double-labeled reconstituted LDL using the agarose gel electrophoresis technique (Apo B was labeled with ^{125}I and the reconstituted cholesteryl linoleate was ^{3}H-labeled). The gel can be cut in slices for determination of radioactivity [38].

Electron Microscopy

An additional method for characterizing reconstituted LDL is electron microscopy. Electron microscopy does not provide data on drug-LDL incorporation or stability but is merely a visual method to compare reconstituted LDL with native LDL.

SDOC (sodium deoxycholate)-depleted LDL reconstituted with egg yolk phosphatidylcholine and cholesteryl oleate showed a circular morphology with a diameter of 219 ± 24 Å, similar to native LDL as determined by negative-stain electron microscopy [43].

In an effort to incorporate the cytotoxic "compound 25" in LDL following the SDOC or sterolhydrolase procedure, Lundberg found cytotoxic m-LDL particles to have a similar electron micrograph in appearance as native LDL [41].

Antibody Precipitation

Antibody precipitation is a simple and fast method to control if the drug is completely incorporated into LDL. The sample is incubated with, for instance, antisera to human LDL (control with antisera to human LDL). After a short incubation, the samples are centrifuged; supernatant and precipitate can be analyzed for drug and lipoprotein [44].

IV. IN VITRO TEST METHODS

More important than physicochemical behavior of the reconstituted LDL is, of course, its biological activity. Clearly, the ultimate test is its ability to reach tumors in vivo, but before that the normal physiological behavior has to be investigated. Cell culture experiments can provide such information.

In vitro experiments with a simple cellular model are therefore a first step in testing the biological activity of reconstituted LDL.

A. Cellular Uptake of Reconstituted LDL

One of the essential features of the drug-LDL complex is its ability to bind to the LDL receptor in a similar way as native LDL. In an early study, Krieger et al. studied the ability of reconstituted LDL (obtained by starch-heptane delipidation) to compete with ^{125}I-LDL for binding, internalization, and hydrolysis in fibroblasts [38]. Krieger et al. also studied saturation kinetics for the uptake and hydrolysis of reconstituted LDL. In this study, the carrier (^{125}I-labeled) as well as the carried compound (^{3}H-cholesteryl linoleate) were measured for their uptake and showed identical uptake and hydrolysis kinetics (Figure 11). This fibroblast system has confirmed the preserved apo B recognition of reconstituted LDL obtained by the Krieger method.

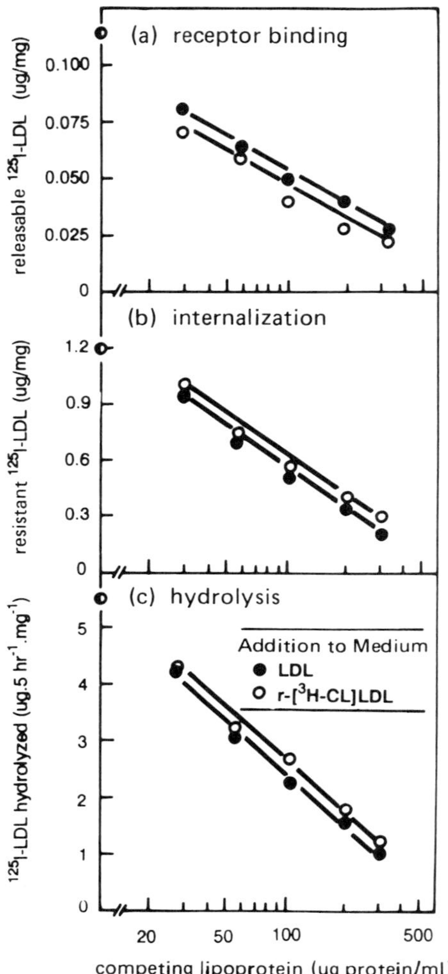

Figure 11 Ability of reconstituted LDL to compete with ^{125}I-LDL for (a) binding, (b) internalization, and (c) proteolytic hydrolysis in fibroblasts at 37°C. On day 7 of cell growth, each monolayer of normal cells received 2 ml of growth medium containing 10% lipoprotein-deficient serum, 25 μg of protein/ml of ^{125}I-LDL (126,000 cpm/μg of protein), and the indicated concentration of either unlabeled LDL (●) or r-[CL-^{3}H] LDL (2590 cpm/nmol of cholesteryl linoleate) (○). After incubation for 5 h at 37°C, the amount of dextran sulfate-releasable ^{125}I-LDL (receptor-bound), dextran sulfate-resistant ^{125}I-LDL (internalized), and ^{125}I-LDL hydrolyzed were determined. (Data from Ref. 38.)

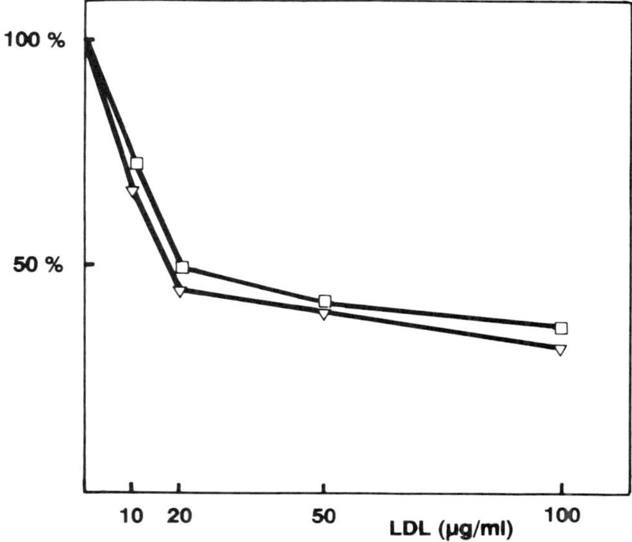

Figure 12 Competition between ^{125}I-LDL and native LDL (▽) or PII-doped LDL (□). Cells were preincubated for 24 h in Ham F10 medium supplemented with 2% Ultroser G, washed, and LDL binding measured with ^{125}I-LDL (10 μg/ml) in the absence or presence of native or PII-doped LDL. Results are expressed in percent of control (^{125}I-LDL alone). 100%: 118 ± 15 ng LDL bound/mg of cell protein; mean of three experimental values. (From Ref. 32.)

In an elegant study, Candide and co-workers have assessed the in vitro biological activity of photofrin II-doped LDL. The apo B recognition can be assessed by competition between ^{125}I-LDL (10 μg/ml) and native LDL or PII-doped LDL (Figure 12).

The PII delivery to MRC 5 fibroblasts was assessed by measuring fluorescence in cells after incubation with various concentrations of PII-doped LDL, PII-doped HDL, and PII-doped albumin. PII-doped LDL delivered PII to the fibroblasts in a saturable way, whereas the fluorescence increased quite linearly with carrier protein concentration for HDL-PII and albumin-PII [32]. PII delivery to the cells could be increased by preincubation with LDL-free media, which upregulates LDL receptor expression.

Sometimes in vitro studies with cultured cells give unexpected results. The well-known carcinogen benzo(a)pyrene, for instance, when added to human plasma in vitro, associated with the plasma lipoproteins, especially the LDL fraction. Remsen and Shireman have treated normal and LDL receptor-negative

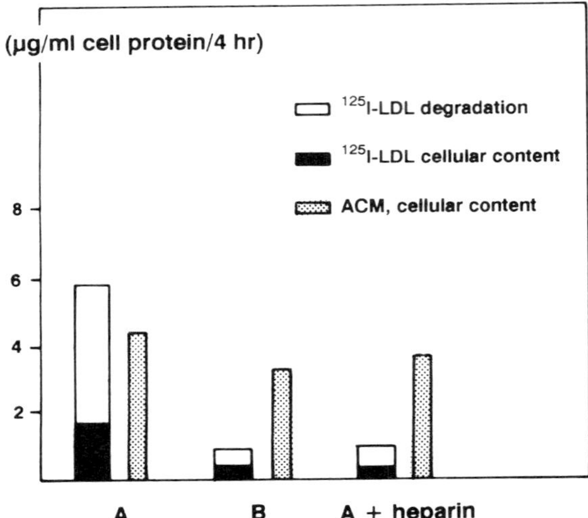

Figure 13a Role of LDL receptor activity in the accumulation and degradation of aclacinomycin A:^{125}I-LDL complex. Glioma cells were preincubated in either LDL-free culture medium (A) or in the presence of 170 µg/ml LDL (B). After 48 h of preincubation, cells were washed and incubated with 12 µg/ml aclacinomycin A:^{125}I-LDL. Heparin (3 mg/ml) was added to some of the group A dishes before incubating with the drug-LDL particle (A + heparin). After 4 h of incubation, medium was assayed for degradation of ^{125}I-LDL and cells were washed and assayed for cell protein, radioactivity, and aclacinomycin A (ACM). (From Ref. 31.)

fibroblasts with media containing free benzo(a)pyrene or the benzo(a)pyrene-LDL complex. Uptake from delipidates or serum-free medium [i.e., free benzo(a)pyrene] was linear with concentration, while cell association of benzo(a)-pyrene bound to LDL was much less and nonlinear at higher protein concentration. While low-density lipoprotein apparently influenced uptake of benzo(a)-pyrene by the cell, no differences were noted in the incorporation of benzo(a)-pyrene by normal and receptor-negative cells. It was concluded that benzo(a)-pyrene diffused into the cells directly from low-density lipoproteins despite the absence of specific receptors, suggesting a rapid redistribution between the LDL and cell membrane [37].

Rudling and co-workers (31) incubated an aclacinomycin A-LDL particle with two groups of human glioma cells. One group was preincubated with lipo-

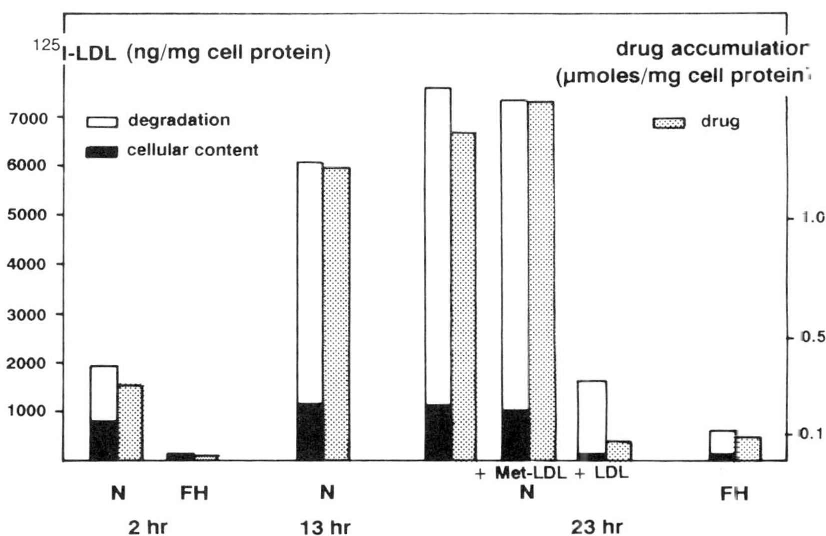

Figure 13b Cellular drug accumulation and cellular content and degradation of ^{125}I-LDL upon incubation at 37°C of normal fibroblasts (N) and receptor-negative fibroblasts from a subject with familial hypercholesterolemia (FH) with r11-Dox-^{125}I-LDL (9 μg/ml of LDL, 1.75 μM of drug) and the effect of adding excess (200 μg/ml) native and methylated LDL (Met-LDL). Each value shows the mean results of duplicate incubations; the range was less than 10%. (From Ref. 45.)

protein-free medium, the other in serum-containing medium. When both drug and ^{125}I-LDL uptake were analyzed, it appeared that although ^{125}I association in the glioma group with low LDL receptor expression was approximately 20% as compared with the other group, drug accumulation was only slightly lower (Figure 13a). In both cases a large part of the cellular drug uptake probably occurred by leakage of drug from the LDL complex to the cells, since the cellular drug accumulation markedly exceeded that which could be explained by uptake and degradation of the ^{125}I-LDL complex.

In a later study, a new lipophilic derivative of doxorubicin (r11-Dox) was synthesized by coupling doxorubicin with a linoleyl and a retinyl anchor. The cellular drug accumulation and cellular content and degradation of ^{125}I-LDL on incubation with normal and receptor-negative fibroblasts was measured at different time intervals (Figure 13b).

The excellent correlation between cellular uptake and degradation of ^{125}I-LDL and cellular drug accumulation as well as the effects of methylated LDL in excess strongly support the hypothesis of a receptor-mediated uptake of r11-Dox-^{125}I-LDL.

B. Cytotoxicity Tests

The ability of reconstituted LDL to kill target cells or inhibit their growth can also be studied with cellular models. In spite of the fact that in vitro cytotoxicity tests are a very simplified model of the much more complex in vivo situation, these studies still are valuable because the effectiveness of the particle is measured directly and cannot be affected by in vivo parameters such as tumor microanatomy and vascularization.

The effect of cytotoxic LDL particles can be studied in many different ways. For instance, the setup for a cytotoxicity study will be different when the drug incorporated is merely growth-inhibiting instead of tumoricidal. Exposure of cells to the cytotoxic LDL (i.e., incubation time, concentration) are other important factors that will influence the outcome of the studies.

A simple way to assess cytotoxicity is cell protein assay. After exposure to the cytotoxic LDL for various lengths of time, cells can be washed and taken up in 0.1 M sodium hydroxide and analyzed for their protein content. With r11-DOX-methyl LDL as a control, Vitols et al. have shown r11-DOX-LDL to inhibit growth of fibroblasts [45].

In an earlier study, Rudling has varied the aclacinomycin A-LDL concentration in the medium and incubated glioma cells for a fixed length of time. After protein assay, the drug-LDL complex appeared to inhibit the growth of the human glioma cells in a concentration-dependent way [31].

Another technique uses trypan blue exclusion for determination of cell viability. Using this method, a toxic "compound-25"-LDL particle was shown to kill human neuroblastoma cells through receptor-mediated endocytosis. Addition of heparin to the medium prevented cell death [41].

The effect of the tumoricidal photosensitizing compound PCO (1-pyrenemethyl 23,24-bisnor-5-cholen-22-oate-3β-yl oleate), when incorporated in LDL and incubated with mammalian cells, could be measured directly by fixing the cells in the dish and staining with crystal violet [33]. (PCO)methyl-LDL was used as a reference and did not influence cell viability.

The toxic effects could be quantified more precisely by pulse-labeling with ^{3}H-thymidine for 2 h to measure the rate of DNA synthesis. ^{3}H-thymidine incorporation varied at varying periods after irradiation, and it was found that DNA synthesis was inhibited for 99% by 18 h after irradiation [33].

V. IN VIVO TESTS

The ultimate goal of LDL-mediated drug targeting is to elicit a selective in vivo effect. Thus far, most research in this field has focused on in vitro studies. Only some tentative in vivo studies have been reported, and most of them indicated a different in vivo behavior than what could be expected from the in vitro data. For this reason, it is essential to include in vivo experiments at an early stage in a study protocol. These early in vivo studies can tell whether the reconstituted LDL has lost or kept its native LDL receptor properties. With rats or mice as models, serum half-life and tissue distribution are important parameters for in vivo validation of the drug-LDL particles.

One of the first attempts to use in vivo studies for validation of a new incorporation procedure was made in 1986. Masquelier and co-workers have studied relative tissue uptake and serum concentrations 90 min after injection of the particle [25]. An additional method to check if the internalization in the in vivo situation follows the LDL pathway is conveniently carried out by pretreating rats with ethinylestradiol, which upregulates liver parenchymal LDL receptors [46]. Liver uptake of the reconstituted LDL should increase significantly, just like liver uptake of incorporated drug.

VI. CONCLUDING REMARKS

The utilization of LDL as a drug carrier may be regarded as an example where endogenous pathways are followed for site-specific drug delivery. In cases where tumor cells express high LDL receptor activity, possibilities are opened up for LDL-mediated antineoplastic therapy. Previous research has indicated that dietary measures can protect healthy body cells (without influencing the tumor) from injected toxic drug-LDL particles. Prerequisite is the development of a functional drug-LDL particle that has kept all in vivo characteristics of native LDL.

By modifying LDL, for instance, by acetylation or lactosylation, the particle can be "steered" to receptors other than the LDL receptor. In these cases, the lipoprotein will by utilized as a drug reservoir, since apo B recognition is lost. The modified particle will now be actively taken up by distinctive liver cell types to open new pathways for curing hepatic diseases.

VII. REFERENCES

1. M. E. Brewster, K. S. Estes, and N. Bodor, *J. Med. Chem. 31*:244 (1988).
2. M. M. Ponpipom, R. L. Bugianesi, J. C. Robbins, T. W. Doebber, and T. Y. Shen, in *Receptor Mediated Targeting of Drugs*, (G. Gregoriadis, G. Poste, J. Senior, and A. Trouet, eds.), NATO ASI Series, Plenum Press, New York and London, 1983, p. 53.
3. S. S. Davis and L. Illum, in *Site-Specific Drug Delivery*, (E. Tomlinson and S. S. Davis, eds.), John Wiley, Chichester, New York, Brisbane, Toronto, and Singapore, 1986, p. 93.
4. M. J. Poznansky and R. L. Juliano, *Pharmacol. Rev. 36*:277 (1984).

5. R. E. Counsell and R. C. Pohland, *J. Med. Chem. 25*:1115 (1982).
6. Th. J. C. van Berkel, J. K. Kruijt, L. Harkes, J. F. Nagelkerke, H. H. Spanjer, and H. J. M. Kempen, in *Site-Specific Drug Delivery*, (E. Tomlinson and S. S. Davis, eds.), John Wiley, Chichester, New York, Brisbane, Toronto, and Singapore, 1986, p. 49.
7. Y. K. Ho, R. G. Smith, M. S. Brown, and J. L. Goldstein, *Blood 52*:1099 (1978).
8. D. Gal, M. Ohashi, P. C. MacDonald, H. J. Buchsbaum, and E. R. Simpson, *Am. J. Obstet. Gynecol. 139*:877 (1981).
9. J. M. Dietschy and D. K. Spady, in *Receptor-Mediated Uptake in the Liver* (H. Greten, E. Windler, and U. Beisiegel, eds.), Springer-Verlag, Berlin, Heidelberg, 1986, p. 56.
10. S. H. Chen, C.-Y. Chang, P.-F. Chen, D. Setzer, M. Tanimura, W.-H. Li, A. M. Gotto, and L. Chan, *J. Biol. Chem. 261*:12918 (1986).
11. L. M. Havekes, D. Schouten, E. C. M. De Wit, L. H. Cohen, M. Griffioen, V. W. M. Van Hinsbergh, and H. M. G. Princen, *Biochem. Biophys. Acta 875*:236 (1986).
12. J. M. Dietschy, *Klin. Wochenschr. 62*:338 (1984).
13. D. Gal, P. C. MacDonald, J. C. Porter, and E. R. Simpson, *Int. J. Cancer 28*:315 (1981).
14. G. Norata, G. Canti, L. Ricci, A. Nicolin, E. Trezzi, and A. L. Catapano, *Cancer Lett. 25*:203 (1984).
15. S. A. Hynds, J. Welsh, J. M. Stewart, A. Jack, M. Soukop, C. S. McArdle, K. C. Calman, C. J. Packard, and J. Shepherd, *Biochim. Biophys. Acta 795*:589 (1984).
16. S. Vitols, M. Björkholm, G. Gahrton, and C. Peterson, *Lancet ii*:1150 (1985).
17. D. K. Spady and J. M. Dietschy, *Proc. Natl. Acad. Sci. USA 82*:4526 (1985).
18. J. F. Nagelkerke, K. P. Barto, and Th. J. C. van Berkel, *J. Biol. Chem. 258*: 12221 (1983).
19. Th. C. J. van Berkel, J. K. Kruijt, H. H. Spanjer, J. F. Nagelkerke, L. Harkes, and H. J. M. Kempen, *J. Biol. Chem. 260*:2694 (1985).
20. A. M. Mommaas-Kienhuis, J. F. Nagelkerke, B. J. Vermeer, W. Th. Daems, and Th. J. C. van Berkel, *Eur. J. Cell Biol. 38*:42 (1985).
21. E. Wisse, *J. Ultrastruct. Res. 31*:125 (1970).
22. H. J. M. Kempen, C. Hoes, J. H. Van Boom, H. H. Spanjer, J. De Lange, A. Langendoen, and Th. J. C. van Berkel, *J. Med. Chem. 27*:1306 (1984).
23. J. Schlepper-Schäfer, D. Hülsmann, A. Djovkar, H. E. Meyer, L. Herbertz, H. Kolb, and V. Kolb-Bachofen, *Exp. Cell Res. 165*:494 (1986).
24. R. A. Firestone, J. M. Pisano, J. R. Falck, M. M. McPhaul, and M. Krieger, *J. Med. Chem. 27*:1037 (1984).
25. M. Masquelier, S. Vitols, and C. Peterson, *Cancer Res. 46*:3842 (1986).
26. R. Burton, C. D. Eck, and B. Lloyd, *Biochem. Soc. Trans. 3*:1251 (1975).

27. J. B. Lloyd, in *Lysosomes and Storage Diseases* (H. G. Hers and F. Van Hoof, eds.), Academic Press, New York and London, 1973, p. 173.
28. M. Krieger, M. J. McPhaul, J. L. Goldstein, and M. S. Brown, *J. Biol. Chem. 254*:3845 (1979).
29. J. M. Shaw, K. V. Shaw, S. Yanovich, M. Iwanik, W. S. Futch, A. Roscwsky, and L. B. Schook, *Ann. N.Y. Acad. Sci. 507*:252 (1987).
30. J. C. Khoo, E. Miller, P. McLoughlin, and D. Steinberg, *Arteriosclerosis 8*: 348 (1988).
31. M. J. Rudling, V. P. Collins, and C. O. Peterson, *Cancer Res. 43*:4600 (1983).
32. C. Candide, J. C. Morlière, J. C. Mazière, S. Goldstein, R. Santus, L. Dubertret, J. P. Reyftmann, and J. Polonovski, *FEBS Lett. 207*:133 (1986).
33. S. T. Mosley, J. L. Goldstein, M. S. Brown, J. R. Falck, and R. G. Anderson, *Proc. Natl. Acad. Sci. U.S.A. 78*:5717 (1981).
34. Th. J. C. van Berkel, J. K. Kruijt, and H. J. M. Kempen, *J. Biol. Chem 260*: 12203 (1985).
35. J. Seki, A. Okita, M. Watanabe, T. Nakagawa, K. Honda, N. Tatewaki, and M. Sugiyama, *J. Pharm.Sci. 74*:12 (1985).
36. S. Yanovich, L. Preston, and J. M. Shaw, *Cancer Res. 44*:3377 (1984).
37. J. F. Remsen and R. B. Shireman, *Cancer Res. 41*:3179 (1981).
38. M. Krieger, M. S. Brown, J. R. Faust, and J. L. Goldstein, *J. Biol. Chem. 253*:4093 (1978).
39. M. Krieger, *Methods Enzymol. 128*:608 (1986).
40. M. T. Walsh and D. Atkinson, *Methods Enzymol. 128*:582 (1986).
41. B. Lundberg, *Cancer Res. 47*:4105 (1987).
42. G. W. Halbert, J. F. B. Stuart, and A. T. Florence, *Cancer Chemother. Pharmacol. 15*:223 (1985).
43. G. S. Ginsburg, M. T. Walsh, D. M. Small, and D. Atkinson, *J. Biol. Chem. 259*:6667 (1984).
44. M. J. Iwanik, K. V. Shaw, B. J. Ledwith, S. Yanovich, and J. M. Shaw, *Cancer Res. 44*:1206 (1984).
45. S. G. Vitols, M. Masquelier, and C. O. Peterson, *J. Med. Chem. 28*:451 (1985).
46. L. Harkes and Th. J. C. van Berkel, *FEBS Lett. 154*:75 (1983).

Index